AF753466

Contemporary Algorithms

Theory and Applications. Volume V

nova
science publishers
www.novapublishers.com

Mathematics Research Developments Series

Mathematics for Optimal Decision-Making and Applications: A Deterministic Approach
Rami A. Maher, PhD
ISBN: 979-8-89530-560-7
eBook ISBN: 979-8-89530-763-2

Numerical Optimization on Time Scales
Svetlin Georgiev
ISBN: 979-8-89530-466-2
eBook ISBN: 979-8-89530-503-4

The Laplace Transform and Its Applications in Science
Victor Martinez-Luaces, PhD (Editor)
ISBN: 979-8-89113-747-9
eBook ISBN: 979-8-89530-293-4

Selected Topics in Fractional Differential Equations
Zahia Drici, PhD, J. Vasundhara Devi, PhD and Farzana A. McRae, PhD (Editors)
ISBN: 979-8-89530-301-6
eBook ISBN: 979-8-89530-428-0

Bernoulli Scheme in Probability Theory
Mohammad Ahsanullah, Valery B. Nevzorov and Sergey M. Ananjeveskii
ISBN: 979-8-89530-003-9
eBook ISBN: 979-8-89530-135-7

Emerging Trends in Mathematical Sciences and Computing
Biswadip Basu Mallik, PhD, Sharmistha Ghosh, PhD, Binay Kumar Pandey, Digvijay Pandey and Vinay Kumar Nassa, PhD (Editors)
ISBN: 979-8-89113-751-6
eBook ISBN: 979-8-89113-818-6

The Direct Integration of Ordinary Differential Equations
A.B Urdaletova, PhD and S.K. Kydyraliev, PhD
eBook ISBN: 979-8-89530-690-1

The Fundamentals of Boundary Element Methods
Olena Sierikova (Editor)
ISBN: 979-8-89113-793-6
eBook ISBN: 979-8-89113-846-9

More information about this series can be found at
https://novapublishers.com/product-category/series/mathematics-research-developments/

Michael I. Argyros
Samundra Regmi
Ioannis K. Argyros
Santhosh George

Contemporary Algorithms

Theory and Applications. Volume V

nova
science publishers
www.novapublishers.com

Copyright © 2025 by Nova Science Publishers, Inc.
DOI: https://doi.org/10.52305/XHZP6026

All rights reserved. No part of this book may be reproduced, stored in a retrieval system or transmitted in any form or by any means: electronic, electrostatic, magnetic, tape, mechanical photocopying, recording or otherwise without the written permission of the Publisher.

We have partnered with Copyright Clearance Center to make it easy for you to obtain permissions to reuse content from this publication. Simply navigate to this publication's page on Nova's website and locate the "Get Permission" button below the title description. This button is linked directly to the title's permission page on copyright.com. Alternatively, you can visit copyright.com and search by title, ISBN, or ISSN.

For further questions about using the service on copyright.com, please contact:
Copyright Clearance Center
Phone: +1-(978) 750-8400 Fax: +1-(978) 750-4470 E-mail: info@copyright.com.

NOTICE TO THE READER

The Publisher has taken reasonable care in the preparation of this book, but makes no expressed or implied warranty of any kind and assumes no responsibility for any errors or omissions. No liability is assumed for incidental or consequential damages in connection with or arising out of information contained in this book. The Publisher shall not be liable for any special, consequential, or exemplary damages resulting, in whole or in part, from the readers' use of, or reliance upon, this material. Any parts of this book based on government reports are so indicated and copyright is claimed for those parts to the extent applicable to compilations of such works.

Independent verification should be sought for any data, advice or recommendations contained in this book. In addition, no responsibility is assumed by the Publisher for any injury and/or damage to persons or property arising from any methods, products, instructions, ideas or otherwise contained in this publication.

The Publisher assumes no responsibility for any statements of fact or opinion expressed in the published contents.

This publication is designed to provide accurate and authoritative information with regard to the subject matter covered herein. It is sold with the clear understanding that the Publisher is not engaged in rendering legal or any other professional services. If legal or any other expert assistance is required, the services of a competent person should be sought. FROM A DECLARATION OF PARTICIPANTS JOINTLY ADOPTED BY A COMMITTEE OF THE AMERICAN BAR ASSOCIATION AND A COMMITTEE OF PUBLISHERS.

Additional color graphics may be available in the e-book version of this book.

Library of Congress Cataloging-in-Publication Data

ISBN: 979-8-89530-724-3 (Hardcover)
ISBN: 979-8-89530-781-6 (eBook)

Published by Nova Science Publishers, Inc. † New York

The first author dedicates this book to his beloved grandparents, Jolanda, Mihallaq, Anastasia and Konstantinos.

The second author dedicates this book to his nieces, Sia Poudel and Shayana Poudel.

The third author dedicates this book to his wonderful children, Christopher, Gus, Michael, and lovely wife, Diana.

The fourth author dedicates this book to all his school teachers.

Contents

Preface

The book provides various avenues for studying algorithms and introduces new techniques and methodologies for problem-solving in Computational Sciences, Engineering, Scientific Computing, and Medicine (e.g., imaging, radiation therapy), among others.

A plethora of universally applicable algorithms is presented in a sound analytical manner. The chapters are written independently, allowing readers to understand them without prior knowledge of earlier chapters. However, some familiarity with Analysis, Linear Algebra, and Computing experience is required.

The organization and content of the book cater to senior undergraduate and graduate students, researchers, practitioners, professionals, and academicians in the aforementioned disciplines. It can also be used as a reference book, including numerous references and open problems.

To avoid repetition, whenever the symbols "ω," "φ," "ψ," "γ," "δ," and "h" are used as functions connected to the local or semi-local convergence analysis of iterative algorithms, they are assumed to be continuous, non-decreasing, and defined on a domain with nonnegative values, with a range in the real number system. Moreover, these functions are related to the operator appearing in the equation to be solved, its derivative, or its divided difference of order one. For x^* denoting a simple solution of the equation $F(x) = 0$, the following conditions hold:

$$\|F'(x^*)^{-1}(F'(x) - F'(x^*))\| \leq \omega_0(\|x - x^*\|)$$

for all $x \in \Omega$,

$$\|F'(x^*)^{-1}(F'(x) - F'(y))\| \leq \omega(\|x - y\|)$$

for all $x, y \in \Omega_0 \subset \Omega$,

$$\|F'(x^*)^{-1}F'(x)\| \leq \omega_1(\|x - x^*\|)$$

for all $x \in \Omega_0$,

$$\|I - a[x, x^*; F]\| \leq \gamma_0(\|x - x^*\|),$$

$$\|I + b[x, x^*; F]\| \leq \gamma(\|x - x^*\|),$$

$$\|I + c[x, x^*; F]\| \leq \gamma_1(\|x - x^*\|),$$

$$\|I + d[x, x^*; F]\| \leq \gamma_2(\|x - x^*\|),$$

$$\|I + L[x, x^*; F]\| \leq \gamma_3(\|x - x^*\|),$$

where $a, b, c, d \in \mathbb{R}$, L is a linear operator, and x^* may also be replaced by x_0 in the conditions involving the "γ" functions. These hold for $x \in \Omega$ or $x \in \Omega_0$.

$$\|F'(x^*)^{-1}[z, x; F]\| \leq \varphi_0(\|z - x^*\|, \|x - x^*\|),$$

for all $x, z \in \Omega$,

$$\|F'(x^*)^{-1}([z,x;F]-[x,x^*;F])\| \leq \varphi(\|z-x^*\|, \|x-x^*\|),$$

$$\|F'(x^*)^{-1}([z,x;F]-[y,x;F])\| \leq \varphi_1(\|z-x^*\|, \|x-x^*\|),$$

$$\|F'(x^*)^{-1}([z,x;F]-[y,z;F])\| \leq \varphi_2(\|z-x^*\|, \|x-x^*\|, \|y-x^*\|),$$

or

$$\|F'(x^*)^{-1}([z,x;F]-[y,v;F])\| \leq \varphi(\|z-x^*\|, \|y-x^*\|, \|x-x^*\|, \|v-x^*\|),$$

for all $x, y, z, v \in \Omega_0$.

$$\|F'(x^*)^{-1}[x,x^*;F]\| \leq \delta(\|x-x^*\|),$$

for all $x \in \Omega_0$,

$$\|F'(x^*)^{-1}([x,x^*;F]-F'(x^*))\| \leq \varphi_3(\|x-x^*\|),$$

for all $x \in \Omega$.

$$\|F'(x_0)^{-1}[x,x_0;F]\| \leq \delta(\|x-x_0\|),$$

for all $x \in \Omega_0$,

$$\|F'(x_0)^{-1}([x,x^*;F]-F'(x_0))\| \leq \psi_0(\|z-x_0\|, \|x-x_0\|),$$

for all $x \in \Omega$.

$$\|F'(x_0)^{-1}([x,x^*;F]-[y,x;F])\| \leq \psi_1(\|z-x_0\|, \|x-x_0\|, \|y-x_0\|),$$

$$\|F'(x_0)^{-1}([z,x;F]-[y,z;F])\| \leq \psi_2(\|z-x_0\|, \|x-x_0\|, \|y-x_0\|),$$

for all $x, y, z \in \Omega_0$.

The "φ" and "ψ" functions are assumed to be symmetric in each variable. Furthermore, the same notation is used to denote these functions in cases where a variable other than x^* or x_0 is missing in the corresponding condition.

Chapter 1

Solution of Generalized Equations in Banach Spaces under Strong Regularity Assumptions

1. Introduction

The letters X, Y denote Banach spaces; $\Omega \subset X$ is a convex and open subset of X, and $F_1 : \Omega \longrightarrow Y$ stands for a continuous operator. Numerous applications from diverse areas of computational science and Engineering can be converted by using mathematical modelling [3,8,14,17,19,20,21,23,26,28,33,35] to finding a solution $s^* \in \Omega$ of the nonlinear equation

$$F_1(x) = 0. \tag{1.1}$$

The closed form of the solution s^* is attainable only in special cases. This forces researchers and practitioners to solve equation (1.1) iteratively. Single-step methods of high convergence order look like the Newton-type defined for each $n = 0, 1, 2, ...$ by

$$x_0 \in \Omega, \quad x_{n+1} = x_n - L_n^{-1} F_1(x_n), \tag{1.2}$$

where $L_n \in \mathcal{L}(X, Y)$ which is the space of continuous operators mapping X into Y, and $L_n^{-1} \in \mathcal{L}(Y, X)$ for each $n = 0, 1, 2, ...$ By L_n, we denote $L(x_n)$. Some choices for the operator L_n can be

$$\begin{aligned} L_n &= F_1'(x_n) \text{ (Newton's method)}, \\ L_n &= [x_n - F_1(x_n), x_n + F_1(x_n); F] \text{ (Steffensen's method)}, \\ L_n &= I \text{ (The Picard method), the identity operator.} \end{aligned}$$

Here F_1', $[.,.;F_1]$ denote Fréchet-derivative and divided differences of order one for the operator F_1, respectively [22,25].

Many other choices are possible. It turns out that the inverse of the operator L_n is costly or impossible to find in general. This concern with the implementation of these methods is the motivation for this paper. Our idea is to replace the inverse with a finite sum of linear operators converging to it. The reasoning is explained as follows. Let $p \in N$ be fixed.

Suppose there exists $\Gamma \in \mathcal{L}(X, Y)$ such that $\Gamma^{-1} \in \mathcal{L}(Y, X)$ and for $A = A(x) = \Gamma^{-1}(\Gamma - L(x))$ the operator $I - A(x)$ is also invertible, i.e. $(I - A(x))^{-1} \in \mathcal{L}(Y, X)$. In this case, the Newton-type method can read as

$$x_0 \in D, \ x_{n+1} = x_n - (I - A)^{-1} \Gamma^{-1} F_1(x_n). \tag{1.3}$$

Note that we have

$$(I-A)^{-1}\Gamma^{-1} = [\Gamma(I-A)]^{-1} = L_n^{-1}. \tag{1.4}$$

However, even if the linear operator Γ^{-1} is known it is still required to find the inverse of $(I-A)$, which is not a fixed operator (in general). But what if we replace this operator with $M = M_p(x) = I + A + \ldots + A^p$. Then, method (1.3) can be written as

$$x_0 \in \Omega, \; x_{n+1} = x_n - M\Gamma^{-1}F_1(x_n). \tag{1.5}$$

It is clear that (1.5) is a useful alternative for (1.3) because of (1.4). By letting $p \longrightarrow +\infty$, we get $\lim_{p\to+\infty} M_p\Gamma^{-1} = L_n^{-1}$ if the limit exists. The condition $\|A\| < 1$ for each $x \in \Omega$ assures the existence of such a limit. If the linear operator M is invertible and the sequence $\{x_n\}$ given by (1.5) converges to some s^*, then by (1.5) we get

$$M^{-1}(x_n - x_{n+1}) = \Gamma^{-1}F_1(x_n)$$

leading to

$$0 = \lim_{n\to+\infty} M^{-1}(x_n - x_{n+1}) = \lim_{n\to+\infty} \Gamma^{-1}F_1(x_n),$$

i.e. $F_1(s^*) = 0$. Thus, the point s^* solves the equation (1.1). The same reasoning leads for $A_1 = (\Gamma - L(x))\Gamma^{-1}$ and $M_1 = I + A_1 + \ldots + A_1^p$ to the method

$$x_0 \in \Omega \;, x_{n+1} = x_n - M_1\Gamma^{-1}F_1(x_n). \tag{1.6}$$

It is clear that the study of the convergence of the method (1.6) is analogous to (1.5). That is why we study only method (1.5) in Section 2. We deal with two kinds of convergence: the semi-local and the local. The first utilizes knowledge in a neighborhood of x_0 and develops estimates related to $\|x_{n+1} - x_n\|$ and $\|s^* - x_n\|$, and the convergence conditions assure that $\lim_{n\to+\infty} x_n = s^*$. In the second kind, knowledge about a neighborhood of s^* is used to provide the same estimates as in the semi-local kind and again $\lim_{n\to+\infty} x_n = s^*$. It is worth noting that the iterates generated by (1.3),(1.5) and (1.6) are not the same in general. But we use the same notation for simplicity. The convergence for both kinds relies on generalized continuity conditions controlling the operators involved [5,6,9,18]. In particular, our semi-local convergence analysis depends on the usage of majorizing sequences [29,30,34]. Notice that the method (1.5) can also be written for $D_n = \Gamma M^{-1}$ as

$$x_0 \in \Omega, \; F_1(x_n) + D_n(x_{n+1} - x_n) = 0. \tag{1.7}$$

In section 3 we also use the developed methodology for solving nonlinear equations to solve generalized equations. That is find $x \in X$ such that

$$F_1(x) + F_2(x) \ni 0. \tag{1.8}$$

Here $F_2 : X \rightrightarrows Y$ is a set-valued operator mapping X into Y with closed graph [1]-[4,13],[17,18,19,21,22,23,26,28,32,35] and operator F is as previously defined. A plethora of applications from mathematical programming, variational inequalities, optimal control, or constrained systems are written in the form (1.8). There is extensive literature on iterative methods solving the generalized equation (1.8)[1,2,3,4,13,17,18,19,21,22,23,26,28,32,35]. Notice that the method used in the literature to solve (1.8) is defined by

$$F_1(x_n) + \bar{D}_n(x_{n+1} - x_n) + F_2(x_{n+1}) \ni 0, \tag{1.9}$$

where $\bar{D}_n$ is a linear operator. It can be chosen as $\bar{D}_n = L_n, \bar{D}_n = F_1'(x_n)$ or $\bar{D}_n \in \partial F_1(x_n)$ or other choices [9,11,23,24]. These methods have the same problems as the ones for solving nonlinear equations. That is why it is justified to consider the analog of (1.7) defined by

$$F_1(x_n) + D_n(x_{n+1} - x_n) + F_2(x_{n+1}) \ni 0 \tag{1.10}$$

The semi-local and local convergence of the method (1.10) is developed in section 3 in an analogous way to section 2 for the method (1.5) or (1.7). In numerical section 4, the examples demonstrate that the number of iterations of the hybrid methods to arrive at a predetermined error tolerance is essentially the same as with the original methods. Moreover, the convergence order is also the same.

In order to achieve all this we redevelop some standard terminology to make the paper as self-contained as possible. More details can be found in [8,11,19]. Let $S(z,\rho)$ and its closure $S[z,\rho]$ denote open and closed balls, respectively of center $z \in X$ and radius $\rho > 0$. Let C be a set in X. Define the distance for $x \in X$ to C by $dist(x,C) = \inf_{x \in C} \|x - y\|$. The generalized set-valued operator G relates with its graph given by $gph(G) = \{(x,y) \in X \times Y, y \in F_2(x)\}$, and its domain $dom(G) = \{x \in X | F_2(x) \neq 0\}$. The inverse of G is given as $G^{-1}(y) = \{x \in X, y \in F_2(x)\}$.

Definition 1. A set-valued operator $H : X \rightrightarrows Y$ is said to be metrically regular at x_0 for y_0 if $y_0 \in H(x_0)$ and there exists neighbourhoods V_1 of x_0 and V_2 of y_0 and $\beta > 0$ such that $gph(H \cap (V_1 \times V_2))$ is closed and for each $(x,y) \in V_1 \times V_2$

$$dist(x, H^{-1}(y)) \leq \beta dist(y, H(x)). \tag{1.11}$$

The regularity modulus of H at x_0 for y_0 is the infimum over all $\beta > 0$ and is denoted by $reg(H; x_0/y_0)$. Additionally if the operator $\Delta : V_2 \to y \to H^{-1}(y) \cap V_1$ is not multivalued on V_2, then we say that H is strongly metrically regular.

Next, we make this concept more precise. The convergence analysis was based using the idea of strong regularity attributed to Robinson [29].

Definition 2. The operator $F_1 + F_2$ is said to be strongly regular at x for y, when $y \in F_1(x) + F_2(x)$, and there exist $\rho_x > 0, \rho_y > 0$, and $\beta > 0$ such that $S(x,\rho_x) \subset \Omega$, the operator $x \to L_{F_1}(x,x)^{-1} \cap S(x,\rho_x)$ is single-valued from $S(y,\rho_y)$ to $S(x,\rho_x)$ with modulus β and satisfies the Lipschitz condition

$$\|L_{F_1}(x,u_2)^{-1} \cap S(x,\rho_x) - L_{F_1}(x,u_1)^{-1} \cap S(x,\rho_x)\| \leq \beta \|u_2 - u_1\|, \tag{1.12}$$

for each $u_1, u_2 \in S(y,\rho_y)$.

The parameter β is the Lipschitz constant in this case.

Finally, Section 5 contains concluding remarks and directions for research.

2. Convergence for the Method (1.5)

We start with the study of the semi-local analysis in this section. Some auxiliary results and definitions are useful.

Lemma 1. (Banach Lemma on Invertible Operators)([14,22,30,34]) If P is a bounded linear operator in X, P^{-1} exists if and only if there is a bounded linear operator P_1 in X such that P_1^{-1} exists and

$$\|I - P_1 P\| < 1.$$

If P^{-1} exists, then

$$P^{-1} = \sum_{n=0}^{\infty}(I - P_1 P)^n P_1$$

and

$$\|P^{-1}\| \leq \frac{\|P_1\|}{1 - \|I - P_1 P\|}.$$

Further, we use majorizing sequences to prove the semi-local convergence.

Definition 3. ([14,22,30,34]) Let $\{x_n\}$ be a sequence in a normed space X. Then a nonnegative scalar sequence $\{u_n\}$ for which

$$\|x_{n+1} - x_n\| \leq u_{n+1} - u_n \quad \forall n \geq 0 \tag{1.13}$$

holds, is a majorizing sequence for $\{x_n\}$. Note that any majorizing sequence is necessarily nondecreasing. Moreover, if the sequence $\{u_n\}$ converges, then $\{x_n\}$ converges too, and for $u^* = \lim_{n \longrightarrow \infty} u_n$

$$\|s^* - x_n\| \leq u^* - u_n.$$

Hence, the study of the convergence of the sequence $\{x_n\}$ reduces to that of $\{u_n\}$.

The analysis requires some conditions. Let $E = [0, +\infty)$.
Suppose that

(H_1) There exists a function $\phi : E \times E \times E \rightarrow [0, +\infty)$ continuous as well nondecreasing in all three variables and invertible operators $M(.)$ and Γ such that for some $x_0 \in \Omega$, and each $x, y \in \Omega$ the following Mysovskii-like condition holds

$$\begin{aligned} & \|M(x)\Gamma^{-1}(F_1(y) - F_1(x) - \Gamma M^{-1}(x))\| \\ \leq \; & \phi(\|x - x_0\|, \|y - x_0\|, \|y - x\|)\|y - x\| \end{aligned}$$

Define the real real sequence $\{\alpha_n\}$ for $\alpha_0 = 0, \alpha_1 \geq \eta := \|M(x_0)\Gamma^{-1}F_1(x_0)\|$ and each $n = 0, 1, 2, ...$ by

$$\alpha_{n+1} = \alpha_n + \phi(\alpha_{n-1}, \alpha_n, \alpha_n - \alpha_{n-1})(\alpha_n - \alpha_{n-1}), \; n = 1, 2, ... \tag{1.14}$$

Notice that the constant η is well defined, since the operator Γ is invertible. Moreover, the sequence $\{\alpha_n\}$ defined by the formula (1.14) is proven to be majorizing for the method (1.5) in Theorem 1. But let us present convergence conditions for it.

(H_2) There exists a parameter $\rho \geq \eta$ such that for each $n = 0, 1, 2, ...$

$$\phi(\alpha_{n-1}, \alpha_n, \alpha_n - \alpha_{n-1}) < 1 \text{ and } \alpha_n \leq \rho.$$

It follows by the condition (H_2) and the formula (1.14) that

$$0 \leq \alpha_{n-1} \leq \alpha_n \leq \rho$$

and there exists $\alpha^* \in [\eta, \rho]$ such that $\lim_{n \rightarrow +\infty} \alpha_n = \alpha^*$. H is well known that this limit is the unique least upper bound of the sequence $\{\alpha_n\}$ and

(H_3) $S[x_0, \alpha^*] \subset \Omega$.

The conditions (H_1)-(H_3) combined with the terminology are utilized for the convergence of the method (1.5).

Theorem 1. Suppose that the conditions (H_1)-(H_3) hold. Then, the sequence $\{x_n\}$ generated by the method (1.5) exists in the ball $S(x_0, \alpha^*)$ remains in the same ball for each $n = 0, 1, 2, ...$ and is convergent to a unique solution $s^* \in S[x_0, \alpha^*]$ of the equation $F_1(x) = 0$ such that

$$\|s^* - x_n\| \leq \alpha^* - \alpha_n,\ n = 0, 1, 2 \ldots. \tag{1.15}$$

Proof. The process of induction is employed to show the assertion

$$\|x_{m+1} - x_m\| \leq \alpha_{m+1} - \alpha_m \text{ for } m = 0, 1, 2, \ldots. \tag{1.16}$$

The choice of η, (1.14) and the method (1.5) give that

$$\|x_1 - x_0\| = \|M(x_0)\Gamma^{-1}F_1(x_0)\| \leq \eta = \alpha_1 - \alpha_0 < \alpha^*,$$

so the assertion (1.16) holds if $m = 0$, and the iterate $x_1 \in S(x_0, \alpha^*)$. Suppose iterates $x_0, x_1, ..., x_m$ exist and (1.16) holds for all integers smaller or equal to $m - 1$. Notice that the iterate x_{m+1} exists by the method (1.5) and the invertability of the operators $M(x_m)$ and Γ. Then, we can write by the method (1.5) the Ostrowski-type representation for $F_1(x_m)$ as

$$F_1(x_m) = F_1(x_m) - F_1(x_{m-1}) - \Gamma M^{-1}(x_m - x_{m-1}). \tag{1.17}$$

Using the condition (H_1), method (1.5), (1.14) and the induction hypothesis on (1.17) we obtain in turn that

$$\begin{aligned}
\|x_{m+1} - x_m\| &= \|M\Gamma^{-1}(F_1(x_m))\| \\
&= \|M\Gamma^{-1}(F_1(x_m) - F_1(x_{m-1}) - \Gamma M^{-1}(x_m - x_{m-1}))\| \\
&\leq \phi(\|x_{m-1} - x_0\|, \|x_m - x_0\|, \|x_m - x_{m-1}\|)\|x_m - x_{m-1}\| \\
&\leq \phi(\alpha_{m-1}, \alpha_m, \alpha_m - \alpha_{m-1})(\alpha_m - \alpha_{m-1}) \\
&= \alpha_{m+1} - \alpha_m,
\end{aligned} \tag{1.18}$$

and

$$\begin{aligned}
\|x_{m+1} - x_0\| &\leq \|x_{m+1} - x_m\| + \|x_m - x_{m-1}\| + \ldots + \|x_1 - x_0\| \\
&\leq \alpha_{m+1} - \alpha_m + \alpha_m - \alpha_{m-1} + \ldots + \alpha_1 - \alpha_0 \\
&= \alpha_{m+1} < \alpha^*,
\end{aligned}$$

which complete the induction for the assertion (1.16) and show that all iterates $\{x_{m+1}\} \subset S(x_0, \alpha^*)$. But the sequence $\{\alpha_m\}$ is complete by the condition (H_2) as convergent. It follows by the estimate (1.16) that $\{x_m\}$ is also complete. But the space X is Banach, so there exists $s^* \in S[x_0, \alpha^*]$ such that $\lim_{m \to +\infty} x_m = s^*$. Next, by letting $m \to +\infty$ in (1.18), the invertibility of the operators $M(.), \Gamma$ and the continuity of the operator F, we deduce that $\lim_{m \to +\infty} M\Gamma^{-1}F_1(s^*) = 0$. By the invertibility of M

$$\begin{aligned}
0 &= M^{-1}(0) = M^{-1}(\lim_{m \to +\infty} M\Gamma^{-1}F_1(x_m)) = \lim_{m \to +\infty} MM^{-1}\Gamma^{-1}F_1(x_m) \\
&= \lim_{m \to +\infty} \Gamma^{-1}F_1(x_m) = \Gamma^{-1} \lim_{m \to +\infty} F_1(x_m) = \Gamma^{-1}F_1(s^*).
\end{aligned}$$

So, $F_1(s^*) = 0$, since $\Gamma(0) = 0$. Let $i = 0,1,2,\ldots$. Then, the triangle inequality and (1.16) give

$$\|x_{m+i} - x_m\| \leq \alpha_{m+i} - \alpha_m. \tag{1.19}$$

Hence, by letting $i \longrightarrow +\infty$ in (1.19) we prove (1.15). Finally, to show the uniqueness part, let $w \in S[x_0, \alpha^*]$ with $F_1(w) = 0$ and $w \neq s^*$. By using the conditions (H_1), (H_2) we can write in turn that

$$\begin{aligned}
\|w - s^*\| &= \|(M\Gamma^{-1})(\Gamma M^{-1})(w - s^*)\| \\
&= \|M\Gamma^{-1}(F_1(w) - F_1(s^*) - \Gamma M^{-1}(w - s^*))\| \\
&\leq \phi(\|s^* - x_0\|, \|w - x_0\|, \|w - s^*\|)\|w - s^*\| \\
&\leq \phi(\alpha^*, \alpha^*, \|w - s^*\|)\|w - s^*\| < \|w - s^*\|,
\end{aligned}$$

which gives a contradiction. Therefore, we conclude that $w = s^*$. □

Remark. The condition Mysoskii-type [22] condition in (H_1) can be replaced as follows:

$(H_1)'$ With operator Γ as in condition (H_1), suppose that there exists $a \in (0, \frac{1}{2})$ such that $\|A\| < a$ and for each $x, y \in \Omega$

$$\begin{aligned}
&\|\Gamma^{-1}(F_1(y) - F_1(x) - \Gamma M^{-1}(y - x))\| \\
\leq\ &\phi_1(\|x - x_0\|, \|y - x_0\|, \|y - x\|)\|y - x\|,
\end{aligned}$$

where the function ϕ_1 is as the function ϕ. Then, the condition $(H_1)'$ implies (H_1) if we take $\phi = a_0\phi_1$, where $a_0 = \frac{1}{1-a}$.

This is the case, since by the definition of the operator M, we have the estimate

$$\begin{aligned}
\|M\| &= \|I + A + \ldots + A^p\| \\
&\leq 1 + a + \ldots + a^p \\
&= \frac{1 - a^{p+1}}{1 - a} = a_0 < \frac{1}{1 - a}.
\end{aligned} \tag{1.20}$$

In this case the invertibility of the operator M is implied by the Banach Lemma 1, since

$$\|I - M\| \leq \|A\| + \ldots + \|A\|^p \leq a + \ldots + a^p = a\frac{1 - a^p}{1 - a} < \frac{a}{1 - a} < 1.$$

So, M is invertible and

$$\|M^{-1}\| \leq b = \frac{1 - a}{1 - 2a}.$$

Next, we develop the local convergence. The role of the initial point x_0 is exchanged by s^* and the function ϕ by ψ.
Suppose:

(H_4) There exists a function $\psi : E \longrightarrow [0, +\infty)$ continuous and nondecreasing such that the equation $\psi(t) - 1 = 0$ has a smallest positive solution denoted by r.

(H_5) There exists a solution $s^* \in \Omega$, and invertible linear operators M, Γ such that for each $x \in \Omega$

$$\begin{aligned} & \|M\Gamma^{-1}(F_1(x) - F_1(s^*) - \Gamma M^{-1}(x - s^*))\| \\ \leq \; & \psi(\|x - s^*\|)\|x - s^*\| \end{aligned}$$

and

(H_6) $S[s^*, r] \subset \Omega$.

Next, the constant r is shown to be a radius of convergence for the method (1.5).

Theorem 2. Suppose that the conditions $(H_4) - (H_6)$ hold. Then, the sequence $\{x_n\}$ for $x_0 \in S(s^*, r) - \{s^*\}$ exists in $S(s^*, r)$, stays in $S(s^*, r)$ and converges to s^* so that

$$\|x_{n+1} - s^*\| \leq \psi(\|x_n - s^*\|)\|x_n - s^*\| \leq \|x_n - s^*\| < r. \tag{1.21}$$

Moreover, x^* is the only solution of the equation $F(x)$ =0 in the ball $U(s^*, r)$.

Proof. The iterates $x_1, x_2, ..., x_{m+1}$ are well defined by the method (1.5), and we can write in turn that

$$\begin{aligned} x_{m+1} - s^* &= x_m - s^* - M\Gamma^{-1}F_1(x_m) \\ &= M\Gamma^{-1}(F_1(x_m) - F_1(s^*) - \Gamma M^{-1}(x_m - s^*)). \end{aligned} \tag{1.22}$$

It follows by the conditions $(H_4), (H_5)$ and (1.22) that

$$\begin{aligned} \|x_{m+1} - s^*\| &\leq \psi(\|x_m - s^*\|)\|x_m - s^*\| \\ &\leq \xi\|x_m - s^*\| \leq \xi^{m+1}\|x_0 - s^*\| < r, \end{aligned} \tag{1.23}$$

where $\xi = \psi(\|x_0 - s^*\|) \in [0, 1)$ showing the assertion (1.21) for each $m = 1, 2, ...,$ since $x_0 \in S(s^*, r) - \{s^*\}$. By letting $m \to +\infty$ in (1.23), we deduce that $\lim_{m \to +\infty} x_m$. In order to show the uniqueness part, suppose there exists a solution $w_1 \in S(s^*, r)$ such that $w_1 \neq s^*$. Then, as in the semi-local case, we can write in turn

$$\begin{aligned} \|w_1 - s^*\| &= \|(M\Gamma^{-1})(\Gamma M^{-1}(w_1 - s^*))\| \\ &= \|M\Gamma^{-1}(F_1(w_1) - F_1(s^*) - \Gamma M^{-1}(w_1 - s^*))\| \\ &\leq \psi(\|w_1 - s^*\|)\|w_1 - s^*\| < \|w_1 - s^*\| \end{aligned} \tag{1.24}$$

by the choice of r. Hence, we conclude that $w_1 = s^*$, since (1.24) contradicts the hypothesis $w_1 \neq s^*$. □

Remark. Comments similar to the ones in Remark 2.4 can be made provided that x_0, ψ are exchanged by s^*, ψ, respectively. It is also worth noting that the condition (H_4) does not necessarily imply the usual condition in local convergence studies that the operator $F_1'(s^*)$ is invertible, i.e. that s^* is a simple solution of the equation $F_1(x) = 0$. Consequently the method (1.5) can be applied to find solutions of multiplicity greater than one.

3. Convergence for the Method (1.10)

Let $\lambda > 0$and $\delta \geq 0$ be given parameters. We use β to denote $\beta = \dfrac{1 - a^{p+1}}{1 - a}$.

Suppose:

(C_1) There exists a function

$$\varphi_1 : [0,\lambda] \times [0,\lambda] \times [0,\lambda] \longrightarrow [0,\delta]$$

which is continuous and nondecreasing. Define the scalar sequence $\{h_n\}$ for $h_0 = 0$, some $h_1 \geq 0, \beta_1 > \beta$ and each $n = 1,2,...$ by

$$h_{n+1} = h_n + \beta_1 \varphi_1 (h_{n-1}, h_n, h_n - h_{n-1})(h_n - h_{n-1}). \tag{1.25}$$

The scalar sequence $\{h_n\}$ is shown to be majorizing for $\{x_n\}$ is generated by the formula in the Theorem 2. However, let us present a convergence criterion for it.

(C_2) There exists $\lambda_0 \in [0, j]$ such that for each $n = 0,1,2,...$ $h_n \leq \lambda_0$.
It follows by this condition and (1.25) that $0 \leq h_{n-1} \leq h_n \leq \lambda_0$ and there exists $h^* \in [0, \lambda_0]$ such that $\lim\limits_{n\to\infty} h_n = h^*$.
The limit point h^* is the unique least upper bound of the sequence $\{h_n\}$.

(C_3) There exists $x_0 \in X$ and $y_0 \in F_1(x_0) + F_2(x_0)$ such that
$\beta\delta < 1$ and $\|y_0\| \leq (1 - \beta\delta)\, min\{\frac{\lambda}{\beta}, \mu\}$.
Choose $h_1 \leq \beta_1 \|y_0\|$.

(C_4) The operator

$$x \longrightarrow Q_{D_n}(x) := F_1(x_0) + D_n(x - x_0) + F_2(x) \tag{1.26}$$

is metrically regular at x_0 for y_0 with constant β and neighborhoods $S(x_0, \lambda)$ and $S(y_0, \mu)$, respectively.
The mapping φ_1 relates to the operators on the method (1.10).

(C_5)

$$\|F_1(x) - F_1(x_n) - D_n(x - x_n)\| \leq \varphi_1(\|x - x_0\|, \|x_n - x_0\|, \|x - x_n\|)\|x - x_n\|,$$

for each $x \in S(x_0, \lambda)$.
Next, the semi-local analysis of convergence is developed using the conditions $(C_1) - (C_5)$.

Theorem 3. Suppose that the conditions $(C_1) - (C_5)$ hold. Then, for each $\gamma \in (\beta\delta, 1)$ there exists a sequence $\{x_n\}$ generated by the method (1.10) which is well defined in $S(x_0, \lambda)$, remains in $S(x_0, \lambda)$ for each $n = 0,1,2,...$ and convergence to a solution $s^* \in S[x_0, \lambda]$ of the generated equation (1.8). Moreover, the following assertion hold

$$\|s^* - x_n\| \leq \gamma^n \lambda \tag{1.27}$$

and

$$dist(0, F_1(x_n) + F_2(x_n)) \leq \gamma^n \|y_0\| \tag{1.28}$$

for each $n = 0,1,2,...$

Thus, the convergence rate is R-linear. Furthermore, If the operator Q_{D_n} is strongly metrically regular with constant β and neighbourhoods $S(x_0, \lambda)$ and $S(y_0, \mu)$, respectively, then the sequence $\{x_n\}$ is the only one satisfying (1.10), and staying in $S(x_0, \lambda)$.

Remark. The proof is similar to the one in [Theorem 2.2, 11]. But it uses (1.10), (C_4) and (C_5), instead of stronger (1.9),

(C_4)' $x \longrightarrow Q_{D_n}(x) := F_1(x_0) + D_n(x - x_0) + F_2(x)$,

(C_5)' $\|F_1(x)-F_1(x_n)-L_n(x-x_n)\| \leq w(\|x-x_n\|)\|x-x_n\|$
for each $x \in S(x_0,\lambda)$, where $w : [0.\lambda] \longrightarrow [0,\delta]$ satisfies $\lim_{n\to+\infty} w(t) = 0$

Proof. Pick $\gamma \in (\beta\delta, 1)$ and β_1 so that $\beta < \beta_1 \leq \frac{\gamma}{\delta}$ and

$$\|y_0\| \leq (1-\gamma)min\left\{\frac{\lambda}{\beta_1},\mu\right\}. \tag{1.29}$$

The choice of β_1 is certainly possible since there are infinitely many numbers between β and $\frac{\gamma}{\delta}$. We shall show the existence of the sequence $\{x_n\}$ using mathematical induction for $n = 1,2,...$ satisfying

(I_n) $\|x_n - x_0\| \leq \frac{1-\gamma^n}{1-\gamma}\beta_1\|y_0\| \leq (1-\gamma^n)\lambda < \lambda.$

(II_n) $\|x_n - x_{n-1}\| \leq h_n - h_{n-1}.$

(III_n) $0 \in F_1(x_{n-1}) + D_{n-1}(x_n - x_{n-1}) + F_2(x_n),$
where $D_{n-1} = D_{n-1}(x_0,...,x_{n-1}).$
By hypothesis $0 \in S(y_0,\mu)$ and $y_0 \in Q_{D_0}(x_0)$. Using the condition (C_4) for Q_{D_0} we get

$$dist(x_0, Q_{D_0}^{-1}(0)) \leq \beta\, dist(0, Q_{D_0}(x_0)) \leq \beta\|y_0\|.$$

If $y_0 = 0$, pick $x_1 = x_0$. Otherwise, It follows that

$$dist(x_0, Q_{D_0}^{-1}(0)) \leq \beta_1\|y_0\|.$$

Thus, there exists $x_1 \in Q_{D_0}^{-1}(0)$ satisfying

$$\|x_1 - x_0\| < \beta_1\|y_0\| < (1-\gamma)\lambda.$$

So, (I_1), (II_1) and (III_1) hold.
Suppose that for some natural number m the element x_m is defined so that the induction hypothesis $(I_m),(II_m)$ and (III_m) hold. We shall show that the iterate x_{m+1} is defined and the assertions $(I_{m+1}),(II_{m+1})$ and (III_{m+1}) hold. It follows by (I_m) that $x_m \in S(x_0,\lambda)$. Set $e_m = F_1(x_0) - F_1(x_m) - D_m(x_0 - x_m)$.
By (C_5) for $x = x_0, (I_m), (II_m)$ and the properties of the function φ_1 we can write

$$\begin{aligned}
\|e_m - y_0\| &\leq \|y_0\| + \|F_1(x_0) - F_1(x_m) - D_m(x_0 - x_m)\| \\
&\leq \|y_0\| + \varphi_1(\|x_0 - x_0\|, \|x_m - x_0\|, \|x_m - x_0\|)\|x_0 - x_m\| \\
&\leq \|y_0\| + \varphi_1(0, h_m, h_m)\|x_0 - x_m\| \\
&\leq \|y_0\| + \delta\|x_0 - x_m\| \leq \|y_0\| + \frac{1-\gamma^m}{1-\gamma}\beta\|y_0\| \\
&\leq (1 + \frac{1-\gamma^m}{1-\gamma}\gamma)\|y_0\| = \frac{1-\gamma^{m+1}}{1-\gamma}\|y_0\| < \mu.
\end{aligned}$$

If $e_m \in Q_{D_m}(x_m)$, set $x_{m+1} = x_m$. Otherwise by the condition (C_4) we have

$$\begin{aligned}
dist(x_m, Q_{D_m}^{-1}(e_m)) &\leq \beta\, dist(e_m, Q_{D_m}(x_n)) \\
&\leq \beta_1\, dist(e_m, Q_{D_m}(x_m)).
\end{aligned}$$

Next, there exists an element $x_{m+1} \in Q_{D_m}^{-1}(e_m)$ satisfying

$$\|x_{m+1} - x_m\| \leq \beta_1 \, dist(e_m, Q_{D_m}(x_m)).$$

By (III_m) it follows

$$\begin{aligned} Q_{D_m}(x_m) &= F_1(x_0) + D_m(x_m - x_0) + F_2(x_m) \\ &\ni F_1(x_0) + D_m(x_m - x_0) - F_1(x_{m-1} - D_{m-1}(x_m - x_{m-1}). \end{aligned}$$

In view of the condition (C_4) with $x = x_m$, (**??**) and (II_m), we obtain in turn that

$$\begin{aligned} \|x_{m+1} - x_m\| &\leq \beta_1 \big\| e_m - \big[F_1(x_0) - F_1(x_{m-1}) + D_m(x_m - x_0) \\ &\quad - D_{m-1}(x_m - x_{m-1})\big] \big\| \\ &= \beta_1 \|F_1(x_m) - F_1(x_{m-1}) - D_{m-1}(x_m - x_{m-1})\| \\ &\leq \beta_1 \varphi_1(\|x_{m-1} - x_0\|, \|x_m - x_0\|, \|x_m - x_{m-1}\| \\ &\leq \beta_1 \varphi_1(h_{m-1}, h_m, h_m - h_{m-1}) = h_{m+1} - h_m, \end{aligned} \tag{1.30}$$

and

$$\begin{aligned} \|x_{m+1} - x_m\| &\leq \beta_1 \delta \|x_m - x_{m-1}\| = \gamma \|x_m - x_{m-1}\| \leq \gamma^n \|x_1 - x_0\| \\ &\leq \gamma^n \beta_1 \|y_0\|. \end{aligned}$$

Thus, the condition (III_m) holds if $m+1$ replaces m. Moreover, by the selection of x_{m+1}, we get

$$e_m \in Q_{D_m}(x_{m+1}) = F_1(x_0) + D_m(x_{m+1-} - x_0) + F_2(x_{m+1}).$$

So, the assertion (III_n) holds if $m+1$ replaces m. Moreover, by (I_m) we get in turn that

$$\begin{aligned} \|x_{m+1} - x_0\| &\leq \|x_{m+1} - x_m\| + \|x_m - x_0\| \\ &\leq \gamma^n \beta_1 \|y_0\| + \frac{1-\gamma^n}{1-\gamma} \beta_1 \|y_0\| \\ &= \frac{1-\gamma^{n+1}}{1-\gamma} \beta_1 \|y_0\|, \end{aligned}$$

which terminates the induction for (I_m). Then, by (C_2) the sequence is complete as convergent. It follows by (II_m) that the sequence $\{x_m\}$ is also complete in a Banach space X and as such it converges to some s^*. By (I_m) we have $s^* \in S(x_0, \lambda)$. We must show that the limit point s^* solves (1.8).

Set $y_m := F_1(x_m) - F_1(x_{m-1}) - D_{m-1}(x_m - x_{m-1})$. It follows by (III_m) that $y_m \in F_1(x_m) + F_2(x_m)$. By using (II_m) and the condition (C_4) for $x = x_m$, we get in turn that

$$\begin{aligned} \|y_m\| &= \|F_1(x_m) - F_1(x_{m-1}) - D_{m-1}(x_m - x_{m-1})\| \\ &\leq \psi_1(\|x_{m-1} - x_0\|, \|x_m - x_0\|, \|x_m - x_{m-1}\|)\|x_m - x_{m-1}\| \\ &\leq \psi_1(h_{m-1}, h_m, h_m - h_{m-1})(\alpha_m - \alpha_{m-1}) = \alpha_{m+1} - \alpha_m \longrightarrow 0 \end{aligned}$$

as $m \to +\infty$.

Consequently, we deduce that $(x_m, y_m) \to (s^*, 0)$ as $m \to +\infty$.

But F is continuous whereas G has a closed graph. Thus, we conclude $0 \in F_1(s^*) + F_2(s^*)$.

Finally, If Q_{D_m} is a strongly metrically regular operator. It follows that the iterate x_{m+1} is unique and is obtained from x_m. □

Next, we develop the local convergence analysis of the method (1.10).

Suppose :

(C_6) There exists a parameter $\beta > 1$ and a function $\varphi_2 : \mathbb{R}_+ \rightarrow \mathbb{R}_+$ which is continuous and nondecreasing such that the equation $\beta_1\varphi_2(t) - 1 = 0$ has a smallest positive solution. Denote such a solution by r_0.

(C_7) There exists a solution $s^* \in X$ of the generalized equation (1.8) and parameters $\beta > 0$ and $\delta \geq 0$ such that $\beta\delta < 1$.

(C_8) The operator $x \longrightarrow T_{D_n}(x) := F_1(s^*) + D_n(x - s^*) + F_2(x)$ is a metrically regular at s^* for 0 with constant β and neighbourhoods $S(s^*, \lambda)$ and $(0, \mu)$ for some $\mu > 0$, respectively.

The function φ_2 relates to the operators on the method (1.10).

(C_9)

$$\|F_1(s^*) - F_1(x_n) - D_n(s^* - x_n)\| \leq \varphi_2(\|s^* - x_n\|)\|s^* - x_n\| \tag{1.31}$$

for $x_0 \in S(s^*, \lambda)$. Next, the local convergence analysis of the method (1.10) is presented using the conditions $(C_6) - (C_4)$.

Theorem 4. Suppose that the conditions $(C_6) - (C_9)$ hold. Then, for each $\gamma \in (\beta\delta, 1)$ the re exists a sequence $\{x_n\}$ generated by the method (1.10) which is well defined in $S(s^*, \lambda)$, remains in $S(s^*, \lambda)$ for each $n = 0, 1, 2, ..$ and converges to s^* so that

$$\|s^* - x_n\| \leq d\|s^* - x_{n-1}\| \leq d^n\|s^* - x_0\| < \lambda,$$

where $d = \beta_1\varphi_2(\|s^* - x_0)\| \in [0, 1)$. Additionally, If the operator T_{D_n} is strongly metrically regular with constant β and neighbourhoods $S(s^*, \lambda)$ and $S(0, \mu)$, respectively, then the sequence $\{x_n\}$ is the only one satisfying (1.8), and satisfying in (s^*, λ).

Proof. Simply follow the proof of Theorem 3 for $\gamma \in (\beta\delta, 1)x_0 = s^*$ and $\beta < \beta_1 \leq \frac{\gamma}{\delta}$ to obtain as in (1.30) but using (C_8) and (1.31) instead of (C_4) and (1.30), respectively to obtain

$$\begin{aligned}
\|s^* - x_m\| &\leq \beta_1\varphi_2(\|s^* - x_{m-1}\|)\|s^* - x_{m-1}\| \\
&\leq d\|s^* - x_{m-1}\| \leq ... \leq d^m\|s^* - x_0\| < \lambda.
\end{aligned}$$

Therefore, we conclude that $\lim_{m \longrightarrow +\infty} x_m = s^*$ and the iterate $x_m \in S(s^*, \lambda)$.

Finally, if the operator T_{D_m} is strongly metrically regular, it follows that the iterate x_m is unique in $S(s^*, \lambda)$ by the way the iterate x_m is derived from x_{m-1}. □

4. Numerical Examples

The examples use $L_n = F_1'(x_n), \Gamma = I$ which is independent of x_0 and s^*.

Example 1. The solution sought for the nonlinear system

$$\begin{aligned}
f_1 &= x - 0.1\sin x - 0.3\cos y + 0.4 \\
f_2 &= y - 0.2\cos x + 0.1\sin y + 0.3
\end{aligned}$$

Let $F_1 = (f_1, f_2)$. Then, the system becomes

$$F_1(s) = 0 \ \ for \ \ s = (x, y)^T.$$

Then

$$F_1'((x,y)) = \begin{bmatrix} 1-0.1\cos(x) & 0.3\sin(y) \\ 0.2\sin(x) & 0.1\cos(y)+1 \end{bmatrix}.$$

Method (1.2)

$$x_{p+1} = x_p - F_1'(x_p)^{-1}F_1(x_p).$$

Method (1.5), $p=1,\ \Gamma = I,$

$$\begin{aligned} M_1(x) &= I+(I-F_1'(x)), \\ Q_1(x) &= x-(I+(I-F_1'(x)))F_1(x), \\ x_{n+1} &= Q_1(x_n). \end{aligned} \tag{1.32}$$

Method (1.5), $p=2,\ \Gamma = I,$

$$\begin{aligned} M_2(x) &= I+(I-F_1'(x))+(I-F_1'(x))^2, \\ Q_2(x) &= x-M_2(x)F_1(x), \\ x_{n+1} &= Q_2(x_n). \end{aligned} \tag{1.33}$$

Method (1.5), $p=3,\ \Gamma = I,$

$$\begin{aligned} M_3(x) &= I+(I-F_1'(x))+(I-F_1'(x))^2+(I-F_1'(x))^3, \\ Q_3(x) &= x-M_3(x)F_1(x), \\ x_{n+1} &= Q_3(x_n). \end{aligned} \tag{1.34}$$

Method (1.5), $p=4,\ \Gamma = I,$

$$\begin{aligned} M_4(x) &= I+(I-F_1'(x))+(I-F_1'(x))^2+(I-F_1'(x))^3+(I-F_1'(x))^4, \\ Q_4(x) &= x-M_4(x)F_1(x), \\ x_{n+1} &= Q_4(x_n). \end{aligned} \tag{1.35}$$

Method (1.5), $p=5,\ \Gamma = I,$

$$\begin{aligned} M_5(x) &= I+(I-F_1'(x))+(I-F_1'(x))^2+(I-F'(x))^3+(I-F_1'(x))^4+(I-F_1'(x))^5, \\ Q_5(x) &= x-M_5(x)F_1(x), \\ x_{n+1} &= Q_5(x_n). \end{aligned} \tag{1.36}$$

Method (1.5), $p=1,5\ ,\ \Gamma = F_1'(x_0),$

$$\begin{aligned} x_{n+1} &= x_n - M\Gamma^{-1}F_1(x_n), \\ A &= B^{-1}(B-F_1'(x)), \\ M &= I+\sum_{i=1}^{p} A^i. \end{aligned} \tag{1.37}$$

Thus, the comparison shows that the behavior of the method (1.5) is essentially the same as Newton's method (1.2). However, the iterates of the method (1.5) are cheaper to obtain than Newton's. As observed in Table 1.1 - Table 1.4, the number of iterations required for the proposed methods with k ranging from 3 to 5 closely aligns with those of Newton's method.

Method	Iterations	Time	Method	Iterations	Time
(1.2) Newton	4	10.82×10^{-6}	(1.2) Newton	4	10.82×10^{-6}
(1.32), $k=1$	6	6.94×10^{-6}	(1.37), $k=1$	8	13.063×10^{-6}
(1.33), $k=2$	5	6.922×10^{-6}	(1.37), $k=2$	6	12.255×10^{-6}
(1.34), $k=3$	4	5.946×10^{-6}	(1.37), $k=3$	5	9.089×10^{-6}
(1.35), $k=4$	4	6.324×10^{-6}	(1.37), $k=4$	5	9.198×10^{-6}
(1.36), $k=5$	4	6.836×10^{-6}	(1.37), $k=5$	4	8.743×10^{-6}

Table 1.1. Number of iterations to achieve tolerance $\varepsilon = 10^{-9}$ with initial guess $x_0 = (1,1)$ and $\|I - F'(x_0)\| = 0.3129 < 1$.

Method	Iterations	Time	Method	Iterations	Time
(1.2) Newton	3	7.215×10^{-6}	(1.2) Newton	3	7.215×10^{-6}
(1.32), $k=1$	5	5.122×10^{-6}	(1.37), $k=1$	3	4.256×10^{-6}
(1.33), $k=2$	4	5.8×10^{-6}	(1.37), $k=2$	3	5.862×10^{-6}
(1.34), $k=3$	3	5.503×10^{-6}	(1.37), $k=3$	3	5.575×10^{-6}
(1.35), $k=4$	3	5.771×10^{-6}	(1.37), $k=4$	3	5.825×10^{-6}
(1.36), $k=5$	3	5.895×10^{-6}	(1.37), $k=5$	3	5.973×10^{-6}

Table 1.2. Number of iterations to achieve tolerance $\varepsilon = 10^{-9}$ with initial guess $x_0 = (0,0)$ and $\|I - F'(x_0)\| = 0.1414 < 1$.

Table 1.5 shows the results of calculations to determine the Computational Order of Convergence (COC) and the Approximated Computational Order of Convergence (ACOC) aiming to compare the convergence order of method (1.5) with the convergence order of Newton's method (1.2).

Definition 4. Computational order of convergence of a sequence $\{x_j\}_{j\geq 0}$ is defined by

$$\overline{\Upsilon}_j = \frac{\ln|e_{j+1}/e_j|}{\ln|e_j/e_{j-1}|},$$

where x_{j-1}, x_j, x_{j+1} are three consecutive iterations near the root α and $e_j = x_j - \alpha$ [6].

Definition 5. The approximated computational order of convergence of a sequence $\{x_j\}_{j\geq 0}$ is defined by

$$\widehat{\Upsilon}_n = \frac{\ln|\widehat{e}_{j+1}/\widehat{e}_j|}{\ln|\widehat{e}_j/\widehat{e}_{j-1}|},$$

where $\widehat{e}_j = x_j - x_{j-1}$. x_j, x_{j-1}, x_{j-2} are three consecutive iterates [6].

Table 1.5 shows that the convergence of the proposed methods closely corresponds with the convergence of Newton's method, particularly for values of k ranging from 4 to 5 with the convergence order closely approximating 2.

Example 2. Let $X = Y = \mathbb{R}^3$ and $\Omega = S[s^*, 1]$. The mapping F is defined on Ω for $a = (a_1, a_2, a_3)^{tr} \in \mathbb{R}^3$ as

$$F_1(a) = (a_1, e^{a_2} - 1, \frac{e-1}{2}a_3^2 + a_3)^{tr}.$$

Then, the definition of the derivative according to Fréchet [22,25] is given for the mapping F_1

$$F_1'(a) = \begin{bmatrix} 1 & 0 & 0 \\ 0 & e^{a_2} & 0 \\ 0 & 0 & (e-1)a_3 + 1 \end{bmatrix}.$$

Method	Iterations	Time	Method	Iterations	Time
(1.2) Newton	5	13.454×10^{-6}	(1.2) Newton	5	13.454×10^{-6}
$(1.32), k = 1$	7	7.184×10^{-6}	$(1.37), k = 1$	9	12.015×10^{-6}
$(1.33), k = 2$	5	10.195×10^{-6}	$(1.37), k = 2$	7	13.245×10^{-6}
$(1.34), k = 3$	5	7.352×10^{-6}	$(1.37), k = 3$	6	9.378×10^{-6}
$(1.35), k = 4$	5	7.883×10^{-6}	$(1.37), k = 4$	6	9.829×10^{-6}
$(1.36), k = 5$	5	8.636×10^{-6}	$(1.37), k = 5$	5	8.936×10^{-6}

Table 1.3. Number of iterations to achieve tolerance $\varepsilon = 10^{-9}$, where $x_0 = (-15, -15)$ and $\|I - F'(x_0)\| = 0.257 < 1$.

Method	Iterations	Time	Method	Iterations	Time
(1.2) Newton	7	16.719×10^{-6}	(1.2) Newton	7	16.719×10^{-6}
$(1.32), k = 1$	8	15.677×10^{-6}	$(1.37), k = 1$	12	24.742×10^{-6}
$(1.33), k = 2$	7	20.052×10^{-6}	$(1.37), k = 2$	8	27.178×10^{-6}
$(1.34), k = 3$	7	14.711×10^{-6}	$(1.37), k = 3$	8	20.43×10^{-6}
$(1.35), k = 4$	7	15.357×10^{-6}	$(1.37), k = 4$	7	19.2×10^{-6}
$(1.36), k = 5$	7	16.49×10^{-6}	$(1.37), k = 5$	7	17.538×10^{-6}

Table 1.4. Number of iterations to achieve tolerance $\varepsilon = 10^{-12}$, where $x_0 = (-15, -15)$ and $\|I - F'(x_0)\| = 0.257 < 1$.

The point $s^* = (0,0,0)^{tr}$ solves the equation $F_1(a) = 0$. Moreover, $F_1'(s^*) = I$. The conditions of the Theorem 2 hold for $p = 1$, if $\varphi(t) = a(t) = (e-1)t$. Then, we can have $r \in (0, 0.2909883534)$.

Example 3. Let $H[0,1]$ stand for the space of continuous functions mapping the interval $[0,1]$ into the real numbers. Let $X = Y = H[0,1]$ and $\Omega = S[s^*, 1]$ with $s^*(v) = 0$. The operator F_1 is defined on $H[0,1]$ as

$$F_1(z)(v) = z(v) - 4\int_0^1 vz(\tau)^3 d\tau.$$

Then, of the derivative according to Fréchet [10,15,1,22,30] is given below for the operator F_1

$$F_1'(z(w))(v) = w(v) - 12\int_0^1 v\tau z(\tau)^2 w(\tau) d\tau$$

for each $w \in H[0,1]$. Therefore, the conditions of the Theorem 2 hold if, since for $s^* = 0$, $F_1'(x^*(v)) = I$ hold for that $p = 1$, if $\varphi(t) = a(t) = 6t$. Then, again by the definition of r, we can choose $r \in (0, 0.8\bar{3})$.

5. References

[1] Adly, S., Cibulka, R., Ngai, H. V., *Newton's method for solving inclusions using set-valued approximations*, *SIAM J. Optim.*, 25 (2015) 159–184.

[2] Adly, S., Ngai, H. V., Vu, N. V., *Newton's method for solving generalized equations: Kantorovich's and Smale's approaches*, *J. Math. Anal. Appl.*, 439 (2016) 396–418.

[3] Allgower, E. L., Georg, K., *Introduction to numerical continuation algorithms*, Springer, Berlin, Heidelberg, New York, (1989).

[4] Aragn Artacho, F. J., Melyakov, A., Dontchev, A. L., Lopez, M., *Local convergence of quasi-Newton methods under metric regularity*, *Comput. Optim. Appl.*, 58 (2014) 225–247.

Method	COC	ACOC
(1.2) Newton	1.8624	1.9697
(1.32), $p=1$	0.863	1
(1.33), $p=2$	0.2695	1.0438
(1.34), $p=3$	1.9714	2.3569
(1.35), $p=4$	1.8354	1.9453
(1.36), $p=5$	1.8642	1.9661
(1.37), $p=1$	0.9065	1.0118
(1.37), $p=2$	0.5912	0.999
(1.37), $p=3$	0.7321	0.9926
(1.37), $p=4$	1.933	2.0151
(1.37), $p=5$	1.8679	1.9578

Table 1.5. The computational Order of Convergence and the Approximated Computational Order of Convergence, where $x_0 = (-15, -15)$, $\varepsilon = 10^{-12}$.

[5] Argyros, I. K., *Convergence and Applications of Newton-Type Iterations*, Springer, Berlin, (2008).

[6] Argyros, I. K., *The theory and application of iterative methods with applications*, Second edition, CRC Press, Taylor and Francis, Boca Raton, Florida, USA, (2002).

[7] Argyros, I. K., George, S., *On the unified convergence analysis for Newton-type algorithms solving generalized equations with the Aubin property, Journal of Complexity*, 81 (2024), 101817.

[8] Aubin, J.-P., Frankowska, H., *Set-Valued Analysis, Systems and Control: Foundations and Applications*, Birkhuser, Boston, (1990).

[9] Catinas, E., *The inexact, inexact perturbed, and quasi-Newton algorithms are equivalent models, Math. Comp.*, 74 (2005), 291–301.

[10] Ciarlet, P. G., Mardare, C., *On the Newton-Kantorovich theorem, Anal. Appl. (Singapore)*, 10 (2012) 249–269.

[11] Cibulka, R., Dontchev, A. L., Preininger, J., Roubal, T., Veliov, V., *Kantorovich-type Theorems for Generalized Equations, Journal of Convex Analysis*, 25 (2018), 459–486.

[12] Deuflhard, P., *Newton algorithms for nonlinear problems. Affine invariance and adaptive algorithms*, Springer Series in Computational Mathematics, 35, Springer-Verlag, Berlin, (2004).

[13] Dontchev, A. L., *Local analysis of a Newton-type method based on partial linearization*, in *The Mathematics of Numerical Analysis*, (Park City, 1995), Lectures Appl. Math. 32, Amer. Math. Soc., Providence, (1996) 295–306.

[14] Dontchev, A. L., Rockafellar, R. T., *Implicit Functions and Solution Mappings. A View from Variational Analysis*, 2nd edition, Springer, Merlin, (2014).

[15] Ezquerro, J. A., Gutirrez, J. M., Hernndez, M. A., Romero, N., Rubio, M. J., *The Newton algorithm: from Newton to Kantorovich. (Spanish), Gac. R. Soc. Mat. Esp.*, 13 (2010), 53–76.

[16] Ezquerro, J. A., Hernndez-Veron, M. A., *Domains of global convergence for Newton's algorithm from auxiliary points, Appl. Math. Lett.*, 85 (2018), 48–56.

[17] Facchinei, F., Pang, J.-S., *Finite-Dimensional Variational Inequalities and Complementarity Problems*, Springer, New York, (2003).

[18] Ferreira, O. P., Silva, G. N., *Inexact Newton's method to nonlinear functions with values in a cone, Applicable Analysis*, (2018).

[19] Izmailov, A. F., Solodov, M. V., *Newton-Type Methods for Optimization and Variational Problems*, Springer, Berlin, (2014).

[20] Kantorovich, L. V., Akilov, G., *Functional analysis in normed spaces*, Moscow: Fizmatgiz, (1959). (German translation, Akademie-Verlag, Merlin, 1964): (English translation (2nd edition)), Pergamon Press, London, 1981.

[21] Kelley, C. T., *Solving Nonlinear Equations with Newton's Method*, Fundamentals of Algorithms, SIAM, Philadelphia, (2003).

[22] Kummer, M., *Newton's method for non-differentiable functions*, in: "Advances in Mathematical Optimization", J. Guddat et al. (eds.), Ser. Math. Res. 45, Akademie-Verlag, Berlin, (1988), 114–125.

[23] Moore, R. H., Nashed, M. Z., *Approximations to generalized inverses of linear operators, SIAM J. Appl. Math.*, 27 (1974), 1–16.

[24] Nashed, M. Z., *Generalized inverses and applications*, Academic Press, New York, (1976).

[25] Padcharoen, A., Kumam, P., Chaipunya, P., Shehu, Y., *Convergence of inertial modified Krasnoselskii-Mann iteration with application to image recovery, Thai J. Mathematics*, 18(1), (2020), 126–142.

[26] Potra, F. A., Ptk, V., *Nondiscrete induction and iterative processes*, Research Notes in Mathematics 103, Pitman, Boston, (1984).

[27] Proinov, P. D., *New general convergence theory for iterative processes and its applications to Newton-Kantorovich type theorems, J. Complexity*, 25 (2010), 3–42.

[28] Qi, L., Sun, J., *A nonsmooth version of Newton's method, Math. Programming A*, 58 (1993) 353–367.

[29] Regmi, S., Argyros, I. K., George, S., Argyros, C. I., *Extended Convergence of Three Step Iterative Methods for Solving Equations in Banach Space with Applications, Symmetry*, 14 (2022), 1484. https://doi.org/10.3390/sym14071484.

[30] Rheinboldt, W. C., *A unified convergence theory for a class of iterative processes, SIAM J. Numer. Anal.*, 5 (1968), 42–63.

[31] Robinson, S. M., *Extension of Newton's method to nonlinear functions with values in a cone, Numer. Math.*, 19 (1972) 341–347.

[32] Silva, G. N., *Kantorovich's theorem on Newton's method for solving generalized equations under the majorant condition, Appl. Math. Comput.*, 286 (2016) 178–188.

[33] Traub, J. F., Wozniakowsi, H., *Convergence and complexity of Newton iteration for operator equations, J. Assoc. Comput. Mach.*, 26 (1979), 250–258.

[34] Ulbrich, M., *Semismooth Newton Methods for Variational Inequalities and Constrained Optimization Problems in Function Spaces*, SIAM, Philadelphia, (2011).

[35] Yamamoto, T., *A convergence theorem for Newton-like algorithms in Banach spaces*, *Numer. Math.*, 51 (1987), 545–557.

Chapter 2

A Non-differentiable Inexact Newton-like Method for Generalized Equations

1. Introduction

We are concerned with the problem of approximating a solution of the generalized equation

$$f(x)+g(x)+G(x) \ni 0, \tag{2.1}$$

where B_1, B_2 are Banach spaces, $f: B_1 \to B_2$ is a differentiable operator in the sense of Fréchet, $g: B_1 \to B_2$ is a continuous operator whose differentiability is not assumed and $G: B_1 \rightrightarrows B_2$ is a set valued operator. The usual case is when $g = 0$[1].

Newton-Secant-type method for generating a sequence approximating a solution $x_* \in B_1$ of the equation (2.1) is

$$x_0 \in B_1, \quad f(x_n)+g(x_n)+D_n(x_{n+1}-x_n)+G(x_{n+1}) \ni 0, \quad n=0,1,2,\ldots \tag{2.2}$$

where D_n is as defined in the previous chapter where the motivation for this paper is justified.

If $G=\{0\}$ and $D(x)=f'(x)$, the method (2.2) reduces to the iteration introduced by Zincenko [43] for solving the equation

$$f(x)+g(x)=0. \tag{2.3}$$

The operator $G: B_1 \rightrightarrows B_2$ is assumed to be closed, non-empty, convex cone in the Banach space B_2 which is denoted by C for simplicity. The semi-local convergence analysis of the method (2.2) is developed using generalized continuity conditions and majorizing sequences. The idea is taken from [1], where we used such conditions to solve the equation (2.3). We also show that even specializations of our results provide finer error analysis than existing ones [1].

2. Mathematical Background

The definition of a convex process was inaugurated by Rockafellar [16] and studied by Robinson [12], Dontchev [4] and others[1]-[42]. In order to make the study as self contained as possible, we present the definition of a convex process. More properties can be found in [15] and the references with in.

Definition 6. An operator P from a linear space $B_1 \times B_1$ into the linear space B_2 is a convex process if

$$P(x,\tilde{x}) + P(y,\tilde{y}) \subset P(x+y,\tilde{x}+\tilde{y}) \quad \text{for each } x,y,\tilde{x},\tilde{y} \in B_1;$$
$$P(a(x,\tilde{x})) = aP(x,\tilde{x}) \quad \text{for each } a > 0 \text{ and } x,\tilde{x} \in B_1$$
$$\text{and} \quad P(0,0) = 0.$$

Then, the domain denoted by *dom*, the range by *rge* and the inverse by P^{-1} are defined as: *domP* is the set of elements from B_1 with $P(x) \neq 0$, *rge* of P is $U\{P(x); x \in dom(P)\}$ and $P^{-1}(y) = \{x; y \in P(x)\}$.

The sets $dom(P)$ and $rge(P)$ are convex set and the inverse is a convex process.

If the norm of P, $\|P\|$ is finite, the convex process is normed.

The following Banach-type perturbation Lemma is needed [34].

Lemma 2. Let P, P_1 be convex processes from B_1 into B_2. Set $dom(P) = D_1$ and $rge(P) = D_2$. Suppose P, P^{-1} and P_1 are normed, and that $\|P^{-1}\| \cdot \|P_1\| < 1$; $C \subset dom(P_1)$, $P_1(P) \subset P_1$, P_1 is closed and $(P - P_1)(x)$ is closed for each $x \in D_1$. Then, $(P - P_1)^{-1}$ satisfies

$$rge(P) \subset rge(P - D_1);$$
$$(P - D_1)^{-1}_{D_2} \text{ is a normed convex process,}$$
$$\text{and} \quad \|P^{-1}(P - P_1)^{-1}_{D_2}\| \leq \frac{1}{1 - \|P_1\|}.$$

We assume familiarity with majorizing sequences and their importance to the study of the convergence of iterative methods [23,26].

3. Algorithm

Let $\Omega \subset B_1$. For each fixed $x \in \Omega$, define

$$P(x) = z = D(x)z - C, \quad z \in B_1.$$

Clearly, P is a normed convex process from B_1 into B_2, with inverse

$$P^{-1}(x)(z) = \{\bar{x} \in B_1, D(x)(z) \in z + C\},$$

which is also a convex process.

For a starting element $x_0 \in \Omega$, $P^{-1} = P^{-1}(x_0)$ such that $P^{-1}(-f(x_0) - g(x_0)) \neq \phi$, consider x_1 as the sum of x_0 and a projection of the origin in B_1 on $P^{-1}(-f(x_0) - g(x_0))$. This process is repeated with x_1 replacing x_0. At the $n-th$ step, we get x_n and define x_{n+1} as the sum of x_n and a projection of the origin in B_1 on $P^{-1}(-f(x_n) - g(x_n))$. Equivalently, the algorithm is, if x_{n-1}, x_n are computed, the iterate x_{n+1} is any solution of the minimization problem:

$$minimize\{\|x - x_n\| : f(x_n) + g(x_n) + D(x_n)(x - x_n) \in C\}. \tag{2.4}$$

Thus, we arrived at:
Algorithm: Newton-Secant-type-Cone $(f, g, C, x_0, \varepsilon)$.

Step 1. If $P^{-1}(-f(x_0)-g(x_0))=\phi$, terminate with failure.

Step 2. If $\varepsilon_1>\varepsilon$ do:

- Pick a solution of the problem

$$minimize\{\|x-x_0\| \| f(x_0)+g(x_0)+D(x_0)(x-x_0)\in C\}.$$

- $e_1=\|x-x_0\|$; $x_0=x$.

Step 3. Return x.

This algorithm reduces to the one in [1] if $A=f'$. If $G=0$, the algorithm reduces to the one considered by us in [3]. Moreover, if $C=\{0\}$, then it reduces to the Zincenko iteration [43].

Remark. The continuity of $A(x_n)$ and since C is closed and convex, imply that the feasible set of (2.4) is a closed convex set for each $n=1,2,\ldots$. Then, the existence of a feasible element $\bar{x}$ implies that any solution of (2.4) is in the intersection of the feasible set of (2.4) with the closed ball of center x_n and radius $\|\bar{x}-x_n\|$. But, B_1 is reflexive and the function $\|x-x_n\|$ is weakly lower semi-continuous, a solution exists (see [1]). Thus, if (2.4) is feasible, then it is solvable and by convexity, this implies that any local solution is global. This important remark is used in Section 4.

4. Convergence

The following conditions are needed in the semi-local convergence of the method (2.2).

Assume:

(H_1) There exists elements $x_0\in\Omega$ so that $P=P(x_0)$ maps B_1 onto B_2.

(H_2) Let $s_0\geq 0$. There exists a continuous and non-decreasing function $v_0:[0,+\infty)\times[0,+\infty)\to(-\infty,+\infty)$ so that for each $x\in\Omega$,

$$\|P^{-1}(D(x)-D(x_0))\|\leq v_0(\|x-x_0\|).$$

(H_3) The equation $v_0(t)-1=0$ has a smallest positive solution denoted by ρ, satisfying $\rho\geq s_o$.

(H_4) There exists a continuous and non-decreasing function $v:[0,\rho-s_0)\times[0,\rho)\to(-\infty,+\infty)$ so that for each $x,y,z\in U(x_0,\rho-s_0)$

$$\|P^{-1}(f(x)-f(y)+g(x)-g(y)+D(x)(y-x))\|\leq v(\|y-x\|).$$

(H_5) $\|x_1-x_0\|\leq s_1\leq\rho$, where x_1 exists by the Remark 3. and is obtained by the Algorithm.

Define the scalar sequence $\{s_n\}$ $s_0=0$, $s_1\geq s_0$, where s_{n+1} is given by

$$s_{n+1}=s_n+\frac{v(s_n-s_{n-1})}{1-v_0(s_n)},\quad n=0,1,2,\ldots \tag{2.5}$$

The sequence $\{s_n\}$ is shown to be majorizing in Theorem 5 for the method (2.2). However, first a convergence condition is needed for it.

(H_6) There exists $\rho_0 \in [s_1, \rho]$ so that for each $n = 0, 1, 2, \dots$

$$v_0(s_n) < 1 \quad \text{and} \quad s_n \leq \rho$$

and the existence of the $\lim_{n \to +\infty} s_n$ denoted by s_*. This limit is the unique least upper bound of the sequence $\{s_n\}$.
and

(H_7) $U[x_0, s_*] \subset D$.

Next, the main semi-local convergence result is developed under the conditions (H_1)-(H_7) and the preceding notation.

Theorem 5. Assume that the conditions (H_1)-(H_7) are valid. Then, the sequence $\{x_n\}$ generated by the Algorithm (2.4) is well defined in $U(x_0, s_*)$, remains in $U(x_0, s_*)$ for each $n = 0, 1, 2, \dots$ and is convergent to some $x_* \in U[x_0, s_*]$ solving the equation (2.1).

Moreover, the following error estimates are valid

$$\|x_* - x_n\| \leq s_* - s_n, \quad n = 0, 1, 2, \dots. \tag{2.6}$$

Proof. Mathematical induction is employed to show the assertion

$$\|x_{n+1} - x_n\| \leq s_{n+1} - s_n, \quad n = 0, 1, 2, \dots. \tag{2.7}$$

The definition of the iterate s_1, s_0 in (H_5) and (2.5) imply the existence of x_1 and that the assertion (2.7) is valid if $n = 0$.

Assume $x_1, \dots x_m, s_0, \dots, s_m$ exist and

$$\|x_m - x_{m-1}\| \leq s_m - s_{m-1} \text{for each } m = 0, 1, 2, \dots, n \tag{2.8}$$

It follows that

$$\begin{aligned} \|x_m - x_0\| &\leq \|x_m - x_{m-1}\| + \cdots + \|x_1 - x_0\| \\ &\leq s_m - s_{m-1} + \cdots + s_1 = s_m. \end{aligned}$$

Thus, the iterates $x_m \in U(x_0, s_*)$. By (H_1), (H_2),(H_3) and (H_7), it follows that

$$\begin{aligned} \|P^{-1}(Dx_m) - D(x_0))\| &\leq v_0(\|x_m x_0\|) \\ &\leq v_0(s_m) \leq v_0(s_*) < 1, \end{aligned} \tag{2.9}$$

by the definition of s_* in (H_6). Moreover, we have

$$\begin{aligned} P(x_m)(x) &= D(x_m)(x) - C \\ &= (D(x_0) + (D(x_m) - D(x_0)))(x) - C \\ &= (P(x_0) - Q_m), \\ Q_m &= D(x_0) - D(x_m). \end{aligned}$$

By Lemma 2 and (2.9), $P(x_m)$ maps B_1 into B_2, $P^{-1}(x_m)$ is normed and

$$\begin{aligned} \|P^{-1}(x_m)P\| &\leq \frac{1}{1 - \|P^{-1}(D(x_m) - Dx_0))\|} \\ &\leq \frac{1}{1 - v_0(s_m}. \end{aligned} \tag{2.10}$$

Moreover, (2.4) is solvable for $m = n$, since $P(x_m)$ maps B_1 into B_2, which establishes the existence of x_{m+1} solving (2.4). Then, consider the problem to find x solving

$$f(x_m) + g(x_m) + D(x_m)(x - x_m) \in f(x_{m-1}) + g(x_{m-1}) + Dx_{m-1})(x_m - x_{m-1}) + C. \quad (2.11)$$

Notice that the right hand side of (2.11) is in the cone C, so x_m solves (2.4). Thus, any x satisfying (2.11) is also feasible for (2.4). We can write (2.11) as

$$x - x_m \in P^{-1}(x_m)(-f(x_m) - g(x_m) + f(x_{m-1}) + g(x_{m-1}) + D(x_{m-1})(x_m - x_{m-1})). \quad (2.12)$$

But the right hand side of (2.12) has an element of least norm. Hence, there exists $\bar{x}$ satisfying (2.12), (2.11) and

$$\begin{aligned} \|\bar{x} - x_m\| &\leq \|P^{-1}(x_m)(-f(x_m) + f(x_{m-1}) + D(x_m)(x_m - x_{m-1}) + (g(x_m) - g(x_{m-1})))\| \\ &\leq \frac{v(\|x_m - x_0\|)}{1 - v_0(s_m)} \\ &\leq \frac{v(s_m)}{1 - v_0(s_m)} \\ &= s_{m+1} - s_m, \end{aligned} \quad (2.13)$$

where we also used (H_4),(2.7)(for $n = m$) and (2.10). Then, by (2.13), it follows

$$\begin{aligned} \|x_{m+1} - x_m\| &\leq \|\bar{x} - x_m\| \leq s_{m+1} - s_m \\ \text{and } \|x_{m+1} - x_0\| &\leq \|x_{m+1} - x_m\| + \|x_m - x_0\| \\ &\leq s_{m+1} - s_m + s_m \\ &= s_{m+1} < s_*. \end{aligned}$$

Thus, the induction for (2.7) is completed and the iterate

$$x_{m+1} \in U(x_0, s_*). \quad (2.14)$$

So, the sequence $\{x_n\}$ is complete in a Banach space B_1 and as such it is convergent to some $x_* \in U[x_0, s_*]$ (since $U[x_0, s_*]$ is a closed set). We can write

$$E_m = (f(x_{m+1}) + g(x_{m+1}) - f(x_*) - g(x_*)) - (f(x_{m+1}) -$$

$$-f(x_m) - D(x_m)(x_{m+1} - x_m)) + (g(x_{m+1}) - g(x_m)).$$

Then, we get

$$E_m \in C - f(x_*) - g(x_*).$$

By the continuity of functions f and g

$$E_m \to 0 \text{ as } m \to +\infty.$$

But, $C - f(x_*) - g(x_*)$ is closed. Thus, we conclude $f(x_*) + g(x_*) \in C$, which implies that x_* solves the equation (2.1). □

5. Specializations and Applications

Application Let us compare our results to the ones in [1]. That is we take $D(x_n) = f'(x)$. The conditions in [1] are non-affine invariant form

(T_1) $\|P^{-1}(x_0)\| \leq b, \quad \|f'(x) - f'(y)\| \leq L_1\|x-y\|.$

(T_2) $\|g(x) - g(y)\| \leq L_2\|x-y\|$
and the majorizing sequence $\{t_n\}$ is defined by

$$t_0 = 0, t_1 \geq 0, \tag{2.15}$$

$$t_{n+1} = t_n + \frac{b}{1-bL_1t_n}\left(\frac{L_1}{2}(t_n - t_{n-1}) + L_2\right)(t_n - t_{n-1}). \tag{2.16}$$

(T_3) $\|x_1 - x_0\| \leq t_1 \leq \min\left(\frac{1}{bL_1}, \frac{(1-bL_2)^2}{2bL_1}\right).$

(T_4) $U[x_0, t_*] \subset D$, where $t_* = \frac{1}{bL_1}\left(1 - bL_1 - \sqrt{(1-bL_2)^2 - 4bL_1t_1}\right).$

In our case, specialize the functions

$$v_0(s,t) = v_0(t) = l_2 t$$
$$\text{and} \quad \text{v(s,t)} = v(t) = \frac{l_1}{2}t^2 + l_2 t.$$

then, the iteration $\{s_n\}$ becomes

$$s_{n+1} = s_n + \frac{1}{l_2 s_n}\left(\frac{l_1}{2}(s_n - s_{n-1}) + l_2\right)(s_n - s_{n-1}). \tag{2.17}$$

Notice that

$$l_1 \leq bL_1 \tag{2.18}$$
$$\text{and} \quad l_2 \leq bL_2. \tag{2.19}$$

Our results are given in affine invariant form. The advantages of affine over non-affine invariant form are well known [1]. It follows by (2.15)-(2.19) and a simple inductive argument that

$$0 \leq s_n \leq t_n \tag{2.20}$$
$$0 \leq s_{n+1} - s_n \leq t_{n+1} - t_n \tag{2.21}$$
$$\text{and} \quad 0 \leq s_* \leq t_*. \tag{2.22}$$

Notice also that our sufficient semi-local convergence conditions (H_5) and (H_6) are weaker than (T_3). That is the conditions (T_1)-(T_4) imply the conditions (H_1)-(H_7) but not necessarily vice versa.

6. References

[1] Adly, S., Van Ngai, H., et al, *Newton's method for solving generalized equations: Kantorovich's and Smale's approaches*, Journal of Mathematical Analysis and Applications, 439(1), (2016), 396-418.

[2] Aragn Artacho, F. J., Belyakov, A., Dontchev, A. L., Lopez, M., *Local convergence of quasi-Newton methods under metric regularity*, Computational Optimization and Applications, 58(1), (2014), 225-247.

[3] Argyros, I. K., *A unifying local-semilocal convergence analysis and applications for two-point Newton-like methods in Banach space*, Journal of Mathematical Analysis and Applications, 298(2), (2004), 374-397.

[4] Argyros, I. K., *Convergence and applications of Newton-type iterations*, Springer Science & Business Media, (2008).

[5] Argyros, I. K., *The Theory and Applications of Iteration Methods*, 2nd edition, CRC Press/Taylor and Francis Publishing Group Inc., Boca Raton, Florida, USA, (2022).

[6] Argyros, I. K., George, S., *Mathematical Modeling for the Solutions of Equations and Systems of Equations with Applications Vol. 4*, Nova Science Publication, New York, (2020).

[7] Argyros, I. K., Magren, . A. (2017). *Iterative Methods and their dynamics with applications: A Contemporary Study*, CRC Press.

[8] Artacho, F. A., Dontchev, A. L., Gaydu, M., Geoffroy, M. H., Veliov, V. M., *Metric regularity of Newton's iteration*, SIAM Journal on Control and Optimization, 49(2), (2011), 339-362.

[9] Aubin, J.-P., *Lipschitz behavior of solutions to convex minimization problems*, Mathematics of Operations Research, 9(1), (1984), 87-111.

[10] Aubin, J.-P., Frankowska, H., *Set-valued analysis*, Springer Science & Business Media, (2009).

[11] Balzs, M., Goldner, G., *On existence of divided differences in linear spaces*, Revue d'analyse numrique et de la thorie de l'approximation, 2, (1973), 5-9.

[12] Cibulka, R., Dontchev, A., Geoffroy, M. H., *Inexact Newton Methods and Dennis-Mor Theorems for Nonsmooth Generalized Equations*, SIAM Journal on Control and Optimization, 53(2), (2015), 1003-1019.

[13] Cibulka, R., Dontchev, A. L., Preininger, J., Roubal, T., Veliov, V., *Kantorovich-type theorems for generalized equations*, J. Convex Anal, 25(2), (2018), 459-486.

[14] Dontchev, A., *Uniform convergence of the Newton method for Aubin continuous maps*, Serdica Mathematical Journal, 22(3), (1996), 385p-398p.

[15] Dontchev, A., Hager, W., *An inverse mapping theorem for set-valued maps*, Proceedings of the American Mathematical Society, 121(2), (1994), 481-489.

[16] Dontchev, A., Hager, W., *Lipschitzian stability for state constrained nonlinear optimal control*, SIAM Journal on Control and Optimization, 36(2), (1998), 698-718.

[17] Dontchev, A. L., Rockafellar, R. T., *Implicit functions and solution mappings: A view from variational analysis*, volume 616, Springer, (2009).

[18] Dontchev, A. L., Rockafellar, R. T., *Newton's method for generalized equations: a sequential implicit function theorem*, Mathematical Programming, 123(1), (2010), 139-159.

[19] Dontchev, A. L., Quincampoix, M., Zlateva, N., *Aubin criterion for metric regularity*, Journal of Convex Analysis, 13(2), (2006), 281.

[20] Levitin, E., *Constrained minimization method*, USSR Computational Mathematics and Mathematical Physics, 6, (1966), 1-50.

[21] Magren, . A., Argyros, I. (2018). *A contemporary study of iterative methods: convergence, dynamics and applications*, Academic Press.

[22] Mordukhovich, B., *Variational Analysis and Generalized Differentiation I, Basic Theory*, (2005).

[23] Nesterov, Y., Nemirovskii, A., *Interior-point polynomial algorithms in convex programming*, SIAM, (1994).

[24] Ortega, J., Rheinboldt, W., *Iterative Solution of Nonlinear Equations in Several Variables*, volume 30. Academic Press, New York, (1970).

[25] Pietrus, A., *Non-Differentiable Perturbed Newton's Method For Functions With Values In A Cone*, Investigacin Operacional, 35(1), (2014).

[26] Pietrus, A., Jean-Alexis, C., *Newton-secant method for functions with values in a cone*, Serdica Mathematical Journal, 39(3-4), (2013), 271-286.

[27] Proinov, P., *Semilocal convergence of two iterative methods for simultaneous computation of polynomial zeros*, C.R. Acad. Bulgare Sci., 59(7), (2006), 705-712.

[28] Robinson, S. M., *Extension of Newton's method to nonlinear functions with values in a cone*, Numerische Mathematik, 19(4), (1972), 341-347.

[29] Robinson, S. M., *Generalized equations and their solutions, Part I: Basic theory*, Springer, (1979).

[30] Robinson, S. M., *Generalized equations and their solutions, Part II: Applications to nonlinear programming*, Springer, (1982).

[31] Robinson, S. M., *Normed convex processes*, Transactions of the American Mathematical Society, 174, (1972), 127-140.

[32] Rockafellar, R. T., *Convex analysis*, volume 11, Princeton University Press, (1997).

[33] Rockafellar, R. T., *Monotone processes of convex and concave type*, Number 77, American Mathematical Soc, (1967).

[34] Rockafellar, R. T., *Lipschitzian properties of multifunctions*, Nonlinear Analysis: Theory, Methods & Applications, 9(8), (1985), 867-885.

[35] Rockafellar, R. T., Wets, R. J.-B., *Variational analysis*, volume 317, Springer Science & Business Media, (2009).

[36] Shakhno, S., *On an iterative algorithm with superquadratic convergence for solving nonlinear operator equations*, Journal of Computational and Applied Mathematics, 231(1), (2009), 222-235.

[37] Shakhno, S., *Convergence of the two-step combined method and uniqueness of the solution of nonlinear operator equations*, Journal of Computational and Applied Mathematics, 261, (2014), 378-386.

[38] Shakhno, S., Gnatyshyn, O. P., *On an iterative algorithm of order 1.839... for solving the nonlinear least squares problems*, Applied Mathematics and Computation, 161(1), (2005), 253-264.

[39] Sharma, J. R., Guha, R. K., Sharma, R., *An efficient fourth order weighted-Newton method for systems of nonlinear equations*, Numerical Algorithms, 62, (2013), 307-323.

[40] Zincenko, A., *Some approximate methods of solving equations with nondifferentiable operators (Ukrainian)*, Dopovidi Akad. Nauk. Ukrain. RSR.9, (1963).

Chapter 3

Generalized Equations in Hilbert Space

Let k be a natural number, H denote a Hilbert space and $M \in \mathbb{L}(H,H)$ be an invertible and positive operator. Define the operators $A = M^{-1}(M - F'(x)), B = 1 + A + ... + A^k$ and $F_1(x) = MB^{-1}$. Given $x_0, x_1, ..., x_n$, define x_{n+1} such that

$$0 \in F(x_n) + F_1(x_n)(x_{n+1} - x_n) + G(x_{n+1}), \tag{3.1}$$

where $F : \Omega \subset H \longrightarrow H$ is a Fréchet-differentiable operator and $G : H \rightrightarrows H$ is a set-valued maximal monotone operator [1,5,7,8,9,10]. The motivation for introducing the hybrid method (3.1) for solving the generalized equation

$$0 \in F(x) + G(x) \tag{3.2}$$

has been given in previous chapter. So for the method used for solving (3.2) is

$$0 \in F(x_n) + F'(x_n)(x_{n+1} - x_n) + G(x_{n+1}). \tag{3.3}$$

We refer the reader to [1,4,5,7,8,9,10] for such studies. The advantages of using (3.2) over (3.3) are also developed in previous chapters in the setting of a Banach space. In this chapter we consider the monotonicity of the operator G to develop convergence and uniqueness of the solution results for the method (3.1) under general conditions on the operator F' used to control it (see also [2,3,6,8]).

1. Mathematical Background

In order to make the chapter as self contained as possible, we introduce some concepts. More properties can be found in [4,10].
The following result gives properties for a positive operator $Q : H \longrightarrow H$ [1,5,7,8,9,10]. Recall that Q is a positive operator if Q is self-conjugate and $< Q(x), x > \geq 0$ for all $x \in H$.

Lemma 3. Let Q stand for a positive operator. Then, the following assertions hold:

(1) $\|Q^2\| = \|Q\|^2$.

(2) If $Q^{-1} \in \mathcal{L}(H,H)$, then Q^{-1} is also positive operator, and

$$< Q(x), x > \geq \frac{\|x\|^2}{\|Q^{-1}\|} \quad \text{for all } x \in H$$

Definition 7. Let $G : H \rightrightarrows H$ be a set valued operator. G is said:

(1) Monotone if for all $x, y \in Dom(G)$
$< u - v, y - x > \geq 0$ for all $u \in G(y)$ and $v \in G(x)$.

(2) Maximal monotone if it is monotone and $< u - v, x - y > \geq 0$ for all $y \in Dom(G)$ and $v \in G(y) \implies x \in Dom(G)$ and $u \in G(x)$.

Let Q^* denote the conjugate operator for Q. Define the operator $\hat{Q} = \frac{1}{2}(Q + Q^*)$. Then, clearly $\hat{Q}$ is a self-conjugate operator. In the rest of the chapter $G : H \rightrightarrows H$ is a set-valued maximal monotone operator and $F : H \longrightarrow H$ a Fréchet-differentiable operator. The solution of (3.2) is defined by x^*.

Remark. Suppose that there exist $d > 0$ such that $< F_1(x_n)y, y > \geq d\|y\|^2$ for all $y \in H$. Then, there exists a unique element x_{n+1} such that (3.2) holds, since G is a maximal monotone operator (see [22, Lemma 2.2]). Consequently, the sequence $\{x_n\}$ given by (3.1) is well defined for all $n = 0, 1, 2, \ldots$.

2. Convergence Analysis

Let $T = [0, +\infty)$. The analysis uses the conditions:
Suppose

(C_1) There exists a continuous and nondecreasing function $h_0 : T \longrightarrow T$ such that the equation $h_0(t) - 1 = 0$ has a smallest positive solution. Denote such solution by ρ and set $T_0 = [0, \rho)$.

(C_2) There exists a continuous and nondecreasing function $h : T_0 \longrightarrow T$. Define the sequence $\{s_n\}$ for $s_0 = 0$, some $s_1 \geq 0$ and each $n = 1, 2, 3, \ldots$ by

$$\begin{aligned} a &= a_n = h_0(s_n),\ b = b_n = a\frac{1 - a^k}{1 - a},\ \bar{b} = \bar{b}_n = \frac{1}{1 - b}, \\ s_{n+1} &= s_n + \bar{b}\int_0^1 h(\theta(s_n - s_{n-1}))d\theta(s_n - s_{n-1}). \end{aligned} \tag{3.4}$$

The sequence $\{s_n\}$ is shown to be majorizing for the method (3.1) in Theorem 3.1. But let us first present a convergence criterion for it.

(C_3) There exists a parameter $\rho_0 \in [0, \rho)$ such that for each $n = 0, 1, 2, \ldots$

$2h_0(s_n) < 1$ and $s_n \leq \rho_0$. It follows by this condition and (3.4) that $0 \leq s_n \leq s_{n+1} \leq \rho_0$ and there exists $x^* \in [0, \rho)$ such that $\lim_{n \to +\infty} s_n = s^*$. The limit point s^* is the unique least upper bound of the sequence $\{s_n\}$. The functions h_0 and h relate to the operators on the method (3.1).

(C_4) There exists $x_0 \in H$ and a positive and invertible operator $M \in \mathcal{L}(H, H)$ such that $\hat{M}$ is also a positive and invertible operator such that for all $x \in H$

$$\|\hat{M}^{-1}\|\|F'(x) - M\| \leq h_0(\|x - x_0\|).$$

If $x - x_0$, we have $\hat{M}^{-1}\|\|F'(x_0) - M\| \leq h_0(0) < 1$. So, $F'(x_0)^{-1} \in \mathcal{L}(B, B)$, and we can choose $s_1 \geq \|x_1 - x_0\|$. Set $\Omega_1 = U(x_0, \rho_0)$.

(C_5)

$$\|\hat{M}^{-1}\|\|F'(x+\theta(y-x))-F'(x)\| \leq h(\theta\|y-x\|) \text{ for each } x,y\in\Omega_1$$

and $\theta \in [0,1]$.

Next,
The semi-local analysis of the method is developed.
Theorem 6. Suppose that the conditions $(C_1)-(C_5)$ hold. Then, the sequence $\{x_n\}$ generated by the method (3.1) is well defined in $U(x_0,s^*)$, remains in $U(x_0,s^*)$ for each $n=0,1,2,...$ and converges to a solution $x^* \in U[x_0,s^*]$ of the generalized equation (3.2) such that

$$\|x^*-x_n\| \leq s^*-s_n. \tag{3.5}$$

Proof. The following assertion shall be established

$$\|x_{m+1}-x_m\| \leq s_{m+!}-s_m \text{ for each } m=0,1,2,... \tag{3.6}$$

If $m=0$, (3.6) holds by the choice of s_1. Suppose (3.6) holds for all integer values of m smaller or equal to $n-1$. It follows by the induction hypothesis in turn that

$$\begin{aligned}\|x_m-x_0\| &\leq \|x_m-x_{m-1}\|+...+\|x_1-x_0\| \\ &\leq s_m-s_{m-1}+s_{m-1}-s_{m-2}+...+s_1-s_0=s_m<s^*.\end{aligned}$$

By hypothesis M is a positive operator and $\hat{M}^{-1}\in \mathcal{L}(H,H)$, so by the computations of the previous chapter

$$\|F_1(x_m)^{-1}\| \leq \|\hat{M}^{-1}\|\bar{b}. \tag{3.7}$$

In view of the Lemma 3 (2), we get in turn that

$$\begin{aligned}\frac{\|x\|^2}{\|\hat{F}_n^{-1}(x_m)\|} &\leq <\hat{F}_1(x_m)x,x> \\ &= <F_1(x_m)x,x> \text{ for all } x\in H.\end{aligned} \tag{3.8}$$

Thus, by Remark 1. there exists a unique x_{m+1} such that

$$0\in F(x_m)+F_1(x_m)(x_{m+1}-x_m)+G(x_{m+1}). \tag{3.9}$$

By the induction hypothesis

$$0\in F(x_{m-1})+F_1(x_{m-1})(x_m-x_{m-1})+G(x_m) \tag{3.10}$$

By the maximality of the operator G, (3.9) and (3.10) we have

$$\begin{aligned}&\langle F(x_{m-1}-F_1(x_{m-1})(x_m-x_{m-1})+F(x_m)+F_1(x_m)(x_{m+1}-x_m), \\ &x_m-x_{m+1}\rangle \geq 0,\end{aligned}$$

so

$$\begin{aligned}<F(x_m)-F(x_{m-1}) &- F_1(x_{m-1})(x_m-x_{m-1}),x_m-x_{m+1}> \\ &\geq <F_1(x_m)(x_m-x_{m+1}),x_m-x_{m-1}>.\end{aligned} \tag{3.11}$$

But by (3.8), notice that

$$\begin{aligned} \frac{\|x_m - x_{m+1}\|^2}{\|\hat{F}_1^{-1}(x_m)\|} &\leq\ < \hat{F}_1(x_m)(x_m - x_{m+1}, x_m - x_{m+1} > \\ &\geq\ < F_1(x_m)(x_m - x_{m+1}), x_m - x_{m+1} > \end{aligned}$$

leading by (3.4),(3.7), (3.11) and (c_5)

$$\begin{aligned} \|x_m - x_{m+1}\| &\leq \|\hat{F}_1(x_m)^{-1}\| \|F(x_m) - F(x_{m-1}) - F_1(x_{m-1})(x_m - x_{m-1})\| \\ &\leq \bar{b}\|\hat{M}^{-1}\| \|F(x_m) - F(x_{m-1}) - F_1(x_{m-1})(x_m - x_{m-1})\| \\ &\leq \bar{b}\int_0^1 h(\theta\|x_m - x_{m-1}\|)d\theta\|x_m - x_{m-1}\| \\ &\leq \bar{b}\int_0^1 h(\theta(s_m - s_{m-1}))d\theta(s_m - s_{m-1}) = s_{m+1} - s_m \end{aligned}$$

and

$$\begin{aligned} \|x_{m+1} - x_0\| &\leq \|x_{m+1} - x_m\| + \|x_m - x_0\| \\ &\leq s_{m+1} - s_m + s_m - s_0 = s_{m+1} < s^*. \end{aligned}$$

The induction for the assertion (3.6) is terminated and $\{x_m\} \subset U(x_0, s^*)$. Therefore, the sequence $\{x_m\}$ is complete in the Hilbert space H and as such it converges to some $x^* \in U[x_0, s^*]$. BY hypothesis G is a maximal monotone operator and F is continuous. Therefore, we conclude that x^* solves (3.2). □

Let us define the sequence $\{t_n\}$ for $t_0 = 0, t_1 \geq \|x_1 - x_0\|$ and each $n = 1, 2, ...$ by

$$t_{n+1} = t_n + \int_0^1 h(\theta(t_n - t_{n-1}))d\theta(t_n - t_{n-1}) \tag{3.12}$$

It follows by this definition that

$$\begin{aligned} &0 \leq t_n \leq t_{n+1} \leq \rho_0 \\ &\text{and} \lim_{n\to+\infty} t_n = s^*. \end{aligned}$$

Next, we show a uniqueness of the solution result for (3.2).

PROPOSITION 1. *Suppose that the conditions $(C_1) - (C_5)$ with F_1 replaced by m hold and there exists $x_1 \in H$ such that*

$$0 \in F(x_0) + M(x_1 - x_0) + G(x_1). \tag{3.13}$$

Then, there exists a unique solution $x^* \in [x_0, s^*]$ of (3.2).

Proof. Let us consider the method

$$0 \in F(x_m) + M(x_{m+1} - x_m) + G(x_{m+1}).$$

The sequence $\{x_m\}$ is well defined by Remark 1. and Lemma 3(2), since M is a positive operator and $\hat{M}^{-1} \in \mathcal{L}(H, H)$. Mathematical induction shall establish the assertion

$$\|x_{m+1} - x_{n+1}\| \leq t_{m+1} - t_m \text{ for all } m = 0, 1, 2, ... \tag{3.14}$$

If $m = 0$, (3.14) holds by the choice of t_1. Suppose that it holds for all integer values smaller or equal to $m-1$. We can write

$$0 \in F(x_{m-1}) + M(x_m - x_{m-1}) + G(x_m)$$

and

$$0 \in F(x_m) + M(x_{m+1} - x_m) + G(x_{m+1}).$$

By the monotone maximality of G, we get

$$< F(x_m) + M(x_{m+1} - x_m) - F(x_{m-1}) - M(x_m - x_{m-1}), x_m - x_{m+1} \geq 0,$$

thus

$$\begin{aligned} < F(x_m) - F(x_{m-1}) - M(x_m - x_{m-1}) \quad &, \quad x_m - x_{m+1} > \\ &\geq \; < M(x_m - x_{m+1}), x_m - x_{m+1} > . \end{aligned}$$

Moreover, by Lemma 2.1 (2), it follows that

$$\begin{aligned} \frac{\|x_m - x_{m+1}\|^2}{\hat{M}^{-1}} &\leq \; < \hat{M}(x_m - x_{m+1}), x_m - x_{m+1} > \\ &= \; < M^{-1}(x_m - x_{m+1}), x_m - x_{m+1} >, \end{aligned}$$

leading to

$$\begin{aligned} \|x_m - x_{m+1}\| &\leq \|\hat{M}^{-1}\| \|F(x_m) - F(x_{m-1}) M(x_m - x_{m-1})\| \\ &\leq \int_0^1 h(\theta \|x_m - x_{m-1}\|) d\theta \|x_m - x_{m-1}\| \\ &\leq \int_0^1 h(\theta(t_m - t_{m-1})) d\theta (t_m - t_{m-1}) = t_{m+1} - t_m. \end{aligned}$$

The induction for the assertion (3.14) is completed. Let $y_0 \in U[x_0, s^*]$ and set $u_0 = \|y_0 - x_0\|$. Define the sequence

$$u_{n+1} = u_n + h(u_n), \;\; n = 0, 1, 2, \ldots$$

Clearly, the sequence $\{u_n\}$ converges to s^*. Let us define the method

$$0 \in F(y_m) + M(y_{m+1} - y_m) + G(y_{m+1}).$$

The sequence $\{y_m\}$ is well defined by the same arguments as $\{x_m\}$. Next, mathematical induction shall establish the assertion

$$\|y_m - x_m\| \leq u_m - t_m. \tag{3.15}$$

If $m = 0$, the assertion (3.15) holds by the initial condition. Suppose that it holds for all values smaller or equal to $m-1$. We can write

$$0 \in F(y_m) + M(y_{m+1} - y_m) + G(y_{m+1})$$

and

$$0 \in F(x_m) + M(x_{m+1} - x_m) + G(x_{m+1}).$$

By the maximal monotonicity of G

$$< F(y_m) + M(y_{m+1} - y_m) - F(x_m) - M(x_{m+1} - x_m), x_{m+1} - y_{m+1} \geq 0,$$

so

$$\begin{aligned} &< F(y_m) - F(x_m) - M(y_m - x_m), x_{m+1} - y_{m+1} > \\ &\geq < M(x_{m+1} - y_{m+1}), x_{m+1} - y_{m+1} > \end{aligned}$$

Then, from Lemma 2.1 (2)

$$\begin{aligned} \frac{\|y_{m+1} - x_{m+1}\|^2}{\|\hat{M}^{-1}\|} &\leq < \hat{M}(x_{m+1} - y_{m+1}), x_{m+1} - y_{m+1} > \\ &= < M(x_{m+1} - y_{m+1}), x_{m+1} - y_{m+1} >, \end{aligned}$$

thus

$$\begin{aligned} \|x_{m+1} - y_{m+1}\| &\leq \|\hat{M}^{-1}\| \|F(y_m) - F(x_m) - M\|y_m - x_m)\| \\ &\leq \int_0^1 h(\theta\|y_m - x_m\|)d\theta\| \|y_m - x_m\| \\ &\leq h(\theta(u_m - t_m))d\theta(u_m - t_m) = u_{m+1} - t_{m+1}. \end{aligned}$$

The induction for the assertion (3.15) is completed. It follows by (3.15) that

$$\lim_{m\to+\infty} y_m = \lim_{m\to+\infty} x_m = x^*.$$

Therefore, we conclude that x^* is the only solution of the equation (3.2) in $U[x_0, s^*]$. Possible choice for the operator M are as given in previous chapters. □

3. References

[1] Browder, F. E., *Multi-valued monotone nonlinear mappings and duality mappings in Banach spaces*, Trans. Am. Math. Soc., 118, (1965), 338-351.

[2] Ezquerro, J. A., Hernández, M. A., *On an application of Newton's method to nonlinear operators with w-conditioned second derivative*, BIT, 42, (2002), 519-530.

[3] Gutiérrez, J. M., Hernández, M. A., *Newton's method under weak Kantorovich conditions*, IMA J. Numer. Anal., 20, (2000), 521-532.

[4] Harker, P. T., Pang, J. S., *Finite-dimensional variational inequality and nonlinear complementarity problems: A survey of theory, algorithms and applications*, Math. Programming, 48, (1990), 161-220.

[5] Josephy, N. H., *Newton's method for generalized equations*, Technical Report, No. 1965, Mathematics Research Center, University of Wisconsin, (1979), Madison, WI.

[6] Kantorovich, L. V., Akilov, G. P., *Functional Analysis*, Oxford, Pergamon, (1982).

[7] Robinson, S. M., *Generalized equations*, in A. Bachem, M. Gretschel and B. Korle, eds., Math Program., the State of the Art, Springer, Berlin, (1982), 346-367.

[8] Uko, L. U., *Generalized equations and the generalized Newton method*, Math. Program., 73, (1996), 251-268.

[9] Wang, J. H., *Convergence ball of Newton's method for inclusion problems and uniqueness of the solution*, J. Nonlinear Convex Anal., (2015).

[10] Zeidler, E., *Nonlinear Functional Analysis and Applications II B, Nonlinear Monotone Operators*, Springer, Berlin, (1990).

Chapter 4

Extended Kantorovich-type Theory for Solving Generalized Equations Using Newton's Method

1. Introduction

We are concerned with the problem of approximating a solution $x^* \in B_1$ of the generalized equation

$$0 \in F(x) + G(x), \tag{4.1}$$

where $F : \Omega \subset B_1 \longrightarrow B_2$ is a Fréchet-differentiable operator and $G : B_1 \rightrightarrows B_2$ is a set-valued operator. The solution is approximated by Newton's method [1] defined as: Given $x_0, x_1, ..., x_n$ find x_{n+1} by

$$0 \in F(x_n) + F'(x_{n+1} - x_n) + G(x_{n+1}). \tag{4.2}$$

There is extensive literature on local as well as semi-local convergence results for the method (4.2) [1,2,3,4,5,6,7,8]. We are motivated by the work in [1], where Kantorovich-type assumption were utilized to establish the convergence of (4.2) under the assumption that F'' exists and Lipchitz-type conditions used to control it. Here, we assume only the existence of F' and use tighter Lipchitz parameters to obtain the following advantages:

(i) Weaker sufficient convergence conditions (semi-local case);

(ii) tighter error bounds on the distances $\|x^* - x_n\|$ (local and semi-local case) and

(iii) A larger radius of convergence.

It is worth noticing that these advantages are obtained under the same conditions, since the new Lipchitz parameters are special cases of the ones used in [1]. Let m be a natural number. Define the operator $B = I + A + A^2 + ... + A^m$, where for an invertible linear operator M, $A = M^{-1}(M - F'(x))$. Define the hybrid Newton method for solving the generalized equation (4.1) as follows: Given $x_0, x_1, ..., x_n$ find x_{n+1} using the method

$$F(x_n) + F_1(x_n)(x_{n+1} - x_n) + G(x_{n+1}), \tag{4.3}$$

where $F_1(x_n) = MB^{-1}(x_n)$. Notice that if $\|A\| < 1$, then $\lim_{n \to +\infty} F_1(x_n) = F'(x_n)$. The motivation for constructing (4.3) has been given in our previous works [2,3]. The results can be extended to (4.3) along the same lines.

2. Preliminaries

We introduce some usefull concepts [1]. If $x \in B_1$ and $S \subset B_1$, by $d(x,S)$ we denote the distance function from x to S, defined by

$$d(x,S) := \inf_{y \in S} \|x-y\|.$$

For subsets S_0, S of B_1, we define the excess of S_0 beyond S by

$$e(S_0,S) := \inf\{t > 0 : S_0 \subset S + tD(x,1)\},$$

where $D(x,1)$ is the unit ball centered at $x \in B_1$. The Hausdorff distance between S_0 and S is

$$d(S_0,S) := \max\{e(S_0,S), e(S,S_0)\}.$$

It is important for the convergence analysis to consider the metric regularity property. An operator $H : B_1 \rightrightarrows B_2$ is metrically regular at $(\tilde{x},\tilde{y}) \in gph(H)$ with modulus $\lambda > 0$ if there exists neighborhoods U of $\tilde{x}$ in B_1 and V of $\tilde{y}$ in B_2 such that

$$d(x, H^{-1}(y)) \leq \lambda d(y, H(y)), \text{ for all } (x,y) \in U \times V.$$

The infimum of all such parameters λ is called the regularity modulus of H at $\tilde{x}$ and $\tilde{y}$. It is usually denoted by $reg(H;(\tilde{x},\tilde{y}))$.
This property is also called Aubin and it is equivalent to the Lipchitz-type property of the operator H^{-1} at $(\tilde{x},\tilde{y})$[]. An operator $Q : B_2 \rightrightarrows B_1$ is Lipchitz-type at $(\tilde{y},\tilde{x}) \in gph(Q)$ with constant λ if there exists a neighbourhood $V \times U$ of $(\tilde{y},\tilde{x})$ in $B_2 \times B_1$ so that

$$e(Q(y) \cap U, Q(y_1)) \leq \lambda \|y - y_1\|, \text{ for all } y_1, y \in V.$$

An operator $Q : B_1 \rightrightarrows B_2$ is called center-Lipchitz continuous on a set $\Omega \subset B_1$ at $x_0 \in \Omega$ iff there exists a parameter $L_0 > 0$ so that

$$d(Q(x), Q(x_0)) \leq L_0 \|x - x_0\|, \text{for all } x \in \Omega.$$

Define the set $\Omega_0 = U \cap D(x, \frac{1}{L_0})$. An operator $Q : B_1 \rightrightarrows B_2$ is called restricted Lipchitz continuous on Ω_0 iff there exists a parameter $L > 0$ so that

$$d(Q(x), Q(\tilde{x})) \leq L\|x - \tilde{x}\|, \text{ for all } x, \tilde{x} \in \Omega_0.$$

An operator $Q : B_1 \rightrightarrows B_2$ is called Lipchitz continuous on Ω iff there exists a parameter $L_1 > 0$ so that

$$d(Q(x), Q(\tilde{x})) \leq L_1 \|x - \tilde{x}\|, \text{ for all } x, \tilde{x} \in \Omega.$$

It follows by these definitions that

$$L_0 \leq L_1 \tag{4.4}$$

and

$$L \leq L_1. \tag{4.5}$$

Theorem 7. Let $P : B_1 \rightrightarrows B_2$ be a set-valued operator with closed graph and let $(\tilde{x}, \tilde{y}) \in gph(P)$. Suppose that operator P is metrically regular at $(\tilde{x}, \tilde{y})$ with modulus $\lambda > 0$ on a neighbourhood $D(\tilde{x}, a) \times D(\tilde{y}, b)$ of $(\tilde{x}, \tilde{y})$ for some $a > 0$ and $b > 0$. Let $\delta > 0, L_0 \in (0, \lambda^{-1})$, and take $\theta = \dfrac{\lambda}{1 - \lambda L_0}$. Let α, β be positive parameters so that

$$2\alpha + \beta\theta < \min\{a, \frac{\delta}{2}\},\ \beta(\theta + \lambda) < \delta,\ 2c\alpha + \beta(1 + c\theta) < b,$$

with $c := \max\{1, \frac{1}{\lambda}\}$. If $\psi : B_1 \longrightarrow B_2$ is Lipchitz continuous on $D(\tilde{x}, \delta)$ with the parameter L, and the sum $P + \psi$ has closed graph, then the operator $P + \psi$ is metrically regular on $D(\tilde{x}, \alpha) \times D(\tilde{y} + \psi(\tilde{x}), \beta)$ with modulus θ.

Proof. Simply replace L_1 by L used in theorem 3.2 in [1]. □

Remark. In view of (4.5) our theorem 2.1 improves theorem 2.1 in [1]

Definition 8. If $P : B_1 \rightrightarrows B_2$ is a set-valued operator, $x_0 \in B_1, r > 0$ and $s > 0$, define

$$V(P, x_0, r, s) = \{(x, y) \in B_1 \times B_2 : x \in D[x_0, r],\ d(y, P(x)) < s\} \tag{4.6}$$

The operator P is matrically regular on the set $V(P, x_0, r, s)$ with modulus $\theta > 0$ if

$$d(x, P^{-1}(y)) \leq \theta d(y, P(x)),\ \text{for all } (x, y) \in V(P, x_0, r, s). \tag{4.7}$$

When the operator P satisfies the Definition 2.3, we use $reg(P, x_0, r, s)$ to denote the infimum of all $\theta > 0$ for which (4.7) holds. If not, we take $reg(P, x_0, r, s) = +\infty$. In the next result we present the global version version of the Theorem 2.1.

Theorem 8. Let P, x_0, r, s, λ be as above so that the operator P is metrically regular on $V(P, x_0, r, s)$ with modulus λ. Take $\theta = \dfrac{\lambda}{1 - KL_0}$ for $L_0 \in (0, \frac{1}{\lambda})$. If $\psi : B_1 \longrightarrow B_2$ is Lipchitz continuous on $D[x_0, r]$ with constant L and $P + \psi$ has closed graph, then the set-valued operator $P + \psi$ is metrically regular on $V(P + \psi, x_0, \frac{r}{4}, R)$ with modulus θ, provided that $L_0 \leq L$, where $R = \min\{s, \frac{r}{s\theta}\}$.

Proof. The proof can be found in Theorem 6.2 in [1]. □

Remark. If $L = L_1$ the results of Theorem 2.4 reduce to the ones in Theorem 6.2 in [1]. But if $L < L_1$, our results extend the previous results.

3. Convergence for the method (4.2)

Let R, ρ and r denote positive numbers and θ be as in the Theorem 7. We need to define some notions of Lipchitz-type continuity.

Definition 9. We say that F' is center Lipchitz continuous on U if there exists a parameter $K_0 = K_0(\theta, x^*, r)$, where $x^* \in U$ is a solution of the generalized equation (4.1) such that for all $x \in D(x^*, r)$

$$\|F'(x) - F'(x^*)\| \leq \frac{K_0}{\theta}\|x - x^*\|. \tag{4.8}$$

Set $\Omega_1 = \Omega_0 \cap D(x^*, \frac{\theta}{K_0})$.

Definition 10. We say that F' is restricted Lipchitz continuous on Ω_1 if there exists a parameter $K = K(\theta, x^*, r)$ such that for all $x, y \in \Omega_1$

$$\|F'(y) - F'(x)\| \leq \frac{K}{\theta}\|y - x\|. \tag{4.9}$$

Definition 11. We say that F' is Lipchitz continuous on U if there exists a parameter

$$K_1 = K_1(\theta_1, x^*, r),\ \theta_1 = \frac{\lambda}{1 - \lambda L},\ \lambda \in (0, \frac{1}{L})$$

such that

$$\|F'(y) - F'(x)\| \leq \frac{K_1}{\theta_1}\|y - x\|. \tag{4.10}$$

Remark. Notice that

$$\theta \leq \theta_1, \tag{4.11}$$

$$K_0 \leq K_1 \tag{4.12}$$

and

$$K \leq K_1. \tag{4.13}$$

In [1] the parameters θ_1 and K_1 are used together with the assumption that F'' exists. Here, we only assume that F' exists and use K, θ for K_1, θ_1, respectively. We also assume

$$K_0 \leq K. \tag{4.14}$$

Otherwise i.e if $K \leq K_0$ the results hold with K_0 replacing K.

Next, the following local convergence result is shown for the method (4.2).

Theorem 9. Let $x^* \in U$ be a solution of the generalized equation (4.1) and take $\Omega = F'(x^*)(x^*) - F(x^*) \in B_2$. Suppose that operator P defined by $P(.) = F'(x^*)(.) + H(.)$ is metrically regular on the neighbourhood $V = D[x^*, r] \times D[\Omega, \rho]$ with modulus θ such that $D[x^*, r] \subset U$. Suppose that (4.8), (4.9) and

$$2Kr < 1 \tag{4.15}$$

hold take $\varepsilon = \min\{r, \rho, \theta\rho\}$.
Then, there exists a sequence $\{x_n\}$ generated by (4.2) which is well defined in $D(x^*, \varepsilon)$, remains in $D(x^*, \varepsilon)$ for all $n = 0, 1, 2, ...$ and converges quadratically to x^* so that

$$\|x^* - x_{n+1}\| < \frac{1}{2^r}\|x^* - x_n\|,\ n = 0, 1, 2, ... \tag{4.16}$$

Proof. Simply replace θ_1, K_1 by θ, K, respectively in the proof of Theorem 3.1 in [1]. □

A semi-local convergence for Newton's method is presented next.

Theorem 10. Let F, H, θ and ε be as in Theorem 2.8. Then, for $D[y,\varepsilon] \subset U$, define

$$\begin{aligned} \gamma &= \gamma(\theta,y) := \theta d(0, F(y)+H(y)), \\ K &= K(\theta,y,\varepsilon) := \theta sup\|F'(z)-F'(x)\|, \text{ for all } z \in D(y,\varepsilon). \end{aligned}$$

Moreover, let $x \in U, \beta \in (0,1]$ and $r > 0, s > 0$ be such that the following conditions are satisfied:

(i) $P = F'(x)+H$ is metrically regular on $V = V(P,x,4r,s)$ with modulus $\theta > reg(P,x,4r,s)$.

(ii) $d(0,H_1(x)) < s$, where $H_1 = F+G$,

(iii) $2\gamma(\theta,x)K \leq \beta$,

(iv) $2\Omega\gamma(\theta,x) \leq r$, with $\Omega = \dfrac{1}{1+\sqrt{1-\beta}}$.

Then, there exists a solution $x^* \in U$ of the generalized equation (4.1) such that

$$\|x^*-x\| \leq 2\Omega\gamma(\theta,x) \leq r. \tag{4.17}$$

Moreover, there exists a sequence $\{x_n\}$ generated by the method (4.2) which is well defined starting at x and converges to x^*. Moreover, for $\mu = \dfrac{1-\sqrt{1-\beta}}{1+\sqrt{1-\beta}}$ the following error estimates hold

$$\|x^*-x\| \leq \frac{4\sqrt{1-\beta}}{\beta}\frac{\mu^{2^n}}{1-\mu^{2^n}}\gamma(\theta,x), \text{ if } \beta < 1, \tag{4.18}$$

$$\|x^*-x\| \leq \|x^*-x_n\| \leq 2^{1-n}\gamma(\theta,x). \tag{4.19}$$

For $w_0(t) = K_0t^2 - 2t + 2\gamma, w(t) = Kt^2 - 2t + 2\gamma$ and $s_0 = 0, s_{n+1} = s_n + \dfrac{w(s_n)}{2(1-K_0s_n)}, n = 0,1,2,..., s^* = \dfrac{1-\sqrt{1-\beta}}{K}$. The following hold

$$0 \leq s_n < s_{n+1} \leq s^*, \quad \lim_{n\to+\infty} s_n = s^*.$$

Furthermore,

$$\begin{aligned} s^*-s_n &\leq \frac{4\sqrt{1-\beta}}{\beta}\frac{\mu^{2^n}}{1-\mu^{2^n}}(s_1-s_0) = \frac{4\sqrt{1-\beta}}{\beta}\frac{\mu^{2^n}}{1-\mu^{2^n}}\gamma, \\ \frac{2(s_{n+1}-s_n)}{1+\sqrt{1+4\mu^{2^n}(1+\mu^{2^n})^{-2}}} &\leq s^*-s_n \leq \mu^{2^{n-1}}(s_n-s_{n-1}), \text{ if } \beta < 1 \end{aligned}$$

and

$$\begin{aligned} s^*-s_n &\leq 2^{-n+1}(s_1-s_0) = 2^{-n+1}\gamma, \\ 2(\sqrt{2}-1)(s_{n+1}-s_n) &\leq s^*-s_n \leq s_n-s_{n-1}, \text{ if } \beta = 1. \end{aligned}$$

Proof. Simply replace K_1 by K in the proof of theorem 3.4 and Lemma 3.5 in [1]. □

Remark. If $K_0 = K = K_1$ the results coincide with the ones in [1]. Otherwise, (i.e. if $K < K_1$) they constitute an improvement with advantages as stated in the introduction. Numerical examples, where $K_0 < K < K_1$ can be found in [2,3].

4. References

[1] Adly, S., Cibulka, R., Ngai, H. V., *Newtons method for solving inclusions using set-valued approximations*, SIAM J. Optim., 25(1), (2015), 159184.

[2] Argyros, I. K., George, S., Regmi, S., Argyros, C. I., *Hybrid Newton-like Inverse Free Algorithms for Solving Nonlinear Equations*, Algorithms, 17, (2024), 154. https://doi.org/10.3390/a17040154.

[3] Argyros, I. K., George, S., Shakhno, S., Regmi, S., Havdiak, M., Argyros, M. I., *Asymptotically Newton-Type Methods without Inverses for Solving Equations*, Mathematics, 12, (2024), 1069. https://doi.org/10.3390/math12071069.

[4] Blum, L., Cucker, F., Shub, M., Smale, S., *Complexity and Real Computation*, Springer, New York, (1998).

[5] Dontchev, A. L., Rockafellar, R. T., *Implicit Functions and Solution Mappings: A View from Variational Analysis*, Springer Monographs in Mathematics, Springer, New York, (2009).

[6] Josephy, N. H., *Newtons method for generalized equations*, Technical summary report, Mathematics Research Center, University of Wisconsin, Madison, (1979).

[7] Kantorovich, L. V., Akilov, G. P., *Analyse Fonctionnelle*, Mir, Moscow, (1981), translated from Russian.

[8] Robinson, S. M., *Strongly regular generalized equations*, Math. Oper. Res., 5(1), (1980), 4362.

Chapter 5

Ball Comparison Between Three Efficient Three-Step Methods under Common Conditions

1. Introduction

In this study, we compare the ball convergence of three efficient three step iterative methods for approximating a solution λ of the nonlinear equation

$$Q(x) = 0, \tag{5.1}$$

where $Q : D \subset X \to Y$ is a continuously differentiable nonlinear operator between the Banach spaces X and Y, and D stand for an open non empty convex compact set of X. The sixth order methods we are interested are:
Cordero et al. [11]:

$$\begin{aligned} y_n &= x_n - \frac{2}{3}Q'(x_n)^{-1}Q(x_n) \\ z_n &= x_n - \frac{1}{2}A_n^{-1}(3Q'(y_n)+Q'(x_n))Q'(x_n)^{-1}Q(x_n) \\ x_{n+1} &= z_n - A_n^{-1}Q(z_n). \end{aligned} \tag{5.2}$$

Cordero et al.[12]:

$$\begin{aligned} y_n &= x_n - \frac{1}{2}Q(x_n)^{-1}Q(x_n) \\ z_n &= \frac{1}{3}(4y_n - x_n) \\ u_n &= y_n + B_n^{-1}Q(x_n) \\ x_{n+1} &= u_n + 2B_n^{-1}Q(z_n). \end{aligned} \tag{5.3}$$

Soleymani et al.[24]:

$$\begin{aligned} y_n &= x_n - \frac{2}{3}Q(x_n)^{-1}Q(x_n) \\ z_n &= x_n - \frac{1}{2}A_n^{-1}(3Q'(y_n)+Q'(x_n))Q'(x_n)^{-1}Q(x_n) \\ x_{n+1} &= z_n - \frac{1}{4}C_nQ'(x_n)^{-1}Q(z_n). \end{aligned} \tag{5.4}$$

where, $A_n = 3Q'(y_n) - Q'(x_n)$, $B_n = Q'(x_n) - 3Q'(z_n)$, and

$$C_n = [A_n^{-1}(3Q'(y_n) + Q'(x_n)]^2.$$

These methods are of order six, convergence of these methods are proved using Taylor expansions involving seventh order derivatives not on these methods. The hypotheses involving seventh order derivatives limit the applicability of these methods. For example: Let $X = Y = \mathbb{R}$, $D = [-\frac{1}{2}, \frac{3}{2}]$. Define f on D by

$$f(s) = \begin{cases} s^3 \log s^2 + s^5 - s^4, & s \neq 0 \\ 0, & s = 0. \end{cases}$$

Then, we get

$$f'(s) = 3s^2 \log s^2 + 5s^4 - 4s^3 + 2s^2,$$
$$f''(s) = 6s \log s^2 + 20s^3 - 12s^2 + 10s,$$
$$f'''(s) = 6 \log s^2 + 60s^2 - 24s + 22,$$

$\lambda = 1$. Obviously $f'''(s)$ is not bounded on D. Hence, the convergence of these methods are not guaranteed by the earlier analyses.

Moreover, no computable convergence radii, upper error estimates on $\|x_n - \lambda\|$ or results on the uniqueness of λ are given in earlier studies. Furthermore, performance comparison is provided between them using numerical examples. Hence, we do not know in advance based on the same set of hypotheses for which method we can obtain larger radius of convergence (i. e., more initial points x_0); tighter error estimates on $\|x_n - \lambda\|$ (i.e. fewer iterates to obtain a desired error tolerance) and best information on the location of the solution.

In this chapter, we address these concerns. The same convergence order is obtained using COC or ACOC (to be precise in Remark 2.) that depend only on the first derivative and the iterates. Hence, we also extend the applicability of these methods. Our technique can be used to compare other methods [1]-[28] along the same lines.

Rest of the chapter is organized as follows. The convergence analysis of these methods are given in Section 2.

2. Ball Convergence Analysis

We shall study the ball convergence analysis of method (5.2), method (5.3) and method (5.4), respectively by developing real functions and parameters. Let $S = [0, \infty)$.

Assume there exists function $\omega_0 : S \longrightarrow S$ continuous and nondecreasing such that equation

$$\omega_0(s) - 1 = 0, \tag{5.5}$$

has a least positive solution r_0. Set $S_0 = [0, r_0)$.

Assume there exist functions $\omega : S_0 \longrightarrow S$ and $\omega_1 : S_0 \longrightarrow S$ continuous and nondecreasing. Define functions g_1 and h_1 on S_0 as

$$g_1(s) = \frac{\int_0^1 \omega((1-\tau)s)d\tau + \frac{1}{3}\int_0^1 \omega_1(\tau s)d\tau}{1 - \omega_0(s)}.$$

and

$$h_1(s) = g_1(s) - 1.$$

Assume equation

$$h_1(s) = 0 \tag{5.6}$$

has a least solution $R_1 \in (0, r_0)$.

Assume equation

$$p(s) - 1 = 0 \tag{5.7}$$

has a least positive solution r_p, where

$$p(s) = \frac{1}{2}(\omega_0(s) + 3\omega_0(g_1(s)s)).$$

Set $r_1 = \min\{r_0, r_p\}$ and $S_1 = [0, r_1)$. Define functions g, a, g_2 and h_2 on S_1 as

$$g(s) = \frac{\int_0^1 \omega((1-\tau)s)d\tau}{1-\omega_0(s)},$$

$$a(s) = \frac{5}{4}\frac{\omega_0(s) + \omega_0(g_1(s)s)}{1-p(s)}$$

$$g_2(s) = g(s) + \frac{a(s)\int_0^1 \omega_1(\tau s)d\tau}{1-\omega_0(s)}$$

and

$$h_2(s) = g_2(s) - 1.$$

Assume

$$h_2(s) = 0 \tag{5.8}$$

has a least solution $R_2 \in (0, r_1)$.

Assume equation

$$\omega_0(g_2(s)s) = 0 \tag{5.9}$$

has a least solution r_2. Set $r = \min\{r_1, r_2\}$ and $S_2 = [0, r_2)$. Define functions b, g_3 and h_3 on S_2 as

$$b(s) = \frac{3\omega_0(g_1(s)s) + \omega_0(s) + 2\omega_0(g_2(s)s)}{2(1-\omega_0(g_2(s)s))(1-p(s))},$$

$$g_3(s) = [g(g_2(s)s) + b(s)\int_0^1 \omega_1(\tau g_2(s)s)d\tau]g_2(s)$$

and

$$h_3(s) = g_3(s) - 1.$$

Assume equation

$$h_3(s) = 0 \tag{5.10}$$

has a least solution $R_3 \in (0, r)$. Consider a radius of convergence R as

$$R = \min\{R_j\},\ j = 1, 2, 3. \tag{5.11}$$

Then, we have

$$0 \leq \omega_0(s) < 1 \tag{5.12}$$

$$0 \leq \omega_0(g_2(s)s) < 1 \tag{5.13}$$

and

$$0 \leq g_j(s) < 1, \tag{5.14}$$

for all $s \in [0, R)$. From now we assume $r_0, r_p, r_1, r_2, r, R_i$ as given above exist. We use $U(x, \gamma) = \{y \in X : \|x - y\| < \gamma\}$ and $\bar{U}(x, \gamma)$ be its closure for $\gamma > 0$ and $e_n = \|x_n - \lambda\|$, for all $n = 0, 1, 2, \ldots$.

The conditions (H) are needed:

(H1) $F : D \longrightarrow Y$ is differentiable in the Fréchet sense and λ is a simple solution of equation (5.1).

(H2) There exists a continuous and nondecreasing function $\omega_0 : S \longrightarrow S$ such that for all $x \in D$

$$\|Q'(\lambda)^{-1}(Q'(x) - Q'(\lambda))\| \leq \omega_0(\|x - \lambda\|).$$

Set $D_0 = D \cap U(\lambda, r_0)$.

(H3) There exist continuous and nondecreasing functions $\omega : S_0 \longrightarrow S, \omega_1 : S_0 \longrightarrow S$ such that for each $x, y \in D_0$

$$\|Q'(\lambda)^{-1}(Q'(y) - Q'(x))\| \leq \omega(\|y - x\|)$$

and

$$\|Q'(\lambda)^{-1}Q'(x) \leq \omega_1(\|x - \lambda\|).$$

(H4) $\bar{U}(\lambda, R) \subset D$, where R is defined by (5.10).

(H5) There exists $R_* \geq R$ such that

$$\int_0^1 \omega_0(\tau R_*)d\tau < 1.$$

Set $D_1 = D \cap \bar{U}(\lambda, R_*)$.

Next, we present the local convergence result for method (5.2).

Theorem 11. Assume conditions (H) hold, and $x_0 \in U(\lambda, R) - \{\lambda\}$. Then, sequence $\{x_n\}$ generated by (5.2) is well defined, remains in $U(\lambda, R)$ and converges to λ. Moreover, the following assertions hold

$$\|y_n - \lambda\| \leq g_1(e_n)e_n \leq e_n < R, \tag{5.15}$$

$$\|z_n - \lambda\| \leq g_2(e_n)e_n \leq e_n, \tag{5.16}$$

and

$$\|x_{k+1} - \lambda\| \leq g_3(e_k)e_k \leq e_k, \tag{5.17}$$

where functions g_j and radius R are given previously. Furthermore, λ is the only solution of equation (5.1) in the set D_1 given in (H5).

Proof. Let $v \in U(\lambda, R) - \{\lambda\}$. Using (5.11), (5.12), (H1) and (H2):

$$\|Q'(\lambda)^{-1}(Q'(v) - Q'(\lambda))\| \leq \omega_0(\|v - \lambda\|) \leq \omega_0(R) < 1,$$

so by a result by Banach on invertible operators [2-8].

$$\|Q'(v)^{-1}Q'(\lambda)\| \leq \frac{1}{1 - \omega_0(\|v - \lambda\|)}, \tag{5.18}$$

and y_0 exists by method (5.2). By (5.11), (5.14) (for $i = 1$), (H1)-(H3) and method (5.2), we get

$$\begin{aligned} \|y_0-\lambda\| &= \|x_0-\lambda-Q'(x_0)^{-1}Q(x_0)+\frac{1}{3}Q'(x_0)^{-1}Q(x_0)\| \\ &\leq \frac{\int_0^1 \omega((1-\tau)\|x_0-\lambda\|)d\tau+\frac{1}{3}\int_0^1 \omega_1(\tau\|x_0-\lambda\|)d\tau}{1-\omega_0(\|x_0-\lambda\|)}\|x_0-\lambda\| \\ &\leq g_1(e_0)e_0 \leq e_0 < R, \end{aligned} \tag{5.19}$$

verifying (5.15) and $y_0 \in U(\lambda, R) - \{\lambda\}$. We can write

$$\begin{aligned} z_0-\lambda &= x_0-\lambda-Q'(x_0)^{-1}Q(x_0) \\ &\quad +[I-\frac{1}{2}A_0^{-1}(3Q'(x_0)+Q'(y_0))]Q'(x_0)^{-1}Q(x_0). \end{aligned} \tag{5.20}$$

We need the estimates

$$\begin{aligned} &\|(2Q'(\lambda))^{-1}(3(Q'(y_0)-Q'(\lambda))+(Q'(\lambda)-Q'(x_0))\| \\ \leq\ &\frac{1}{2}(3\omega_0(\|y_0-\lambda\|)+\omega_0(e_0) \leq p(e_0) < 1, \end{aligned}$$

so

$$\|A_0^{-1}Q'(\lambda)\| \leq \frac{1}{2(1-p(e_0))}. \tag{5.21}$$

Then, we get

$$\begin{aligned} &\|I-\frac{1}{2}A_0^{-1}(3Q'(x_0)+Q'(y_0))\| \\ &= \|\frac{1}{2}A_0^{-1}(2A_0-3Q'(x_0)-Q'(y_0))\| \\ &= \|\frac{5}{2}A_0^{-1}(Q'(y_0)-Q'(x_0))\| \\ &\leq \frac{5}{4}\frac{\omega_0(\|y_0-\lambda\|)+\omega_0(e_0)}{1-p(e_0)} \\ &= a(e_0), \end{aligned} \tag{5.22}$$

leading together with (5.20) to

$$\begin{aligned} \|z_0-\lambda\| &\leq [g(e_0)+\frac{a(e_0)\int_0^1 \omega_1(\tau e_0)d\tau}{1-\omega_0(e_0)}]e_0 \\ &\leq g_2(e_0)e_0 \leq e_0, \end{aligned} \tag{5.23}$$

verifying (5.16) and $z_0 \in U(\lambda, R)$. We can write by method (5.2)

$$x_1-\lambda = z_0-\lambda-Q'(z_0)^{-1}Q(z_0)+[Q'(z_0)^{-1}-2A_0^{-1}]Q(z_0). \tag{5.24}$$

So

$$\begin{aligned} \|x_1-\lambda\| &\leq [g(\|z_0-\lambda\|)+b(e_0)\int_0^1 \omega_1(\tau\|z_0-\lambda\|)d\tau]\|z_0-\lambda\| \\ &\leq g_3(e_0)e_0 \leq e_0, \end{aligned} \tag{5.25}$$

verifying (5.18) and $x_1 \in U(\lambda, R)$, where we used

$$\begin{aligned} \|Q'(z_0)^{-1} - 2A_0^{-1}\| &= \|Q'(z_0)^{-1}[(3(Q'(y_0) - Q'(\lambda)) \\ &\quad -(Q'(x_0) - Q'(\lambda)) - 2(Q'(z_0) - Q'(\lambda))]A_0^{-1} \\ &\leq \frac{3\omega_0(\|y_0 - \lambda\|) + \omega_0(e_0) + 2\omega_0(\|z_0 - \lambda\|)}{2(1 - \omega_0(\|z_0 - \lambda\|))(1 - p(e_0))} \\ &= b(e_0). \end{aligned} \tag{5.26}$$

The induction is complete for $n = 0$. Repeat the preceding calculations but replace x_0, y_0, z_0, x_1 in the above arguments by x_k, y_k, z_k, x_{k+1}. Then, the induction is completed. Thus, we get

$$\|x_{k+1} - \lambda\| \leq \lambda \|x_k - \lambda\| < R, \tag{5.27}$$

where $\lambda = g_3(e_0) \in [0, 1)$, we deduce $x_{k+1} \in U(\lambda, R)$ and $\lim_{k \longrightarrow \infty} x_k = \lambda$. Let $d \in D_1$ with $Q(d) = 0$. Set $M = \int_0^1 Q'(d + \tau(\lambda - d))d\tau$. In view of (H2) and (H5), we get

$$\|Q'(\lambda)^{-1}(M - Q'(\lambda))\| \leq \int_0^1 \omega_0(\tau\|\lambda - d\|)d\tau \leq \int_0^1 \omega_0(\tau R_*)d\tau < 1,$$

so $\lambda = d$, follows since M^{-1} exists and $0 = Q(\lambda) - Q(d) = M(\lambda - d)$.

□

Next, in an analogous way for method (5.3) but using the estimates

$$\begin{aligned} \|y_0 - \lambda\| &\leq \frac{[\int_0^1 \omega((1-\tau)e_0)d\tau + \frac{1}{2}\int_0^1 \omega_1(\tau e_0)d\tau]}{1 - \omega_0(e_0)} e_0 \\ &= \bar{g}_1(e_0)e_0 \leq e_0 < \bar{R}, \end{aligned} \tag{5.28}$$

$$\begin{aligned} \|z_n - \lambda\| &= \| - \frac{1}{3}(4(y_n - \lambda))\| \\ &\leq \frac{1}{3}(4\bar{g}_1(e_n) + 1)e_n \\ &= \bar{g}_2(e_n)e_n \leq e_n, \end{aligned} \tag{5.29}$$

$$\begin{aligned} \|u_n - \lambda\| &= \|(x_n - \lambda - Q'(x_n)^{-1}Q(x_n) \\ &\quad + [\frac{1}{2}Q'(x_n)^{-1} + B_n^{-1}]F(x_n)\| \end{aligned} \tag{5.30}$$

$$\begin{aligned} &\leq [g(e_n) + c(e_n)\int_0^1 \omega_1(\tau e_n)d\tau]e_n \\ &= \bar{g}_3(e_n)e_n \leq e_n, \end{aligned} \tag{5.31}$$

$$\begin{aligned} \|\frac{1}{2}Q'(x_n)^{-1} + B_n^{-1}\| &= \|Q'(x_n)^{-1}(\frac{1}{2}B_n + Q'(x_n))B_n^{-1}\| \\ &= \frac{1}{2}\|Q'(x_n)^{-1}(-3Q'(z_n) + 2Q'(x_n))B_n^{-1}\| \\ &\leq \frac{3}{4}\frac{3\omega_0(\|z_n - \lambda\|) + 2\omega_0(e_n) + 1}{(1 - \omega_0(e_n))(1 - q(e_n))} \leq c(e_n) \end{aligned}$$

and

$$\|B_n^{-1}Q'(x_n)\| \leq \frac{1}{2(1-q(e_n))},$$

$$\begin{aligned}
\|x_{n+1}-\lambda\| &= \|(x_n-\lambda-Q'(x_n)^{-1}Q(x_n)) \\
&\quad +(\frac{1}{2}Q'(x_n)^{-1}+B_n^{-1})Q(x_n)+2B_n^{-1}Q(z_n)\| \\
&\leq [g(e_n)+c(e_n)\int_0^1 \omega_1(\tau e_n)d\tau \\
&\quad +\frac{\int_0^1 \omega_1(\tau\bar{g}_2(e_n)e_n)d\tau}{1-q(e_n)}]e_n \\
&= \bar{g}_4(e_n)e_n \leq e_n, \qquad (5.32)
\end{aligned}$$

where $\bar{R}_1, \bar{R}_2, \bar{R}_3, \bar{R}_4$ are the least positive solutions of equations $h_j(s) := g_j(s)-1=0$, respectively,

$$\bar{R} = \min\{\bar{R}_j\},\ j=1,2,3,4 \qquad (5.33)$$

and equation

$$q(s)-1=0$$

has a least positive solution r_q. Clearly, here

$$\bar{g}_1(s) = \frac{\int_0^1 \omega((1-\tau)s)\tau + \frac{1}{2}\int_0^1 \omega_1(\tau s)d\tau}{1-\omega_0(s)},$$

$$\bar{g}_2(s) = \frac{1}{3}(1+4\bar{g}_1(s)),$$

$$\bar{g}_3(s) = g(s)+c(s)\int_0^1 \omega_1(\tau s)d\tau,$$

$$\bar{g}_4(s) = g(s)+c(s)\int_0^1 \omega_1(\tau s)d\tau + \frac{\int_0^1 \omega_1(\tau\bar{g}_2(s)s)d\tau}{1-q(s)},$$

$$q(s) = \frac{1}{2}(3\omega_0(\bar{g}_2(s)s)+\omega_0(s))$$

and

$$c(s) = \frac{3}{4}\frac{3\omega_0(\bar{g}_2(s)s)+2\omega_0(s)+1}{(1-\omega_0(s))(1-q(s))}.$$

Hence, we arrive at the following result for method (5.3).

Theorem 12. Under the conditions (H) but with $\bar{R}, \bar{g}_1, \bar{g}_2, \bar{g}_3$ replacing R, g_1, g_2, g_3, respectively the conclusion of Theorem 11 hold for method (5.3).

□

Next, we study method (5.4) to obtain

$$\|y_n-\lambda\| \leq g^1(e_n)e_n \leq e_n < \tilde{R},$$

$$\|z_n-\lambda\| \leq g^2(e_n)e_n$$

and

$$\begin{aligned}\|x_{n+1}-\lambda\| &= \|(z_n-\lambda)+\frac{1}{4}[A_n^{-1}(3Q'(y_n)+Q'(x_n))]^2Q'(x_n)Q(z_n)\| \\ &\leq [1+\frac{1}{16}(\frac{(3\omega_1(\|y_n-\lambda\|)+\omega_1(e_n))\int_0^1\omega_1(\tau g^2(e_n))d\tau}{(1-p(e_n))(1-\omega_0(e_n))})^2]g_2(e_n)e_n \\ &\leq \tilde{g}_3(e_n)e_n\leq e_n, \end{aligned} \tag{5.34}$$

where

$$\tilde{g}_3(s)=[1+\frac{1}{16}(\frac{(3\omega_1(g^1(s)s)+\omega_1(s))\int_0^1\omega_1(\tau g_2(s)s)d\tau}{(1-p(s))(1-\omega_0(s))})^2]g_2(s)$$

where equation

$$\tilde{h}_3 9s)=\tilde{g}_3(s)-1=0$$

has a least solution $\tilde{R}_3\in(0,r_p)$. A radius of convergence $\tilde{R}$ for method (5.4) is

$$\tilde{R}=\min\{R_1,R_2,\tilde{R}_3\}. \tag{5.35}$$

Hence, we arrive at the following result for method (5.4).

Theorem 13. Under the conditions (H) but with $\tilde{R},\tilde{g}_3$ replacing R,g_3, respectively, the conclusions of Theorem 11 hold for method (5.4).

Remark. We can find the convergence order by resorting to the computational order of convergence (COC) defined by

$$\xi=\ln\left(\frac{\|x_{n+1}-x^*\|}{\|x_n-x^*\|}\right)/\ln\left(\frac{\|x_n-x^*\|}{\|x_{n-1}-x^*\|}\right)$$

or the approximate computational order of convergence

$$\xi_1=\ln\left(\frac{\|x_{n+1}-x_n\|}{\|x_n-x_{n-1}\|}\right)/\ln\left(\frac{\|x_n-x_{n-1}\|}{\|x_{n-1}-x_{n-2}\|}\right).$$

This way we obtain in practice the order of convergence without resorting to the computation of higher order derivatives appearing in the method or in the sufficient convergence criteria usually appearing in the Taylor expansions for the proofs of these results.

3. References

[1] Alzahrani, A. K. H., Bhel, R., Alshomrani, A., *Some higher order iteration functions for solving nonlinear models*, Appl. Math. Comput., 334, (2018), 80-93.

[2] Amat, S., Busquier, S., Gutirrez, J. M., *Geometrical constructions of iterative functions to solve nonlinear equations*, J. Comput. Appl. Math., 157, (2003), 197-205.

[3] Argyros, I. K., *Computational Theory of Iterative Methods, Series: Studies in Computational Mathematics*, 15, Editors: Chui C.K. and Wuytack L., Elsevier Publ. Company, New York, (2007).

[4] Argyros, I. K., George, S., *Mathematical modeling for the solution of equations and systems of equations with applications*, Volume-III, Nova Publishes, NY, (2019).

[5] Argyros, I. K., George, S., *Mathematical modeling for the solution of equations and systems of equations with applications*, Volume-IV, Nova Publishes, NY, (2019).

[6] Argyros, I. K., George, S., Magren, A. A., *Local convergence for multi-point-parametric Chebyshev-Halley-type method of higher convergence order*, J. Comput. Appl. Math., 282, (2015), 215-224.

[7] Argyros, I. K., Magren, A. A., *A study on the local convergence and the dynamics of Chebyshev-Halley-type methods free from second derivative*, Numer. Algorithms, 71, (2015), 1-23.

[8] Argyros, I. K., Magren, A. A., *Iterative methods and their dynamics with applications*, CRC Press, New York, USA, (2017).

[9] Cordero, A., Martnez, E., Torregrosa, J. R., *Iterative methods of order four and five for systems of nonlinear equations*, Appl. Math. Comput., 231, (2009), 541–551.

[10] Cordero, A., Torregrosa, J. R., *Variants of Newton's method for functions of several variables*, Appl. Math. Comput., 183, (2006), 199–208.

[11] Cordero, A., Torregrosa, J. R., *Variants of Newton's method using fifth-order quadrature formulas*, Appl. Math. Comput., 190, (2007), 686–698.

[12] Darvishi, M. T., Barati, A., *A fourth-order method from quadrature formulae to solve systems of nonlinear equations*, Appl. Math. Comput., 188, (2007), 257–261.

[13] Frontini, M., Sormani, E., *Some variant of Newton's method with third-order convergence*, Appl. Math. Comput., 140, (2003), 419–426.

[14] Frontini, M., Sormani, E., *Third-order methods from quadrature formulae for solving systems of nonlinear equations*, Appl. Math. Comput., 149, (2004), 771–782.

[15] Gautschi, W., *Numerical Analysis: An Introduction*, Birkhuser, Boston, (1997).

[16] Grau-Snchez, M., Grau, ., Noguera, M., *On the computational efficiency index and some iterative methods for solving systems of nonlinear equations*, J. Comput. Appl. Math., 236, (2011), 1259–1266.

[17] Gutirrez, J. M., Hernndez, M. A., *A family of ChebyshevHalley type methods in Banach spaces*, Bull. Aust. Math. Soc., 55, (1997), 113–130.

[18] Hommeier, H. H. H., *A modified Newton method for root finding with cubic convergence*, J. Comput. Appl. Math., 157, (2003), 227–230.

[19] Hommeier, H. H. H., *A modified Newton method with cubic convergence: the multivariable case*, J. Comput. Appl. Math., 169, (2004), 161–169.

[20] Lofti, T., Bakhtiari, P., Cordero, A., Mahdiani, K., Torregrosa, J. J., *Some new efficient iterative methods for solving systems of equations*, Intern. J. Computer Mathematics, 92, 9, (2015), 1921-1934.

[21] Noor, M. A., Wassem, M., *Some iterative methods for solving a system of nonlinear equations*, Appl. Math. Comput., 57, (2009), 101–106.

[22] Ortega, J. M., Rheinboldt, W. C., *Iterative Solution of Nonlinear Equations in Several Variables*, Academic Press, New York, (1970).

[23] Ostrowski, A. M., *Solution of Equations and Systems of Equations*, Academic Press, New York, (1966).

[24] Ozban, A. Y., *Some new variants of Newton's method*, Appl. Math. Lett., 17, (2004), 677–682.

[25] Sharma, J. R., Guha, R. K., Sharma, R., *An efficient fourth order weighted-Newton method for systems of nonlinear equations*, Numer. Algor., 62, (2013), 307–323.

[26] Traub, J. F., *Iterative Methods for the Solution of Equations*, Prentice-Hall, Englewood Cliffs, (1964).

[27] Weerakoon, S., Fernando, T. G. I., *A variant of Newton's method with accelerated third-order convergence*, Appl. Math. Lett., 13, (2000), 87–93.

[28] Wolfram, S., *The Mathematica Book, 5th edn*, Wolfram Media, (2003).

Chapter 6

An Extended Derivative-Free Method of Convergence Order Seven for Equations under Generalized Conditions

1. Introduction

Let $\mathcal{B}_1, \mathcal{B}_2$ be Banach spaces, $\Omega \subset \mathcal{B}_1$ be a nonempty open set and $F : \Omega \longrightarrow \mathcal{B}_2$ be a Fréchet-differentiable operator. Applications from different disciplines of sciences can be formulated like nonlinear equation

$$\mathcal{G}(x) = 0. \tag{6.1}$$

A solution $x_* \in \Omega$ in closed form for (6.1) is obtained only in special cases. So, iterative procedures are mainly utilized to obtain iterates that approximate x_*. More details can be found in [1,2,3,5] and the references there in (see also [4,5,6,9,10,11]).

We consider the following seventh-order iterative procedure defined for $x_0 \in \Omega$ and each $n = 0, 1, 2, \ldots$ by

$$\begin{aligned}
u_n &= x_n + \mathcal{G}(x_n), \\
s_n &= x_n - \mathcal{G}(x_n), \\
y_n &= x_n - [u_n, s_n; \mathcal{G}]^{-1}\mathcal{G}(x_k) \\
z_n &= y_n - (3I - 2[u_n, s_n; \mathcal{G}]^{-1}[y_n, x_n; \mathcal{G}])[u_n, s_n; \mathcal{G}]^{-1}\mathcal{G}(y_n) \\
x_{n+1} &= z_n - (\frac{13}{4}I - [u_n, s_n; \mathcal{G}]^{-1}[z_n, y_n; \mathcal{G}]) \\
&\quad \times (\frac{7}{2}I - \frac{5}{4}[u_n, s_n; \mathcal{G}]^{-1}[z_n, y_n; \mathcal{G}])[u_n, s_n; \mathcal{G}]^{-1}\mathcal{G}(z_n),
\end{aligned} \tag{6.2}$$

where $[.,.;\mathcal{G}] : \Omega \times \Omega \longrightarrow \mathcal{B}_2$ is a divided difference of order one [1,2]. The motivation, derivation and benefits out of using procedure (6.2) have been well explained in [5,15,16].

The major concern in [15,16] is the necessity for the existence of derivatives of F up to order eight, reducing its applicability in problems involving operator whose higher order derivatives are not bounded. For example, let $f : [-2, 2] \longrightarrow \mathbb{R}$ be defined by

$$f(x) = \begin{cases} \frac{1}{20}(x^4 logx^2 + x^6 - x^5) & if\, x \neq 0 \\ 0 & if\, x = 0. \end{cases}$$

Then, we get by this definition

$$\begin{aligned} f'(x) &= \frac{1}{20}(2x^3+4x^3 logx^2+6x^5-5x^4) \\ f''(x) &= \frac{1}{20}(14x^2+12x^2 logx^2+30x^4-20x^3) \\ f'''(x) &= \frac{1}{20}(52x+24x logx^2+120x^3-60x^2) \\ f^{IV}(x) &= \frac{1}{20}(24 logx^2+360x^2-120x+100). \end{aligned}$$

Note that, the fourth derivative of the function f is not bounded.

The ball convergence analysis was based on generalized continuity conditions. Motivated by optimization considerations we present a finer ball convergence analysis under weaker conditions. Relevant work can be found in [6]-[14].

The new local convergence analysis can be found in Section 2. The examples appear in Section 3 followed by the conclusion in Section 4.

2. Ball Analysis

Functions and parameters are provided for $S=[0,\infty)$, $a\geq 0$, $b\geq 0$ and $c\geq 0$.

Assume:

(a) Function $w_0(at,bt)-1=0$ has a least zero $r_0\in S-\{0\}$ with $w_0: S\times S\longrightarrow S$ a nondecreasing and continuous function (NCF).

Set $S_0=[0,r_0)$.

(b) Function $g_1(t)-1=0$ has a least zero $\rho_1\in S_0-\{0\}$ with $w: S_0\times S_0\longrightarrow S$ a NCF and $g_1: S_0\longrightarrow S$ defined as

$$g_1(t)=\frac{w(ct,bt)}{1-w_0(at,bt)}.$$

(c) Function $g_2(t)-1=0$ has a least zero $\rho_2\in S_0-\{0\}$ with $g_2: S_0\longrightarrow S$, $h: S_0\longrightarrow S$ defined as

$$g_2(t) = \left[\frac{w(h(t),bt)}{1-w_0(at,bt)}+\frac{2cw_1(h(t),ct)}{(1-w_0(at,bt))^2}\right]g_1(t),$$

where

$$h(t)=(a+g_1(t))t.$$

(d) Function $g_3(t)-1=0$ has a least zero $\rho_3\in S_0-\{0\}$ with $g_3: S_0\longrightarrow S$ defined as

$$\begin{aligned} g_3(t) &= \Bigg[\frac{w((a+g_2(t))t,bt)}{1-w_0(at,bt)}+\frac{c}{16(1-w_0(at,bt))} \\ &\quad \times(65+\frac{81w_1((a+g_2(t))t,(b+g_1(t))t)}{1-w_0(at,bt)} \\ &\quad + 20\left(\frac{w_1((a+g_1(t))t,(b+g_1(t))t)}{1-w_0(at,bt)}\right)^2\Bigg)\Bigg]g_2(t). \end{aligned}$$

Set

$$\rho = \min\{\rho_i\},\ i = 1,2,3. \tag{6.3}$$

Parameter ρ shall be shown to be a convergence radius for method (6.2). Set $S_1 = [0,\rho)$. Notice that for each $t \in S_1$

$$0 \le w_0(at,bt) < 1 \tag{6.4}$$

and

$$0 \le g_i(t) < 1. \tag{6.5}$$

We denote by $\bar{D}(x_*,q)$ the closure of the open ball $D(x_*,q)$ with center $x_* \in \mathcal{B}_1$ and of radius $q > 0$.

The conditions (A) shall be utilized provided functions "w" are as given previously, and x_* is a simple zero of $\mathcal{G}$.

Assume:

(i) $\|I + [x,x_*;\mathcal{G}]\| \le a$,
$\|I - [x,x_*,\mathcal{G}]\| \le b$ and

$$\|\mathcal{G}'(x_*)^{-1}([x,y;\mathcal{G}] - \mathcal{G}'(x_*))\| \le w_0(\|x-x_*\|,\|y-x_*\|)$$

hold for each $x,y \in \Omega$. Set $\Omega_0 = \Omega \cap D(x_*,r_0)$.

(ii) $\|\mathcal{G}(x)\| \le c\|x-x_*\|$,

$$\|\mathcal{G}'(x_*)^{-1}([x,y;\mathcal{G}] - [z,x_*;\mathcal{G}])\| \le w(\|x-z\|,\|y-x_*\|)$$

and

$$\|\mathcal{G}'(x_*)^{-1}([x,y;\mathcal{G}] - [u,v;\mathcal{G}])\| \le w_1(\|x-u\|,\|y-v\|)$$

hold for each $x,y,u,v \in \Omega_0$.

(iii) $\bar{D}(x_*,R) \subset \Omega$ with $R = \max\{\rho, a\rho, b\rho, c\rho\}$.

(iv) There exists $\bar{\rho} \ge \rho$ satisfying

$$w_0(0,\bar{\rho}) < 1 \text{ or } w_0(\bar{\rho},0) < 1.$$

Set $\Omega_1 = \Omega \cap \bar{D}(x_*,\bar{\rho})$.

The main ball convergence result for method (6.2) is presented next utilizing conditions (A).

Lemma 4. Assume conditions (A) hold. Then, $\lim_{n\longrightarrow\infty} x_n = x_*$ provided that $x_0 \in D(x_*,\rho) - \{x_*\}$. Moreover, the only zero of function $\mathcal{G}$ on the domain Ω_1 is x_*.

Proof. The following assertions shall be established using induction.

$$\|y_n - x_*\| \le g_1(\|x_n - x_*\|)\|x_n - x_*\| \le \|x_n - x_*\| < \rho, \tag{6.6}$$

$$\|z_n - x_*\| \le g_2(\|x_n - x_*\|)\|x_n - x_*\| \le \|x_n - x_*\| \tag{6.7}$$

and

$$\|x_{n+1} - x_*\| \le g_3(\|x_n - x_*\|)\|x_n - x_*\| \le \|x_n - x_*\|, \tag{6.8}$$

where the radius r is defined in (6.3) and the functions g_i are as given previously.

We have in turn by (A1) and (A3)

$$\begin{aligned} \|u_0 - x_*\| &\leq \|x_0 - x_* + G(x_0)\| \\ &= \|(I + [x_0, x_*; G])(x_0 - x_*)\| \\ &\leq \|I + [x_0, x_*; G]\| \|x_0 - x_*\| \\ &\leq a\|x_0 - x_*\| < R, \\ \|s_0 - x_*\| &= \|(I - [x_0, x_*; V])(x_0 - x_*)\| \\ &\leq b\|x_0 - x_*\| < R \end{aligned}$$

and

$$\begin{aligned} \|G'(x_*)^{-1}([u_0, s_0; G] - G'(x_*))\| &\leq w_0(u_0 - x_*\|, \|s_0 - x_*\|) \\ &\leq w_0(a\rho, b\rho) < 1. \end{aligned}$$

The last estimate and a lemma of Banach on invertible linear operators [1,2,14] imply that $[u_0, s_0; G]^{-1} \in L(Y, X)$ and

$$\|[u_0, s_0; G]^{-1} F'(x_*)\| \leq \frac{1}{1 - w_0(a\|x_0 - x_*\|, b\|x_0 - x_*\|)}. \tag{6.9}$$

Notice also that y_0 is well defined by the first substep of (6.2) from which, we can also write

$$\begin{aligned} y_0 - x_* &= x_0 - x_* - [u_0, s_0; G]^{-1} G(x_0) \\ &= [u_0, s_0; G]^{-1}([u_0, s_0; G] - [x_0, x_*; G])(x_0 - x_*). \end{aligned} \tag{6.10}$$

By (6.3), (6.5) (for $i = 1$), (A2), (6.9) and (6.10), we get in turn

$$\begin{aligned} \|y_0 - x_*\| &\leq \frac{w(\|u_0 - x_*\|, \|s_0 - x_*\|)\|x_0 - x_*\|}{1 - w_0(a\|x_0 - x_*\|, b\|x_0 - x_*\|)} \\ &\leq g_1(\|x_0 - x_*\|)\|x_0 - x_*\| \leq \|x_0 - x_*\| < \rho, \end{aligned} \tag{6.11}$$

proving (6.6) if $n = 0$ and $y_0 \in D(x_*, \rho)$. Iterate z_0 is also well defined by (6.9) and the second substep of method (6.2) from which we can also write

$$\begin{aligned} z_0 - x_* &= y_0 - x_* - [u_0, x_0; G]^{-1} G(y_0) \\ &\quad + 2(I - [u_0, s_0; G]^{-1}[y_0, x_0; G])[u_0, s_0; G]^{-1} G(y_0) \\ &= y_0 - x_* - [u_0, s_0; G]^{-1} G(y_0) \\ &\quad + 2[u_0, s_0; G]^{-1}([u_0, s_0; G] - [y_0, x_0; G]) \\ &\quad \times [u_0, s_0; G]^{-1} G(y_0). \end{aligned} \tag{6.12}$$

In view of (6.3), (6.5) (for $i = 2$), (6.9)-(6.12), we first obtain the estimates

$$\begin{aligned} &\|y_0 - x_* - [u_0, s_0; G]^{-1} G(y_0)\| \\ = &\|[u_0, s_0; G]^{-1}([u_0, s_0; G] - [y_0, x_*; G])(y_0 - x_*)\| \\ \leq &\frac{w(\|u_0 - y_0\|, \|s_0 - x_*\|)\|y_0 - x_*\|}{1 - w_0(a\|x_0 - x_*\|, b\|x_0 - x_*\|)} \\ \leq &\frac{w(h(\|x_0 - x_*\|), b\|x_0 - x_*\|)\|y_0 - x_*\|}{1 - w_0(a\|x_0 - x_*\|, b\|x_0 - x_*\|)}, \end{aligned} \tag{6.13}$$

since

$$\begin{aligned}\|u_0-y_0\| &= \|(u_0-x_*)+(x_*-y_0)\|\\ &\le (a+g_1(\|x_0-x_*\|))\|x_0-x_*\|\\ &= h(\|x_0-x_*\|)\end{aligned}$$

and

$$\begin{aligned}&2\|[u_0,s_0;G]^{-1}([u_0,s_0;G]-[y_0,x_0;G])[u_0,s_0;G]^{-1}G(y_0)\|\\ \le\ & \frac{2w_1(\|u_0-y_0\|,\|s_0-x_*\|)c\|y_0-x_*\|}{(1-w_0(a\|x_0-x_*\|,b\|x_0-x_*\|))^2}. \qquad (6.14)\end{aligned}$$

By summing up (6.13) and (6.14) and using (6.12), we get

$$\begin{aligned}\|z_0-x_*\| &\le \left[\frac{w(h(\|x_0-x_*\|),b\|x_0-x_*\|)}{1-w_0(a\|x_0-x_*\|,b\|x_0-x_*\|)}\right.\\ &\quad\left.+\frac{2w_1(h(\|x_0-x_*\|),c\|x_0-x_*\|)}{(1-w_0(a\|x_0-x_*\|,b\|x_0-x_*\|))^2}\right]\|y_0-x_*\| \qquad (6.15)\\ &\le g_2(\|x_0-x_*\|)\|x_0-x_*\|\le\|x_0-x_*\|, \qquad (6.16)\end{aligned}$$

proving (6.7) in $n=0$ and $z_0\in D(x_*,\rho)$. It follows from (6.9) and the third substep of method (6.2) that iterate x_1 is well defined, and we can write

$$\begin{aligned}x_1-x_* &= z_0-x_*-[u_0,s_0;G]^{-1}([u_0,s_0;G]-[z_0,y_0;G])\\ &\quad+(\frac{65}{16}I+\frac{81}{16}[u_0,s_0;G]^{-1}([u_0,s_0;G]-[z_0,y_0;G])\\ &\quad+\frac{5}{4}([u_0,s_0;G]^{-1}([u_0,s_0;G]-[z_0,y_0;G])^2)\\ &\quad\times[u_0,s_0;G]^{-1}G(z_0). \qquad (6.17)\end{aligned}$$

We need the estimates

$$\begin{aligned}&\|z_0-x_*-[u_0,s_0;G]^{-1}G(z_0)\|\\ =\ &\|[u_0,s_0;G]^{-1}([u_0,s_0;G]-[z_0,x_*;G])(z_0-x_*)\|\\ \le\ &\frac{w(\|(u_0-x_*)+(x_*-z_0)\|,\|s_0-x_*\|)\|z_0-x_*\|}{(1-w_0(a\|x_0-x_*\|,b\|x_0-x_*\|)}\\ \le\ &\frac{w((a+g_2(\|x_0-x_*\|))\|x_0-x_*\|,b\|x_0-x_*\|)\|z_0-x_*\|}{1-w_0(a\|x_0-x_*\|,b\|x_0-x_*\|)} \qquad (6.18)\end{aligned}$$

and

$$\begin{aligned}&\|(\frac{65}{16}I+\frac{81}{16}[u_0,s_0;G]^{-1}([u_0,s_0;G]-[z_0,y_0;G])\\ &+\frac{5}{4}([u_0,s_0;G]^{-1}([u_0,s_0;G]-[z_0,y_0;G])^2\\ &\times[u_0,s_0;G]^{-1}G(z_0)\|\\ \le\ &\frac{1}{16}\left[65+\frac{81w_1(\|u_0-z_0\|,\|s_0-y_0\|)}{1-w_0(a\|x_0-x_*\|,b\|x_0-x_*\|)}\right.\\ &\left.+20\left(\frac{w_1(\|u_0-z_0\|,\|s_0-y_0\|)}{1-w_0(a\|x_0-x_*\|,b\|x_0-x_*\|)}\right)^2\right]\\ &\times\frac{c\|z_0-x_*\|}{1-w_0(a\|x_0-x_*\|,b\|x_0-x_*\|)}. \qquad (6.19)\end{aligned}$$

Summing up (6.18) and (6.19), we obtain from (6.17), (6.3), (6.6) (for $i = 3$), (6.9), (6.11) and (6.16)

$$\|x_1 - x_*\| \leq g_3(\|x_0 - x_*\|)\|x_0 - x_*\| \leq \|x_0 - x_*\|, \tag{6.20}$$

proving (6.8) in $n = 0$ and $x_1 \in D(x_*, \rho)$, where we also used

$$\begin{aligned} \|u_0 - z_0\| &= \|(u_0 - x_*) + (x_* - z_0)\| \\ &\leq \|u_0 - x_*\| + \|x_* - z_0\| \\ &\leq (a + g_2(\|x_0 - x_*\|))\|x_0 - x_*\| \end{aligned}$$

and

$$\begin{aligned} \|s_0 - y_0\| &= \|(s_0 - x_*) + (x_* - y_0)\| \\ &\leq \|s_0 - x_*\| + \|x_* - y_0\| \\ &\leq (b + g_1(\|x_0 - x_*\|))\|x_0 - x_*\|. \end{aligned}$$

Simply switch u_0, x_0, z_0, x_1 by u_i, x_i, z_i, x_{i+1} in the previous calculations to terminate the induction for assertions (6.6)-(6.8). Then, from the estimation

$$\|x_{i+1} - x_*\| \leq p\|x_i - x_*\| < \rho, \tag{6.21}$$

where $p = g_3(\|x_0 - x_*\|) \in [0, 1)$, we conclude $\lim_{i \longrightarrow \infty} x_i = x_*$ and $x_{i+1} \in D(x_*, \rho)$.

Set $M = [x_*, \gamma; \mathcal{G}]$ for $\gamma \in \Omega_1$ with $\mathcal{G}(\gamma) = 0$. Then, using (A1) and (A4), we obtain in turn

$$\begin{aligned} \|\mathcal{G}'(x_*)^{-1}(M - \mathcal{G}'(x_*))\| &\leq w_0(0, \|x_* - \gamma\|) \\ &\leq w_0(0, \bar{\rho}) < 1, \end{aligned} \tag{6.22}$$

thus $\gamma = x_*$ follows from the invertability of M and the approximation $0 = \mathcal{G}(x_*) - \mathcal{G}(\gamma) = M(x_* - \gamma)$. □

3. Numerical Experiments

We compute the radius of convergence in this section.

Example 4. Let $X = Y = \mathbb{R}^3$, $\Omega = B[0, 1]$. Define function $\mathcal{G}(w)$ on Ω for $w = (a_1, a_2, a_3)^T$ by

$$\mathcal{G}(w) = \left(e^{a_1} - 1, a_2^2 \frac{e-1}{2} + a_2, a_3\right)^T.$$

Here, $x^* = (0, 0, 0)^T$. We have

$$\mathcal{G}'(w) = \begin{pmatrix} e^{a_1} & 0 & 0 \\ 0 & (e-1)a_2 + 1 & 0 \\ 0 & 0 & 1 \end{pmatrix}.$$

The conditions (A) are verified for

$$a = \frac{e+3}{2}, b = \frac{e-1}{2}, c = \frac{1}{2}e^{\frac{1}{e-1}},$$

$$w_0(s,t) = \frac{1}{2}(e-1)(s+t), w(s,t) = \frac{1}{2}(e^{\frac{1}{e-1}}s + (e-1)t)$$

and

$$w_1(s,t) = \frac{1}{2}e^{\frac{1}{e-1}}(s+t).$$

Then,

$$\rho_1 = 0.21126938353173808335627952129031,$$

$$\rho_2 = 0.12620865641391252801833323241707 = \rho$$

and

$$\rho_3 = 0.065489747227670754547307476220085$$

4. References

[1] Argyros, I. K., *Convergence and Applications of Newton-type Iterations*, Springer Verlag, Berlin, Germany, (2008).

[2] Argyros, I. K., *The theory and applications of iteration methods*, 2nd Edition, CRC Press, Engineering Series, Taylor and Francis Group, (2022).

[3] Argyros, I. K., *Unified Convergence Criteria for Iterative Banach Space Valued Methods with Applications*, Mathematics, (2021), 9(16), 1942; https://doi.org/10.3390/math9161942.

[4] Grau-Snchez, M., Noguera, M., Amat, S., *On the approximation of derivatives using divided difference operators preserving the local convergence order of iterative methods*, J. Comput. Appl. Math., 237, (2013), 263-272.

[5] Hernndez, M. A., Romero, N., *General study of iterative processes of R-order at least three under weak convergence conditions*, Journal of Optimization Theory and Applications, 133(2), (2007), 163-177.

[6] Liu, Z., Zheng, Q., Zhao, P., *A variant of Steffensen's method of fourth-order convergence and its applications*, Appl. Math. Comput., 216, (2012), 1978-1983.

[7] Magren, A. A., Gutirrez, J. M., *Real dynamics for damped Newton's method applied to cubic polynomials*, J. Comput. Appl. Math., 275, (2015), 527–538.

[8] Ren, H., Wu, Q., Bi, W., *A class of two-step Steffensen type methods with fourth-order convergence*, Appl. Math. Comput., 209, (2009), 206-210.

[9] Shakhno, S. M., Iakymchuk, R. P., Yarmola, H. P., *Convergence analysis of a two-step method for the nonlinear squares problem with decomposition of operator*, J. Numer. Appl. Math., 128, (2018), 82-95.

[10] Shakhno, S. M., Gnatyshyn, O. P., *On an iterative algorithm of order 1.839... for solving nonlinear operator equations*, Appl. Math. Appl., 161, (2005), 253-264, https://doi.org/10.1016/j.amc.2003.12.025.

[11] Sharma, J. R., Arora, H., *An efficient derivative-free numerical method for solving systems of nonlinear equations*, Appl. Anal. Discrete. Math., 7, (2013), 390-403.

[12] Sharma, J. R., Arora, H., *Efficient derivative-free numerical methods for solving systems of nonlinear equations*, Comp. Appl. Math., 35(1), (2016), 269-284.

[13] Steffensen, J. F., *Remarks on iteration*, Skand. Aktuar Tidskr., 16, (1933), 64-72.

[14] Traub, J. F., *Iterative methods for the solution of equations*, Prentice Hall, New Jersey, U.S.A, (1964).

[16] Wang, X., Zhang, T., *A family of Steffensen type method with seventh-order convergence*, Numer. Algor., 62, (2013), 429-444.

[17] Wang, X., Zhang, T., Qian, W., Teng, M., *Seventh-order derivative-free iterative method for solving nonlinear systems*, Numer. Algor., 70, (2015), 545-558.

Chapter 7

Semi-Local Convergence of a Seventh-Order Method for Nonlinear Equations

1. Introduction

Let $\mathcal{B}_1, \mathcal{B}_2$ be Banach spaces, $\Omega \subset \mathcal{B}_1$ be a nonempty open set and $F : \Omega \longrightarrow \mathcal{B}_2$ be a $C^1(\Omega)$ operator. A plethora of applications from computational sciences can be formulated like nonlinear equation

$$\mathcal{G}(x) = 0. \tag{7.1}$$

The task of finding a solution $x_* \in \Omega$ in closed form is achieved only in special cases. That is why iterative procedures are mostly utilize to generate iterates approximating x_*. A survey of iterative procedures can be found in [1] and the references there in (see also [2,3,4]).

In particular, consider the following third-order Newton-like procedure defined for $x_0 \in \Omega$ and each $n = 0, 1, 2, \ldots$ by

$$\begin{aligned} y_n &= x_n + \mathcal{G}'(x_k)^{-1}\mathcal{G}(x_k) \\ x_{n+1} &= x_n - A_n^{-1}\mathcal{G}(x_n), \end{aligned} \tag{7.2}$$

has been used [1]], where $A_n = \mathcal{G}'(x_n)[x_n, y; \mathcal{G}]^{-1}\mathcal{G}(x_n)$, $[.,.;\mathcal{G}] : \Omega \times \Omega \longrightarrow L(\mathcal{B}_1, \mathcal{B}_2)$ is a divided difference of order one [2,3,4]. The motivation, derivation and benefits out of using procedure (7.2) over third-order ones such as e.g. Chebyshev's (containing a bilinear operator) have been also well explained in [2,3,4].

The semi-local convergence analysis was based on generalized continuity conditions, and the results are given in non-affine invariant form. Motivated by optimization considerations we present a finer semi-local convergence analysis with benefits: results are presented in affine invariant form and under weaker conditions. Relevant work can be found in [6]-[12].

The results in [1] are presented in the next Section to make the article as self contained as possible and for comparison. More details can be found in [1]. The new semi-local convergence analysis can be found in Section 3. The examples appear in Section 4 followed by the conclusion in Section 5.

2. History of the Procedure

Throughout the article $U(x_0,\rho)=\{x\in\mathcal{B}_1:\|x-x_0\|<\rho\}$ and $U[x_0,\rho]=\{x\in\mathcal{B}_1:\|x-x_0\|\leq\rho\}$ for some $\rho>0$.

The following semi-local convergence result for procedure (7.2) was shown in [1, Theorem 3].

Theorem 14. Suppose:

(1)
$$\|[x,y;\mathcal{G}]-[\bar{x},\bar{y};\mathcal{G}]\|\leq f(\|x-\bar{x}\|,\|y-\bar{y}\|) \tag{7.3}$$
for all $x,\bar{x},y,\bar{y}\in\Omega$, where $f:S\times S\longrightarrow S$ is a non-decreasing and continuous function satisfying
$$f(0,t)=f(t,0)=\frac{1}{2}f(t,t). \tag{7.4}$$

(2) $\|\mathcal{G}'(x_0)^{-1}\|\leq\beta$ for some $x_0\in\Omega$ with $\mathcal{G}'(x_0)^{-1}\in L(\mathcal{B}_2,\mathcal{B}_1)$

(3) $Max\{\|\mathcal{G}'(x_0)^{-1}\mathcal{G}(x_0)\|,\|A_0^{-1}\mathcal{G}(x_0)\|\}\leq a$

(4) Equation $\left(1-\dfrac{M_0}{1-2\beta f(t,t)}\right)t-1=0$ has a smallest solution T, where $M_0=\beta f(a,a)$.

(5) If $\beta f(T,T)<\dfrac{1}{3}$ and $U[x_0,T]\subset\Omega$, then $M=\dfrac{M_0}{1-2\beta f(T,T)}\in(0,1)$. Then, iteration $\{x_n\}$ given by method (7.2) exists in $U(x_0,T)$, stays in $U(x_0,T)$ for each $n=0,1,2,\ldots$, $\lim_{n\longrightarrow\infty}x_n=x_*\in U[x_0,T]$ and $F(x_*)=0$. Moreover, the limit point x_* is the only solution of equation $F(x)=0$ in $U[x_0,T]$.

Remark. Conditions (7.3) and (7.4) are in general restrictive and may not hold although method can converge. Results are given in non-affine invariant form. The disadvantages over affine invariant results are well explained in the literature [2,3,4,8,9]. The error bounds on $\|x_{n+1}-x_n\|,\|x_*-x_n\|$ can be tighter, if function f is replaced by a tighter one. The uniqueness of solution may be extended. We positively answer to all these concerns in Section 3.

3. Convergence Analysis

The semi-local convergence analysis is based on some constants and real functions. Let λ and μ be non-negative constants. Set $S=[0,\infty)$.

Suppose there exists:

(i) Function $g_0:S\longrightarrow S$ non-decreasing and continuous such that equation
$$g_0(t)-1=0$$
has a smallest solution $\rho_0\in S-\{0\}$ with $\rho_0\geq\lambda$. Set $S_0=[0,\rho_0)$.

(ii) Function $h_0:S_0\times S_0\longrightarrow S$ non-decreasing and continuous in both variables such that equation
$$h_0(t,t)-1=0$$
has a smallest solution $\rho_1\in S_0-\{0\}$ with $\rho_1\geq\mu$. Set $\rho_2=\min\{\rho_0,\rho_1\}$, $\rho_3=\max\{\rho_0,\rho_1\}$ and $S_1=[0,\rho_2)$.

(iii) Equation

$$p(t) - 1 = 0$$

has a smallest solution $\rho \in S_1 - \{0\}$ with $\rho \geq \rho_3$. Set $S_2 = [0, \rho)$. Define function $q : S_2 \longrightarrow S$ by

$$q(t) = \frac{1}{1 - p(t)},$$

where

$$p(t) = \frac{g_0(t)^2 + 2g_0(t) + h_0(t,t)}{1 - h_0(t,t)}.$$

Consider function $h : S_2 \longrightarrow S, h_1 : S_2 \longrightarrow S$ to be non-decreasing and continuous. Define constants

$$\alpha_1 = h(\lambda) + \frac{(1 + g_0(0))h_1(\mu)}{1 - h_0(0,\mu)},$$

$$\gamma_1 = \frac{\alpha_1}{1 - g_0(\lambda)}, \; p_1 = \frac{g_0(\lambda)^2 + 2g_0(\lambda) + h_0(0,\mu)}{1 - h_0(0,\mu)},$$

$$\delta_1 = q_1\alpha_1, \; q_1 = \frac{1}{1 - p_1},$$

functions on the interval S_2 with values in S by

$$\alpha(t) = h(\lambda) + \frac{(1 + g_0(t))h_1(\gamma_1\mu)}{1 - h_0(t,t)},$$

$$\gamma(t) = \frac{\alpha(t)}{1 - g_0(t)}, \; \delta(t) = q(t)\alpha(t),$$

$$\varphi(t) = (\gamma_1 + \frac{\delta^2(t)}{1 - \delta(t)} + \delta_1 + 1)\lambda - 1.$$

(iv) Equations

$$\gamma(t) - 1 = 0$$

and

$$\delta(t) - 1 = 0$$

have smallest solutions $\rho_4, \rho_5 \in S_1 - \{0\}$, respectively with $\rho_4 \geq \rho_3$ and $\rho_5 \geq \rho_3$ and

(v) Equation

$$\varphi(t) = 0$$

has a smallest solution $r \in S_2 - \{0\}$. Set $S_3 = [0, r)$. Then, it follows from these definitions that

$$0 \leq g_0(\lambda) < 1, \; 0 \leq g_0(t) < 1, \tag{7.5}$$

$$0 \leq h_0(0,\mu) < 1, \tag{7.6}$$

$$0 \leq p_1 < 1, \tag{7.7}$$

$$0 \leq p(t) < 1, \tag{7.8}$$

$$0 \leq \gamma(t) < 1, \tag{7.9}$$

$$0 \leq \delta(t) < 1, \tag{7.10}$$

and

$$0 \leq \varphi(t) < 1, \tag{7.11}$$

hold for all $t \in S_3$.

Additionally the following conditions are used:

(vi) There exists $x_0 \in \Omega$, $\lambda \geq 0, \mu \geq 0$ such that $\mathcal{G}'(x_0)^{-1}, A_0^{-1} \in L(\mathcal{B}_2, \mathcal{B}_1)$,

$$\|\mathcal{G}'(x_0)^{-1}\mathcal{G}(x_0)\| \leq \mu,$$

$$A_0^{-1}\mathcal{G}(x_0)\| \leq \lambda.$$

and

$$\|\mathcal{G}'(x_0)^{-1}([x,y;\mathcal{G}] - \mathcal{G}'(x_0))\| \leq h_0(\|x-x_0\|, \|y-x_0\|)$$

(vii) $\|\mathcal{G}'(x_0)^{-1}(\mathcal{G}'(x) - \mathcal{G}'(x_0))\| \leq g_0(\|x-x_0\|)$ for each $x \in \Omega$. Set $\Omega_0 = U(x_0, \rho_0) \cap \Omega$.

(viii) $\|\mathcal{G}'(x_0)^{-1}([y,x;\mathcal{G}] - \mathcal{G}'(x))\| \leq h(\|y-x\|)$ and

$$\|\mathcal{G}'(x_0)^{-1}([x,y;\mathcal{G}] - \mathcal{G}'(x))\| \leq h_1(\|y-x\|)$$

for all $x, y \in \Omega_0$ and

(ix) $U[x_0, r] \subset \Omega$.

Next, we need some relations connecting the iterates $\{x_n\}, \{y_n\}$.
Lemma 5. Suppose iterates $\{x_n\}, \{y_n\}$ exist for each $n = 0, 1, 2, \ldots$. Then, the following assertions hold:

$$\begin{aligned} B_{n+1} := [x_{n+1}, x_n; \mathcal{G}] - A_n & \\ = \; & [x_{n+1}, x_n; \mathcal{G}] - \mathcal{G}'(x_n) \\ & + \mathcal{G}'(x_n)[x_n, y_n; \mathcal{G}]^{-1}([x_n, y_n; \mathcal{G}] - \mathcal{G}'(x_n)), \end{aligned} \tag{7.12}$$

$$\begin{aligned} \mathcal{G}'(x_0)^{-1}(A_{n+1} - \mathcal{G}'(x_0)) &= \mathcal{G}'(x_0)^{-1}(\mathcal{G}(x_{n+1}) - \mathcal{G}'(x_0)) \\ & \quad \times [x_n, y_n; \mathcal{G}]^{-1}\mathcal{G}'(x_0)\mathcal{G}'(x_0)^{-1}(\mathcal{G}'(x_{n+1}) - \mathcal{G}'(x_0)) \\ & \quad + \mathcal{G}'(x_0)^{-1}(\mathcal{G}'(x_{n+1}) - \mathcal{G}'(x_0))[x_n, y_n; \mathcal{G}]^{-1}\mathcal{G}'(x_0) \\ & \quad + [x_n, y_n; \mathcal{G}]^{-1}\mathcal{G}'(x_0)\mathcal{G}'(x_0)^{-1} \\ & \quad \times (\mathcal{G}'(x_{n+1}) - [x_n, y_n; \mathcal{G}]) \end{aligned} \tag{7.13}$$

$$\mathcal{G}'(x_0)^{-1}\mathcal{G}(x_{n+1}) = \mathcal{G}'(x_0)^{-1}B_{n+1}(x_{n+1} - x_n) \tag{7.14}$$

and

$$\begin{aligned} x_{n+2} - x_{n+1} &= A_{n+1}^{-1}\mathcal{G}(x_{n+1}) \\ &= A_{n+1}^{-1}B_{n+1}(x_{n+1} - x_n) \\ &= A_{n+1}^{-1}\mathcal{G}'(x_0)\mathcal{G}'(x_0)^{-1}B_{n+1}(x_{n+1} - x_n). \end{aligned} \tag{7.15}$$

Proof. By the definition of A_n and B_{n+1}, we have in turn that

$$\begin{aligned} B_{n+1} &= [x_{n+1}, x_n; \mathcal{G}] - \mathcal{G}'(x_n) \\ & \quad + \mathcal{G}'(x_n) - \mathcal{G}'(x_n)[x_n, y_n; \mathcal{G}]^{-1}\mathcal{G}'(x_n) \\ &= [x_{n+1}, x_n; \mathcal{G}] - \mathcal{G}'(x_n) + \mathcal{G}'(x_n)(I - [x_n, y_n; \mathcal{G}]^{-1}\mathcal{G}'(x_n)) \\ &= [x_{n+1}, x_n; \mathcal{G}] - \mathcal{G}'(x_n) \\ & \quad + \mathcal{G}'(x_n)[x_n, y_n; \mathcal{G}]^{-1}([x_n, y_n; \mathcal{G}] - \mathcal{G}'(x_n)), \end{aligned}$$

showing (7.12). Using the definition of A_{n+1}, we get in turn that

$$\begin{aligned} A_{n+1}-G'(x_0) &= (G'(x_{n+1})-G'(x_0))[x_n,y_n;G]^{-1}G'(x_{n+1}) \\ &\quad +G'(x_0)[x_n,y_n;G]^{'}G'(x_{n+1})-G'(x_0) \\ &= (G'(x_{n+1})-G'(x_0))[x_n,y_n;G]^{-1}(G'(x_{n+1})-G'(x_0)) \\ &\quad +(G'(x_{n+1})-G'(x_0))[x_n,y_n;G]^{'}(G'(x_{n+1})-G'(x_0)) \\ &\quad +(G'(x_{n+1})-G'(x_0))[x_n,y_n;G]^{-1}G'(x_0) \\ &\quad +G'(x_0)[x_n,y_n;G]^{'}G'(x_{n+1})-G'(x_0), \end{aligned}$$

so

$$\begin{aligned} G'(x_0)^{-1}(A_{n+1}-G'(x_0)) &= G'(x_0)^{-1}(G'(x_{n+1})-G'(x_0)) \\ &\quad \times[x_n,y_n;G]^{'}G'(x_0)G'(x_0)^{-1}(G'(x_{n+1})-G'(x_0)) \\ &\quad +G'(x_0)^{-1}(G'(x_{n+1})-G'(x_0))[x_n,y_n;G]^{'}G'(x_0) \\ &\quad +[x_n,y_n;G]G'(x_0)G'(x_0)^{-1}(G'(x_{n+1})-[x_n,y_n;G], \end{aligned}$$

showing (7.13). Moreover, by the definition of method (7.2) we can write

$$\begin{aligned} G(x_{n+1}) &= G(x_{n+1})-G(x_n)+G(x_n) \\ &= ([x_{n+1},x_n;G]-A_{n+1})(x_{n+1}-x_n) \\ &= B_{n+1}(x_{n+1}-x_n), \end{aligned}$$

so

$$G'(x_0)^{-1}G(x_{n+1})=G'(x_0)^{-1}B_{n+1}(x_{n+1}-x_n),$$

showing (7.14). Furthermore, by the second substep of method (7.2) and (7.14) we obtain

$$\begin{aligned} x_{n+2}-x_{n+1} &= A_{n+1}^{-1}G(x_{n+1}) \\ &= A_{n+1}^{-1}B_{n+1}(x_{n+1}-x_n) \\ &= A_{n+1}^{-1}G'(x_0)G'(x_0)^{-1}B_{n+1}(x_{n+1}-x_n). \end{aligned}$$

□

Next, we show the semi-local convergence result for method (7.2).

Theorem 15. Suppose conditions (i)-(ix) hold. Then, iterations $\{x_n\},\{y_n\}$ initiated at $x_0\in\Omega$ and produced by method (7.2) exist in $U(x_0,r)$, stay in $U(x_0,r)$ for each $n=0,1,2,\ldots,$ $\lim_{n\longrightarrow\infty} x_n=x_*$ and $G(x_*)=0$.

Proof. It follows from (vi), the definition of r that y_0 is well defined, and

$$\|y_0-x_0\|=\|G'(x_0)^{-1}G(x_0)\|\leq\mu<r,$$

so $y_0\in U(x_0,r)$. Iterate x_1 is well defined by method (7.2) for $n=0$ and $\|x_1-x_0\|\leq\mu<r$, so $x_1\in U(x_0,r)$. Let $w\in U(x_0,r)$. Then, by (vii)

$$\|G'9x_0)^{-1}(G'(w)-G'(x_0))\|\leq g_0(\|w-x_0\|)\leq g_0(r)<1. \tag{7.16}$$

So, $G'(w)^{-1}\in L(\mathcal{B}_2,\mathcal{B}_1)$, and

$$\|G'(w)^{-1}G'(x_0)\|\leq\frac{1}{1-g_0(\|w-x_0\|)} \tag{7.17}$$

follow by a Lemma due to Banach for invertible linear operators [3,9,11]. By (7.13) and (viii), we get

$$\begin{aligned}\|\mathcal{G}'(x_0)(A_{k+1}-\mathcal{G}'(x_0))\| &\leq \frac{g_0(\|x_{k+1}-x_0\|)^2}{1-h_0(\|x_k-x_0\|,\|y_k-x_0\|)}\\ &+\frac{g_0(\|x_{k+1}-x_0\|}{1-h_0(\|x_k-x_0\|,\|y_k-x_0\|}\\ &+\frac{g_0(\|x_{k+1}-x_0\|)+h_0(\|x_k-x_0\|,\|y_k-x_0\|)}{1-h_0(\|x_k-x_0\|,\|y_k-x_0\|)}\\ &= p_{k+1}<1,\end{aligned}$$

so

$$\|A_{k+1}^{-1}\mathcal{G}'(x_0)\| \leq q_{k+1}, \tag{7.18}$$

and iterate x_{k+2} is well defined by the second substep of method (7.2). By the definition of B_{k+1} in (7.12) and α_{k+1} we get in turn that

$$\begin{aligned}\|\mathcal{G}'(x_0)^{-1}B_{k+1}\| &\leq h(\|x_{k+1}-x_k\|\\ &+\frac{(1+g_0(\|x_k-x_0\|))h_1(\|y_k-x_k\|)}{1-g_0(\|x_k-x_0\|,\|y_k-x_0\|)}=\alpha_{k+1}.\end{aligned} \tag{7.19}$$

It then follows from (7.15), (7.18) and (7.19)

$$\begin{aligned}\|x_{k+2}-x_{k+1}\| &\leq \|A_{k+1}^{-1}\mathcal{G}'(x_0)\|\|\mathcal{G}'(x_0)^{-1}B_{k+1}\|\|x_{k+1}-x_k\|\\ &\leq q_{k+1}\alpha_{k+1}\|x_{k+1}-x_k\|\\ &\leq \bar{\delta}(r)\|x_{k+1}-x_k\|,\end{aligned} \tag{7.20}$$

where

$$\bar{\delta}(r)=\begin{cases}\delta_1 & \text{if } k=0\\ \delta(r) & \text{if } k=1,2,\ldots.\end{cases}$$

Similarly, by the first substep of method (7.2), (7.17) (for $w=x_{k+1}$) and (7.12), we have the estimate

$$\begin{aligned}\|y_k-x_k\| &\leq \|\mathcal{G}'(x_0)^{-1}\mathcal{G}'(x_0)\|\|\mathcal{G}'(x_0)^{-1}\mathcal{G}(x_k)\|\\ &\leq \frac{\alpha_k}{1-g_0(\|x_k-x_0\|)}\|x_k-x_{k-1}\|\\ &\leq \bar{\gamma}(r)\|x_k-x_{k-1}\|,\end{aligned} \tag{7.21}$$

where

$$\bar{\gamma}(r)=\begin{cases}\gamma_1 & \text{if } k=1\\ \gamma(r) & \text{if } k=2,\ldots.\end{cases}$$

Notice that we also used

$$\begin{aligned}\|x_{k+1}-x_0\| &\leq \|x_{k+1}-x_k\|+\|x_k-x_{k-1}\|+\ldots+\|x_2-x_1\|+\|x_1-x_0\|\\ &\leq \delta^k\lambda+\delta^{k-1}\lambda+\ldots+\delta^2\lambda+\delta_1\lambda+\lambda\\ &= (\delta^2\frac{1-\delta^{k-1}}{1-\delta}+\delta_1+1)\lambda\\ &\leq (\frac{\delta^2}{1-\delta}+\delta_1+1)\lambda<r\end{aligned} \tag{7.22}$$

(by (v)), so $x_{k+1} \in U(x_0, r)$. Similarly,

$$\begin{aligned} \|y_k - x_0\| &\leq \|y_k - x_k\| + \|x_k - x_0\| \\ &\leq \gamma^{k-1}\gamma_1\lambda + \delta^{k-1}\lambda + \ldots + \delta^2\lambda + (\frac{\delta^2}{1-\delta} + \delta_1 + 1)\lambda \\ &\leq (\gamma_1 + \frac{\delta^2}{1-\delta} + \delta_1 + 1)\lambda < r. \end{aligned} \tag{7.23}$$

Hence, iterate $y_k \in U(x_0, r)$. Similarly, we have $x_{k+2}, y_{k+1} \in U(x_0, r)$. It follows from (7.20) that

$$\|x_{k+2} - x_{k+1}\| \leq \delta^{k+1}\lambda. \tag{7.24}$$

Let $m \geq 1$ be a positive integer. Then, it follows by (7.24) that

$$\begin{aligned} \|x_{k+m} - x_{k+1}\| &\leq \|x_{k+m} - x_{k+m-1}\| \\ &\quad + \|x_{k+m-1} - x_{k+m-2}\| + \ldots + \|x_{k+2} - x_{k+1}\| \\ &\leq (\delta^{k+m-1} + \delta^{k+m-2} + \ldots + \delta^{k+1})\lambda \\ &= (\delta^{m-2} + \delta^{m-3} + \ldots + 1)\delta^{k+1}\lambda \\ &= \frac{1-\delta^{m-1}}{1-\delta}\delta^{k+1}\lambda \leq \delta^{k+1}\lambda. \end{aligned} \tag{7.25}$$

It follows that sequence $\{x_k\}$ is complete in a Banach space $\mathcal{B}_1$, and such it converges to some $x_* \in U[x_0, r]$, (since $U[x_0, r]$ is a closed set.). Moreover, using (7.14) and (7.19), we obtain

$$\|\mathcal{G}'(x_0)^{-1}\mathcal{G}(x_{k+1})\| \leq \alpha(r)\|x_{k+1} - x_k\|. \tag{7.26}$$

By letting $k \longrightarrow \infty$ in (7.26) and using the continuity of $\mathcal{G}$ we deduce $\mathcal{G}(x_*) = 0$. Furthermore, by letting $m \longrightarrow \infty$ in (7.25), we conclude

$$\|x_* - x_k\| \leq \delta^k\lambda. \tag{7.27}$$

□

Next, we present a uniqueness of the solution result by not necessarily using all the conditions of Theorem 15..

PROPOSITION 2. *Suppose*

$$\|\mathcal{G}^{-1}(x_0)(\mathcal{G}'(x) - \mathcal{G}'(x_0))\| \leq g_0(\|x - x_0\|)$$

for all $x \in \Omega$, the point x_* is a solution of equation $\mathcal{G}(x) = 0$ with $x_* \in U[x_0, b] \subset \Omega$ for some $b > 0$ and these exists $c \geq b$ such that

$$\int_0^1 g_0((1-\theta)b + \theta c)d\theta < 1.$$

Define $\Omega_1 = U[x_0, c] \cap \Omega$. Then, the point x_* is the only solution of equation $\mathcal{G}(x) = 0$ in the set Ω_1.

Proof. Let $y_* \in \Omega_1$ with $\mathcal{G}(y_*) = 0$. Define operator $Q = \int_0^1 \mathcal{G}'(x_* + \theta(y_* - x_*))d\theta$. Then, we obtain in turn that

$$\begin{aligned}\|\mathcal{G}'(x_0)^{-1}(Q - \mathcal{G}'(x_0))\| &\leq \int_0^1 g_0((1-\theta)\|x_* - x_0\| + \theta\|y_* - x_)\|]d\theta \\ &\leq \int_0^1 g_0((1-\theta)b + \theta c)d\theta < 1.\end{aligned}$$

So, we conclude $y_* = x_*$ by the using the identity $0 = \mathcal{G}(y_*) - \mathcal{G}(x_*) = Q(y_* - x_*)$, and the invertiblility of linear operator Q.

□

Remark. The concerns in Remark 2. were addressed by Theorem 15. In particular, notice that majorant functions are tighter than f, and the latter implies the new functions but not necessarily vice versa.

4. Numerical Experiments

We compute the radius of convergence in this section.

Example 5. Let us consider a scalar function $\mathcal{G}$ defined on the set $D = U[u_0, 1-s]$ for $s \in (0,1)$ by

$$\mathcal{G}(x) = x^3 - s.$$

Choose $x_0 = 1$. The conditions of Theorem 15 are satisfied provided that $\mu = \frac{1-s}{3}, g_0(t) = (3-s)t, h_0(u,t) = \frac{1}{2}(3-s)(u+t), h(t) = h_1(t) = (1 + \frac{1}{3-s})t, \lambda = \frac{1-s}{y_0^2 + x_0y_0 + x_0^2}, y_0 = x_0 - \mathcal{G}'(x_0)^{-1}\mathcal{G}(x_0)$. Then, we have

$$\rho_0 = \rho_1 = \rho_2 = \rho_3 = 0.4878, \rho = 0.1152, \rho_4 = 0.3565, \rho_5 = 0.1079$$

$$\text{and } r = 0.0962.$$

5. References

[1] Amat, S., Bermudez, C., Busquier, S., Plaza, S., *On a third-order Newton-type method free of bilinear operators*, Numer. Linear Algebra Appl., 17, (2010), 639-653.

[2] Argyros, I. K., *Convergence and Applications of Newton-type Iterations*, Springer Verlag, Berlin, Germany, (2008).

[3] Argyros, I. K., *The theory and applications of iteration methods*, 2nd Edition, CRC Press, Engineering Series, Taylor and Francis Group, (2022).

[4] Argyros, I. K., *Unified convergence criteria for Banach space valued methods with applications*, Mathematics, 9, 16, (2021), 1942.

[5] Behl, R., Maroju, P., Martinez, E., Singh, S., *A study of the local convergence of a fifth-order iterative method*, Indian J. Pure Appl. Math., 51, 2, (2020), 439-455.

[6] Ezquerro, J. A., Hernndez, M. A., *Halleys method for operators with unbounded second derivative*, Applied Numerical Mathematics, 57(3), (2007), 354-360.

[7] Gutirrez, J. M., *A new semi-local convergence theorem for Newtons method*, Journal of Computational and Applied Mathematics, 79, (1997), 131-145.

[8] Hernndez, M. A., Romero, N., *General study of iterative processes of R-order at least three under weak convergence conditions*, Journal of Optimization Theory and Applications, 133(2), (2007), 163-177.

[9] Hernndez, M. A., Rubio, M. J., *Semilocal convergence of the secant method under mild convergence conditions of differentiability*, Computers and Mathematics with Applications, 44(3-4), (2002), 277-285.

[10] Kantorovich, L. V., Akilov, G. P., *Functional Analysis*, Pergamon Press: Oxford, (1982).

[11] Kou, J., Li, Y., Wang, X., *A modification of Newtons method with third-order convergence*, Applied Mathematics and Computation, 181, (2007), 1106-1111.

[12] Traub, J. F., *Iterative methods for the solution of equations*, Prentice Hall, New Jersey, U.S.A, (1964).

Chapter 8

Semi-Local Convergence of a Fifth-Order Method for Solving Nonlinear Equations

1. Introduction

The local as well as the semi-local convergence analysis of two-step and three-step methods is presented under $\omega-$ continuity conditions. The methods are defined for $x_0 \in D$, and all $n = 0, 1, 2, \ldots$ by

$$\begin{aligned} y_n &= x_n - F'(x_n)^{-1}F(x_n) \\ B_n &= 2F'\left(\frac{3x_n+y_n}{4}\right) - F'\left(\frac{x_n+y_n}{2}\right) + 2F'\left(\frac{x_n+3y_n}{4}\right) \\ x_{n+1} &= x_n - 3B_n^{-1}F(x_n), \end{aligned} \tag{8.1}$$

$$\begin{aligned} y_n &= x_n - F'(x_n)^{-1}F(x_n) \\ z_n &= y_n - [\frac{1}{4}I + \frac{1}{2}(F'(y_n)^{-1}F'(x_n) \\ &\quad + \frac{1}{4}(F'(y_n^{-1}(F'(x_n))^2]F'(x_n)^{-1}(F(y_n) \\ x_{n+1} &= z_n - [\frac{1}{2}I + \frac{1}{2}(F'(y_n)^{-1}F'(x_n))^2]F'(x_n)^{-1}F(z_n), \end{aligned} \tag{8.2}$$

$$\begin{aligned} y_n &= x_n - \frac{2}{3}F'(x_n)^{-1}F(x_n) \\ x_{n+1} &= x_n - \frac{1}{2}[-I + \frac{9}{2}F'(y_n)^{-1}F'(x_n) \\ &\quad + \frac{3}{4}F'(x_n)^{-1}F'(y_n)]F'(x_n)^{-1}F(x_n), \end{aligned} \tag{8.3}$$

$$\begin{aligned} y_n &= x_n - \frac{2}{3}F'(x_n)^{-1}F(x_n) \\ z_n &= x_n - \frac{1}{2}[(3F'(y_n) - F'(x_n))^{-1}(3F'(y_n) + F'(x_n))] \\ &\quad \times F'(x_n)^{-1}F(x_n) \\ x_{n+1} &= z_n - 2(3F'(y_n) - F'(x_n))^{-1}F(z_n) \end{aligned} \tag{8.4}$$

and

$$\begin{aligned} y_n &= x_n - F'(x_n)^{-1}F(x_n) \\ A_n &= F'(x_n) + 4F'\left(\frac{x_n + y_n}{2}\right) + F'(y_n) \\ x_{n+1} &= x_n - A_n^{-1}F(x_n), \end{aligned} \tag{8.5}$$

The two-step methods are of convergence order four whereas the three step are of convergence order six [5]-[11], [13]. Similar work can be done on the methods in [1]-[4], [12].

2. Local convergence

The majorant functions are provided for each method.

Method (8.1)

$$g_1(t) = \frac{\int_0^1 \omega((1-\theta)t)d\theta}{1-\omega_0(t)},$$

$$\begin{aligned} g_2(t) &= \frac{\int_0^1 \omega((1-\theta)t)d\theta}{1-\omega_0(t)} \\ &+ \frac{\bar{\omega}(t)(1+\int_0^1 \omega_0(\theta t)d\theta)}{3(1-\omega_0(t))(1-q(t))}, \\ q(t) &= 2\omega_0\left(\frac{(3+g_1(t))t}{4}\right) + \omega_0\left(\frac{(1+g_1(t))t}{2}\right) \\ &+2\omega_0\left(\frac{(1+3g_1(t))t}{4}\right) \end{aligned}$$

$$\bar{\omega}(t) = \begin{cases} v_1(t) \\ \text{or} \\ v_2(t), \end{cases}$$

$$\begin{aligned} v_1(t) &= 2\left[\omega_0\left(\frac{(1+g_1(t))t}{4}\right) + \omega_0\left(\frac{(1+g_1(t))t}{2}\right)\right] \\ &+2\omega_0\left(\frac{(1+g_1(t))t}{4}\right) \end{aligned}$$

and

$$\begin{aligned} v_2(t) &= 2\left[\omega_0\left(\frac{(3+g_1(t))t}{4}\right) + 3\omega_0(t) + \omega_0\left(\frac{(1+g_1(t))t}{2}\right)\right. \\ &\left. + \omega_0\left(\frac{(1+3g_1(t))t}{4}\right)\right] \end{aligned}$$

Method (8.2)

$$g_1(t) = \frac{\int_0^1 \omega((1-\theta)t)d\theta}{1-\omega_0(t)},$$

$$
\begin{aligned}
g_2(t) &= \Bigg[\frac{\int_0^1 \omega((1-\theta)g_1(t)t)d\theta}{1-\omega_0(g_1(t)t)} \\
&+\frac{\bar{\omega}(t)(1+\int_0^1 \omega_0(\theta g_1(t)t)d\theta)}{(1-\omega_0(t))(1-\omega_0(g_1(t)t))} \\
&+\frac{1}{4}\frac{\bar{\omega}}{1-\omega_0(g_1(t)t)}(4+\frac{\bar{\omega}(t)}{1-\omega_0(g_1(t)t)}) \\
&(\frac{(1+\int_0^1 \omega_0(\theta g_1(t)t)d\theta}{1-\omega_0(t)}\Bigg] g_1(t)
\end{aligned}
$$

and

$$
\begin{aligned}
g_3(t) &= \left[1+\frac{1}{2}\left(2+\frac{2\bar{\omega}(t)}{1-\omega_0(g_1(t)t)}+\left(\frac{2\bar{\omega}(t)}{1-\omega_0(g_1(t)t)}\right)^2\right)\right. \\
&\left.\times\frac{(1+\int_0^1 \omega_0(\theta g_2(t)t)d\theta)}{1-\omega_0(t)}\right] g_2(t).
\end{aligned}
$$

where

$$
\bar{\omega}(t)=\begin{cases} v_1(t) \\ \text{or} \\ v_2(t), \end{cases}
$$

$$
v_1(t)=\omega((1+g_1(t))t)
$$

and

$$
v_2(t)=\omega_0(t)+\omega_0(g_1(t)t).
$$

Method (8.3)

$$
g_1(t)=\frac{\int_0^1 \omega((1-\theta)t)d\theta+\frac{1}{3}(1+\int_0^1 \omega_0(\theta t)d\theta}{1-\omega_0(t)},
$$

and

$$
\begin{aligned}
g_2(t) &= \frac{\int_0^1 \omega((1-\theta)t)d\theta}{1-\omega_0(t)} \\
&+3\bar{\omega}(t)\left(\frac{3}{1-\omega_0(g_1(t)t)}+\frac{1}{1-\omega_0(t)}\right) \\
&\times\frac{(1+\int_0^1 \omega_0(\theta t)d\theta)}{1-\omega_0(t)},
\end{aligned}
$$

where $\bar{\omega}(t)$ is given in method (8.2).

Method (8.4)

$$
g_1(t)=\frac{\int_0^1 \omega((1-\theta)t)d\theta+\frac{1}{3}(1+\int_0^1 \omega_0(\theta t)d\theta}{1-\omega_0(t)},
$$

$$
q(t)=\frac{1}{3}(3\omega_0(g_1(t)t)+\omega_0(t)),
$$

$$g_2(t) = \frac{\int_0^1 \omega((1-\theta)t)d\theta}{1-\omega_0(t)} + 3\bar{\omega}(t)\frac{(1+\int_0^1 \omega_0(\theta t)d\theta)}{4(1-\omega_0(t))(1-q(t))}$$

and

$$g_3(t) = \left[1 + \frac{1+\int_0^1 \omega_0(\theta g_2(t)t)d\theta}{1-q(t)}\right] g_2(t),$$

where $\bar{\omega}(t)$ is given in method (8.2).

Method (8.5)

$$g_1(t) = \frac{\int_0^1 \omega((1-\theta)t)d\theta}{1-\omega_0(t)},$$

$$q(t) = \frac{1}{6}\left[\omega_0(t) + \omega_0\left(\frac{(1+g_1(t))t}{2}\right) + \omega_0(g_1(t)t)\right],$$

$$g_2(t) = \frac{\int_0^1 \omega((1-\theta)t)d\theta}{1-\omega_0(t)} + \bar{\omega}(t)\frac{(1+\int_0^1 \omega_0(\theta t)d\theta)}{6(1-\omega_0(t))(1-q(t))},$$

where

$$\bar{\omega}(t) = \begin{cases} v_1(t) \\ \text{or} \\ v_2(t), \end{cases}$$

$$v_1(t) = 4\omega\left(\frac{(1+g_1(t))t}{2}\right) + \omega((1+g_1(t))t)$$

and

$$v_2(t) = 5\omega_0(t) + 4\omega_0\left(\frac{(1+g_1(t))t}{2}\right) + \omega_0(g_1(t)t).$$

Next, we present the estimates that lead to the functions “$g$for each method.

Method (8.1)

$$\begin{aligned} y_n - x^* &= x_n - x^* - F'(x_n)^{-1}F(x_n), \\ \|y_n - x^*\| &\leq g_1(\|x_n - x^*\|)\|x_n - x^*\| < \|x_n - x^*\| < r, \end{aligned}$$

$$\begin{aligned} x_{n+1} - x^* &= x_n - x^* - F'(x_n)^{-1}F(x_n) + (F'(x_n)^{-1} - 3B_n^{-1})F(x_n) \\ &= y_n - x^* + F'(x_n)^{-1}(B_n - 3F'(x_n))B_n^{-1}F(x_n), \end{aligned}$$

$$\begin{aligned} \|(3F'(x^*))^{-1}(B_n - 3F'(x^*))\| &\leq 2\|F'(x^*)^{-1}(F'(\frac{3x_n+y_n}{4}) - F'(x^*))\| \\ &+ \|F'(x^*)^{-1}(F'(\frac{x_n+y_n}{2}) - F'(x^*))\| \\ &+ 2\|F'(x^*)^{-1}(F'(\frac{x_n+3y_n}{4}) - F'(x^*))\| \\ &\leq q_n, \end{aligned}$$

$$
\begin{aligned}
B_n - 3F'(x_n) &= 2(F'(\frac{3x_n+y_n}{4}) - F'(x_n)) \\
&\quad -(F'(\frac{x_n+y_n}{2}) - F'(x_n)) \\
&\quad +2(F'(\frac{x_n+3y_n}{4}) - F'(x_n)), \\
\|F'(x^*)^{-1}(B_n - 3F'(x_n))\| &\leq q_n < 1, \\
\|B_n^{-1}F'(x^*)\| \leq \frac{1}{3(1-q_n)}
\end{aligned}
$$

and

$$\|x_{n+1} - x^*\| \leq g_2(\|x_n - x^*\|)\|x_n - x^*\| \leq \|x_n - x^*\|.$$

Method (8.2)

$$
\begin{aligned}
\|y_n - x^*\| &\leq g_1(\|x_n - x^*\|)\|x_n - x^*\| \leq \|x_n - x^*\|, \\
z_n - x^* &= y_n - x^* - F'(y_n)^{-1}F(y_n) \\
&\quad +(F'(y_n)^{-1} - F'(x_n)^{-1})F(y_n) \\
&\quad +[I - \frac{1}{4}I - \frac{1}{2}F'(y_n)^{-1}F'(x_n) \\
&\quad -\frac{1}{4}(F'(y_n)^{-1}F'(x_n))^2]F'(x_n)^{-1}F(y_n) \\
&= y_n - x^* - F'(y_n)^{-1}F(y_n) \\
&\quad +F'(y_n)^{-1}(F'(x_n) - F'(y_n))F'(x_n)^{-1}F(y_n) \\
&\quad +[\frac{1}{2}(I - F'(y_n)^{-1}F'(x_n)) + \frac{1}{4}(I - F'(y_n)F'(x_n)^2)]F'(x_n)^{-1}F(y_n),
\end{aligned}
$$

$$
\begin{aligned}
\|z_n - x^*\| &\leq \left[\frac{\int_0^1 \omega((1-\theta)\|y_n - x^*\|)d\theta}{1 - \omega_0(\|y_n - x^*\|)} + \frac{\bar{\omega}_n(1 + \int_0^1 \omega_0(\theta\|y_n - x^*\|)d\theta}{(1 - \omega_0(\|x_n - x^*\|))(1 - \omega_0(\|y_n - x^*\|))}\right. \\
&\quad +\frac{1}{4}\frac{\bar{\omega}_n}{1 - \omega_0(\|y_n - x^*\|)}\left(4 + \frac{\bar{\omega}_n}{1 - \omega_0(\|y_n - x^*\|)}\right) \\
&\quad \left.\times \frac{(1 + \int_0^1 \omega_0(\theta\|y_n - x^*\|)d\theta)}{1 - \omega_0(\|x_n - x^*\|)}\right]\|y_n - x^*\| \\
&\leq g_2(\|x_n - x^*\|)\|x_n - x^*\| \leq \|x_n - x^*\|,
\end{aligned}
$$

$$x_{n+1} - x^* = z_n - x^* - \frac{1}{2}[2I - 2(I - F'(y_n)^{-1}F'(x_n)) + (I - F'(y_n)^{-1}F'(x_n))^2]F'(x_n)^{-1}F(z_n)$$

and

$$
\begin{aligned}
\|x_{n+1} - x^*\| &\leq \left[1 + \frac{1}{2}\left(2 + \frac{\bar{\omega}_n}{1 - \omega_0(\|y_n - x^*\|)}\right.\right. \\
&\quad \left.\left. + \left(\frac{\bar{\omega}_n}{1 - \omega_0(\|y_n - x^*\|)}\right)^2\right)\frac{(1 + \int_0^1 \omega_0(\theta\|z_n - x^*\|)}{1 - \omega_0(\|x_n - x^*\|)}\right]\|z_n - x^*\| \\
&\leq g_3(\|x_n - x^*\|)\|x_n - x^*\| \leq \|x_n - x^*\|.
\end{aligned}
$$

Method (8.3)

$$\begin{aligned}
\|y_n - x^*\| &\leq g_1(\|x_n - x^*\|)\|x_n - x^*\| \leq \|x_n - x^*\|,\\
x_{n+1} - x^* &= x_n - x^* - F'(x_n)^{-1}F(x_n) + [\frac{3}{2}I - \frac{9}{8}F'(y_n)^{-1}F'(x_n)\\
&\quad -\frac{3}{8}F'(x_n)^{-1}F'(y_n)]F'(x_n)^{-1}F(x_n)\\
&= x_n - x^* - F'(x_n)^{-1}F(x_n) + \frac{1}{8}[9(I - F'(y_n)^{-1}F'(x_n)\\
&\quad +3(I - F'(x_n)^{-1}F'(y_n))]F'(x_n)^{-1}F(x_n),
\end{aligned}$$

so

$$\begin{aligned}
\|x_{n+1} - x^*\| &\leq \Bigg[\frac{\int_0^1 \omega((1-\theta)\|x_n - x^*\|)d\theta}{1 - \omega_0(\|x_n - x^*\|)}\\
&\quad +3\bar{\omega}_n\left(\frac{3}{1 - \omega_0(\|y_n - x^*\|)} + \frac{1}{1 - \omega_0(\|x_n - x^*\|)}\right)\\
&\quad \times\frac{(1 + \int_0^1 \omega_0(\theta\|x_n - x^*\|)}{1 - \omega_0(\|x_n - x^*\|)}\Bigg]\|x_n - x^*\|\\
&\leq g_2(\|x_n - x^*\|)\|x_n - x^*\| \leq \|x_n - x^*\|.
\end{aligned}$$

Method (8.4)

$$\|y_n - x^*\| \leq g_1(\|x_n - x^*\|)\|x_n - x^*\| \leq \|x_n - x^*\|,$$

$$\begin{aligned}
&\|(2F'(x^*))^{-1}(3(F'(y_n) - F'(x^*)) - (F'(x_n) - F'(x^*))\|\\
&\leq \frac{1}{2}(3\omega_0(\|y_n - x^*\|) + \omega_0(\|x_n - x^*\|)) \leq q_n,\\
A_n &= 3F'(y_n) - F'(x_n),\\
\|A_n^{-1}F'(x^*)\| &\leq \frac{1}{2(1 - q_n)},
\end{aligned}$$

$$\begin{aligned}
z_n - x^* &= x_n - x^* - F'(x_n)^{-1}F(x_n)\\
&\quad +[I - \frac{1}{2}A_n^{-1}(3F'(y_n) + F'(x_n))]F'(x_n)^{-1}F(x_n)\\
&= x_n - x^* - F'(x_n)^{-1}F(x_n)\\
&\quad +\frac{3}{2}A_n^{-1}(F'(y_n) - F'(x_n))F'(x_n)^{-1}F(x_n),
\end{aligned}$$

$$\begin{aligned}
\|z_n - x^*\| &\leq \Bigg[\frac{\int_0^1 \omega((1-\theta)\|x_n - x^*\|)d\theta}{1 - \omega_0(\|x_n - x^*\|)}\\
&\quad + \frac{3\bar{\omega}_n(1 + \int_0^1 \omega_0(\theta\|x_n - x^*\|)d\theta}{4(1 - \omega_0(\|x_n - x^*\|))(1 - q_n)}\Bigg]\|x_n - x^*\|\\
&\leq g_2(\|x_n - x^*\|)\|x_n - x^*\| \leq \|x_n - x^*\|.
\end{aligned}$$

Method (8.5)

$$\|y_n - x^*\| \leq g_1(\|x_n - x^*\|)\|x_n - x^*\| \leq \|x_n - x^*\|,$$

$$\begin{aligned}
\|(6F'(x^*))^{-1}(A_n - F'(x^*))\| &\leq \frac{1}{6}[\|F'(x^*)^{-1}(F'(x_n) - F'(x^*))]\\
&\quad +4\|F'(x^*)^{-1}(F'(\frac{x_n+y_n}{2}) - F'(x^*))\|\\
&\quad +\|F'(x^*)^{-1}(F'(y_n) - F'(x^*))\|] \leq q_n,\\
\|A_n^{-1}F'(x^*)\| &\leq \frac{1}{6(1-q_n)},
\end{aligned}$$

$$\begin{aligned}
x_{n+1} - x^* &= x_n - x^* - F'(x_n)^{-1}F(x_n)\\
&\quad +(F'(x_n)^{-1} - 6A_n^{-1})F(x_n)\\
&= x_n - x^* - F'(x_n)^{-1}F(x_n)\\
&\quad +F'(x_n)^{-1}(A_n - 6F'(x_n))A_n^{-1})F(x_n),
\end{aligned}$$

so

$$\begin{aligned}
\|x_{n+1} - x^*\| &\leq \Bigg[\frac{\int_0^1 \omega((1-\theta)\|x_n - x^*\|)d\theta}{1 - \omega_0(\|x_n - x^*\|)}\\
&\quad +\frac{\bar{\omega}_n(1 + \int_0^1 \omega_0(\theta\|x_n - x^*\|)d\theta)}{6(1 - \omega_0(\|x_n - x^*\|))(1 - q_n)}\Bigg]\|x_n - x^*\|\\
&\leq g_3(\|x_n - x^*\|)\|x_n - x^*\| \leq \|x_n - x^*\|.
\end{aligned}$$

Notice that as in previous chapters the radius of convergence r is determined as the smallest solution of the equations $g_i(t) - 1 = 0$.

Next, the semi-local convergence for these methods is developed based on the concept of majorizing sequences. We use the same notation for the functions ω_0 and ω although they relate to x_0 (not x^*) as indicated in the preface.

Method (8.1)

The sequence $\{t_n\}$, $\{s_n\}$ majorizing this method for $t_0 = 0$ and $s_0 \geq 0$ are defined for all $n = 0, 1, 2, \ldots$ by

$$\begin{aligned}
q_n &= 2\omega_0\left(\frac{3t_n+s_n}{4}\right)+\omega_0\left(\frac{t_n+s_n}{2}\right)+2\omega_0\left(\frac{t_n+3s_n}{4}\right),\\
\bar{\omega}_n &= \begin{cases} v_n^1 \\ v_n^2, \end{cases}\\
v_n^1 &= 2\omega\left(\frac{s_n-t_n}{4}\right)+\omega\left(\frac{s_n-t_n}{3}\right)+2\omega\left(\frac{s_n-t_n}{4}\right),\\
v_n^2 &= 2\omega_0\left(\frac{3t_n+s_n}{4}\right)+5\omega_0(t_n)+2\omega_0\left(\frac{3s_n+t_n}{4}\right),\\
t_{n+1} &= s_n+\frac{\bar{\omega}_n(s_n-t_n)}{3(1-q_n)},\\
\alpha_{n+1} &= \int_0^1 \omega((1-\theta)(t_{n+1}-t_n))d\theta)(t_{n+1}-t_n)\\
&\quad +(1+\omega_0(t_n))(t_{n+1}-s_n),\\
s_{n+1} &= t_{n+1}+\frac{\alpha_{n+1}}{1-\omega_0(t_{n+1})}.
\end{aligned}$$

Method (8.2)

$$\begin{aligned}
u_n &= s_n+\frac{1}{4}\left(2+\frac{\bar{\omega}_n}{1-\omega_0(s_n)}\right)^2\frac{\int_0^1\omega((1-\theta)(s_n-t_n))d\theta(s_n-t_n)}{1-\omega_0(t_n)},\\
t_{n+1} &= u_n+\frac{1}{2(1-\omega_0(t_n))}\left(2+2\frac{\bar{\omega}_n}{1-\omega_0(s_n)}+\left(\frac{\bar{\omega}_n}{1-\omega_0(s_n)}\right)^2\right)\\
&\quad \times((1+\int_0^1\omega_0(s_n+\theta(u_n-s_n))d\theta)(u_n-s_n)\\
&\quad +\int_0^1\omega((1-\theta)(s_n-t_n))d\theta(s_n-t_n)),\\
s_{n+1} &= t_{n+1}+\frac{\alpha_{n+1}}{1-\omega_0(t_{n+1})},
\end{aligned}$$

where the sequence $\{\alpha_{n+1}\}$ is given in method (8.2).

Method (8.3) (for s_0 being $\frac{2}{3}\eta$).

$$\begin{aligned}
t_{n+1} &= s_n+\frac{3}{2}\left(\frac{1}{3}+9\frac{\bar{\omega}_n}{8(1-\omega_0(s_n))}+3\frac{\bar{\omega}_n}{8(1-\omega_0(s_n))}\right)(s_n-t_n),\\
s_{n+1} &= t_{n+1}+\frac{2\beta_{n+1}}{3(1-\omega_0(t_{n+1}))},
\end{aligned}$$

where

$$\beta_{n+1} = (1+\int_0^1\omega_0(t_n+\theta(t_{n+1}-t_n))d\theta)(t_{n+1}-t_n)+\frac{3}{2}(1+\omega_0(t_n))(s_n-t_n).$$

Method (8.4) (for s_0 being $\frac{2}{3}\eta$).

$$\begin{aligned}
u_n &= s_n + \frac{(3\omega_n + 4 + 4\omega_0(t_n))(s_n - t_n)}{8(1-q_n)}, \\
t_{n+1} &= u_n + \frac{(1 + \int_0^1 \omega_0(s_n + \theta(s_n - t_n))d\theta(u_n - s_n) + \int_0^1 \omega(\theta(s_n - t_n))d\theta(s_n - t_n)}{1 - q_n}, \\
s_{n+1} &= t_{n+1} + \frac{2\beta_{n+1}}{3(1 - \omega_0(t_{n+1}))},
\end{aligned}$$

where the sequence $\{\beta_{n+1}\}$ is given in the Method (8.3).

Method (8.5)

$$\begin{aligned}
t_{n+1} &= s_n + \frac{\bar{\omega}_n(s_n - t_n)}{6(1-q_n)} \\
s_{n+1} &= t_{n+1} + \frac{\alpha_{n+1}}{1 - \omega_0(t_{n+1})},
\end{aligned}$$

where the sequence $\{\alpha_{n+1}\}$ is given in the Method (8.1).

Notice that convergence criteria for these sequences have been already given in earlier Chapters for similar methods.

Next, we provide the estimates justifying the choice of the majorizing sequences. In order to avoid repetitions some estimates already be given in previous Chapters are skipped.

Method (8.1)

$$\begin{aligned}
x_{n+1} - y_n &= (F'(x_n)^{-1} - 3B_n^{-1})F(x_n) \\
&= -(3B_n^{-1} - F'(x_n)^{-1})F(x_n) \\
&= -b_n^{-1}(3F'(x_n) - B_n)F'(x_n)^{-1}F(x_n) \\
&= B_n^{-1}(3F'(x_n) - B_n)(y_n - x_n),
\end{aligned}$$

so

$$\|x_{n+1} - y_n\| \leq \frac{\bar{\omega}_n(s_n - t_n)}{3(1-q_n)} = t_{n+1} - s_n$$

and

$$\|y_{n+1} - x_{n+1}\| \leq \|F'(x_{n+1})^{-1}F'(x_0)\| \|F'(x_0)^{-1}F(x_{n+1})\| \leq s_{n+1} - t_{n+1}.$$

Method (8.2)

$$\begin{aligned}
z_n - y_n &= \frac{1}{4}(I + F'(y_n)^{-1}F'(x_n))F'(x_n)^{-1}F(y_n),\\
\|z_n - y_n\| &\leq \frac{1}{4}\left(2 + \frac{\bar{\omega}_n}{1-\omega_0(s_n)}\right)^2\\
&\quad \times \frac{\int_0^1 \omega((1-\theta)(s_n - t_n))d\theta(s_n - t_n)}{1-\omega_0(t_n)}\\
&= u_n - s_n,\\
\|x_{n+1} - z_n\| &\leq \frac{1}{2(1-\omega_0(t_n))}\left(2 + \frac{2\omega_n}{1-\omega_0(s_n)} + \left(\frac{\omega_n}{1-\omega_0(s_n)}\right)^2\right)\\
&\quad \times ((1 + \int_0^1 \omega_0(s_n + \theta(u_n - s_n))d\theta)(u_n - s_n)\\
&\quad + \int_0^1 \omega((1-\theta)(s_n - t_n))d\theta(s_n - t_n))\\
&= t_{n+1} - u_n
\end{aligned}$$

and

$$\|y_{n+1} - x_{n+1}\| \leq s_{n+1} - t_{n+1}.$$

Method (8.3)

$$\begin{aligned}
x_{n+1} - y_n &= \left[\frac{2}{3}I + \frac{1}{2}I - \frac{9}{4}F'(y_n)^{-1}F'(x_n) - \frac{3}{4}F'(x_n)^{-1}F'(y_n)\right](-\frac{3}{2}(y_n - x_n))\\
&= -\frac{3}{2}\left[-\frac{1}{3}I + \frac{1}{8}(9(I - F'(y_n)^{-1}F'(x_n)) + 3(I - F'(x_n)^{-1}F'(y_n))\right](y_n - x_n),\\
\|x_{n+1} - y_n\| &\leq \frac{3}{2}(\frac{1}{3} + \frac{9\bar{\omega}_n}{8(1-\omega_0(s_n))} + \frac{3\bar{\omega}_n}{8(1-\omega_0(t_n))})(s_n - t_n)\\
&= t_{n+1} - s_n,
\end{aligned}$$

$$\begin{aligned}
F(x_{n+1}) &= F(x_{n+1}) - F(x_n) - \frac{3}{2}F'(x_n)(y_n - x_n)\\
&= \int_0^1 [F'(x_n + \theta(x_{n+1} - x_n))d\theta - F'(x_0) + F'(x_0)](x_{n+1} - x_n)\\
&\quad -\frac{3}{2}(F'(x_n) - F'(x_0) + F'(x_0))(y_n - x_n),\\
\|F'(x_0)^{-1}F(x_{n+1})\| &\leq (1 + \int_0^1 \omega_0(t_n + \theta(t_{n+1} - t_n))(t_{n+1} - t_n)\\
&\quad + \frac{3}{2}(1 + \omega_0(t_n))(s_n - t_n)
\end{aligned}$$

and

$$\|y_{n+1} - x_{n+1}\| \leq \frac{2\beta_{n+1}}{3(1-\omega_0(t_{n+1}))} = s_{n+1} - t_{n+1}.$$

Method (8.4)

$$
\begin{aligned}
z_n - y_n &= \left[\frac{2}{3}I - \frac{1}{2}A_n^{-1}(3F'(y_n)+F'(x_n))\right](-\frac{3}{2}(y_n - x_n)) \\
&= \frac{1}{6}(4A_n - 3(3F'(y_n)+F'(x_n))(-\frac{3}{2}(y_n - x_n)), \\
&= \frac{1}{4}A_n^{-1}((3(F'(y_n)-F'(x_n)) - 4F'(x_n)))(y_n - x_n), \\
\|z_n - y_n\| &\le \frac{(4+3\bar{\omega}_n + 4\omega_0(t_n))(s_n - t_n)}{8(1-q_n)} \\
&= u_n - s_n, \\
\|x_{n+1} - z_n\| &\le \frac{1}{1-q_n}(1+\int_0^1 \omega_0(s_n + \theta(u_n - s_n))d\theta)(u_n - s_n) \\
&\quad + \int_0^1 \omega((1-\theta)(s_n - t_n))d\theta(s_n - t_n) \\
&= t_{n+1} - u_n,
\end{aligned}
$$

and

$$\|y_{n+1} - x_{n+1}\| \le \frac{2\beta_{n+1}}{1-\omega_0(t_{n+1})} = s_{n+1} - t_{n+1}.$$

Method (8.4)

$$
\begin{aligned}
x_{n+1} - y_n &= (F'(x_n)^{-1} - 6A_n^{-1})F(x_n) \\
&\quad -(6A_n^{-1} - F'(x_n)^{-1})F(x_n) \\
&= -A_n^{-1}(6F'(x_n) - A_n)F'(x_n)^{-1}F(x_n) \\
&= A_n^{-1}(6F'(x_n) - A_n)(y_n - x_n),
\end{aligned}
$$

so

$$\|x_{n+1} - y_n\| \le \frac{\bar{\omega}_n(s_n - t_n)}{6(1-q_n)} = t_{n+1} - s_n$$

and

$$\|y_{n+1} - x_{n+1}\| \le \frac{\alpha_{n+1}}{1-\omega_0(t_{n+1})} = s_{n+1} - t_{n+1}.$$

3. References

[1] Argyros, I. K., *On the Secant method*, Publ. Math. Debrecen, 43, (1993), 223-238.

[2] Argyros, I. K., *Convergence and Application of Newton-type Iterations*, Springer, (2008).

[3] Argyros, I. K., George, S., *Ball convergence of a novel Newton-Traub composition for solving equations*, Cogent Math., 3(1), (2016), Article ID 1155333.

[4] Argyros, I. K., George, S., *On the complexity of extending the convergence region for Traubs method*, Journal of Complexity, 56, (2020), 101423.

[5] Babajee, D. K. R., *On a two-parameter Chebyshev-Halley-like family of optimal two-point fourth-order methods free from second derivatives*, Afr. Mat., 26, (2015), 689-697.

[6] Cordero, A., Hueso, J. L., Martinez, E., Torregrosa, J. R., *Increasing the convergence order of an iterative method for nonlinear systems*, Appl. Math. Lett., 25, (2012), 2369-2374.

[7] Cordero, A., Hueso, J. L., Martinez, E., Torregrosa, J. R., *A modified Newton-Jarratts composition*, Numer. Algor., 55, (2010), 87-99.

[8] Cordero, V., Torregrosa, J. R., *Variants of Newtons method using fifth-order quadrature formulas*, Appl. Math. Comput., 190, (2007), 686-698.

[9] Frontini, M., Sormani, E., *Third-order methods from Quadrature Formulae for solving systems of nonlinear equations*, Appl. Math. Comput., 140, (2004), 771-782.

[10] Homeier, H. H., *On Newton-type methods with cubic convergence*, Comp. Appl. Math., 176, (2005), 425-432.

[11] Noor, M. A., Waseem, M., *Some iterative methods for solving a system of nonlinear equations*, Comput. Math. Appl., 57, (2009), 101-106.

[12] Ostrowski, A. M., *Solutions of Equations and System of Equations*, Academic Press: New York, NY, USA, (1960).

[13] Sharma, J. R., Guha, R. K., Sharma, R., *An efficient fourth-order weighted-Newton method for systems of nonlinear equations*, Numer. Algor., 62, (2013), 307-323.

Chapter 9

Iterative Derivative-Free Methods of Order Seven and Eight

1. Introduction

We study the local convergence of the three step method defined for each $n = 0, 1, 2, \cdots$ by

$$\begin{aligned} u_n &= x_n + \alpha F(x_n) \\ v_n &= x_n + \beta F(x_n) \\ A_n &= [u_n, v_n; F] \\ y_n &= x_n - A_n^{-1} F(x_n) \\ z_n &= T_4(x_n, y_n) \\ B_n &= [z_n, x_n; F] + [z_n, y_n; F] - [y_n, x_n; F] \\ x_{n+1} &= z_n - B_n^{-1} F(z_n), \end{aligned} \tag{9.1}$$

where α, $\beta \in \mathbb{R}$, $[.,.;F] : \Omega \to L(X,Y)$ is a divided difference of order one [3] and $T_4 : X \times X \to X$ is an iterative function of order four. The convergence order seven and eight was established in [3, 4, 5] using Taylor series expansions and hypotheses reaching up to the eighth derivative of the operator F. But these derivatives do not appear on the method which may converge even if these derivatives do not exist. Other problems exist with this approach. There are not computable error estimates on $\|x_n - x^*\|$ or uniqueness of the solution x^* results. Theses problems are addressed in this chapter and under hypotheses only on the operators on the method (9.1). The motivation, construction and efficiency of the method can be found in [3, 4, 5]. Our technique can be used to extend the applicability of other methods in a similar way [1, 2].

2. Convergence

The local convergence analysis uses scalar majorant functions and real parameters. Let $M = [0, \infty)$ and $a, b, c \geq 0$ be given parameters.
Suppose:

(1) There exists a continuous and non-decreasing function $w_0 : M \times M \to R$ such that the equation $w_0(at, bt) - 1 = 0$ has a smallest positive solution ρ. Set $M_0 = [0, \rho)$.

(2) There exist continuous and nondecreasing functions $w : M_0 \to R$, $w_1 : M_1 \times M_1 \to R$, $M_1 \leq M_0$ such that the equations $g_1(t) - 1 = 0$ and $g_2(t) - 1 = 0$ have smallest positive solutions in M_0 and M_1, denoted by r_1 and r_2, where

$$g_1(t) = \frac{w(c|\alpha|t, bt)}{1 - w_0(at, bt)}$$

and $g_2(t) = w_1(t, g_1(t)t)$.

(3) There exist continuous and nondecreasing function $w_2 : M_1 \times M_1 \times M_1 \to R$, such that the equation $q(t) - 1 = 0$ has a smallest positive zero $\rho_0 \in M_1$, where

$$q(t) = w_0(g_2(t)t, g_1(t)t) + w_2(g_2(t)t, g_1(t)t, t).$$

Set $\rho_1 = min\{\rho_0, \rho\}$ and $M_2 = [0, \rho_1]$.

(4) There exists a continuous and nondecreasing function $w_3 : M_2 \times M_2 \to R$, such that the equation $g_3(t) - 1 = 0$ has a smallest positive zero $r_3 \in M_2$, where

$$g_3(t) = \frac{(w_2(g_2(t)t, g_1(t)t, t) + w_3(g_2(t)t, g_1(t)t))g_2(t)}{1 - q(t)}.$$

Then, the parameter

$$r = \min\{r_i\}, i = 1, 2, 3 \tag{9.2}$$

is shown to be a radius of convergence for the method (9.1) in Theorem 2.1. Set $M_3 = [0, r)$. It follows by these definitions that for each $t \in M_3$

$$0 \leq w_0(ab, bt) < 1 \tag{9.3}$$

$$0 \leq q(t) < 1 \tag{9.4}$$

and

$$0 \leq g_i(t) < 1. \tag{9.5}$$

The functions w_0, w_1, w_2, w_3 relate to the operators on the method (9.1) as follows. Suppose:

(H_1) There exist an invertible operator M and a solution $x^* \in \Omega$ of the equation $F(x) = 0$ and parameters $a, b, c \geq 0$ such that $\|I + \alpha[x, x^*; F]\| \leq a$, $\|I + \beta[x, x^*; F]\| \leq b$ and $\|[x, x^*; F]\| \leq c$, for each $x \in \Omega$.

(H_2) $\|M^{-1}([x, y; F] - M)\| \leq w_0(\|x - x^*\|, \|y - x^*\|)$, for each $x, y \in \Omega$. Set $\Omega_0 = \Omega \cap U(x^*, \rho)$.

(H_3)

$$\begin{aligned} &\|M^{-1}([x + \alpha F(x), x + \beta F(x); F] - [x, x^*; F])\| \\ \leq\ & w(|\alpha|\|F(x)\|, \|\|x - x^* + \beta F(x)\|), \end{aligned}$$

$$\begin{aligned} \|T_4(x, y) - x^*\| &\leq w_1(\|x - x^*\|)\|x - x^*\|, \\ \|M^{-1}([z, x; F] - [y, x; F])\| &\leq w_2(\|z - x^*\|, \|y - x^*\|, \|x - x^*\|), \\ \|M^{-1}([z, y; F] - [z, x^*; F])\| &\leq w_3(\|z - x^*\|, \|y - x^*\|) \end{aligned}$$

for each $x, y, z \in \Omega_0$.

(H_4) $U[x^*,R] \subset \Omega$, where $R = \max\{r, ar, br, cr\}$. Under this notation and conditions we present the local convergence analysis of the method (9.1).

Theorem 16. Suppose that the conditions $(H_1)-(H_4)$ hold and pick $x_0 \in U(x_*, r) - \{x_*\}$. Then, the sequence $\{x_n\}$ generated by method (9.1) is convergent to x^* so that the following error estimates hold:

$$\|y_n - x^*\| \leq g_1(\|x_n - x^*\|)\|x_n - x^*\| \leq \|x_n - x^*\| < r \tag{9.6}$$

$$\|z_n - x^*\| \leq g_2(\|x_n - x^*\|)\|x_n - x^*\| \leq \|x_n - x^*\| \tag{9.7}$$

and

$$\|x_{n+1} - x^*\| \leq g_3(\|x_n - x^*\|)\|x_n - x^*\| \leq \|x_n - x^*\|, \tag{9.8}$$

where the radius is defined in (9.2) and the function g_i are as given previously.

Proof. Mathematical inductions is used to show the estimates (9.6)-(9.8) together with the series of the estimates

$$\begin{aligned}
& \|M^{-1}([x_n + \alpha F(x_n), x_n + \beta F(x_n); F] - F'(x_*)) \\
\| \leq \; & w_0(\|(I + \alpha[x_n, x^*; F])(x_n - x^*)\|, \\
& \|(I + \beta[x_n, x^*; F])(x_n - x^*)\|) \\
\leq \; & w_0(a\|x_n - x^*\|, b\|x_n - x^*\|) \\
\leq \; & w_0(a\rho, b\rho) < 1
\end{aligned}$$

$$\|[x_n + \alpha F(x_n), x_n + \beta F(x_n); F]^{-1} M\| \leq \frac{1}{1 - w_0(a\|x_n - x^*\|, b\|x_n - x^*\|)},$$

$$\begin{aligned}
& \|M^{-1}([x_n + \alpha F(x_n), x_n + \beta F(x_n); F] - [x_n, x^*; F])\| \\
\leq \; & w(|\alpha|\|F(x_n)\|, \\
+ \; & \|x_n + \beta F(x_n) - x^*\|) \\
\leq \; & w(|\alpha|c|\|x_n - x^*\|, b\|x_n - x^*\|).
\end{aligned}$$

So from

$$y_n - x^* = x_n - x^* - A_n^{-1} F(x_n) = A_n^{-1}(A_n - [x_n, x^*; F])(x_n - x^*),$$

we obtain

$$\begin{aligned}
\|y_n - x^*\| & \leq \frac{w(c|\alpha|\|x_n - x^*\|, b\|x_n - x^*\|)\|x_n - x^*\|}{1 - w_0(a\|x_n - x^*\|, b\|x_n - x^*\|)} \\
& \leq g_1(\|x_n - x^*\|)\|x_n - x^*\| \leq \|x_n - x^*\| < r.
\end{aligned}$$

Thus, the iterates $y_n \in U(x^*, r) - \{x^*\}$, and the estimate (9.6) holds.
Notice also that

$$\begin{aligned}
\|x_n + \alpha F(x_n) - x^*\| & = \|(I + \alpha[x_n, x^*; F])(x_n - x^*)\| \\
& \leq a\|x_n - x^*\| \leq ar \leq R
\end{aligned}$$

and similarly

$$\|x_n + \beta F(x_n) - x^*\| \leq br \leq R,$$

where we also used

$$\begin{aligned}
\|F(x_n)\| = \|F(x_n) - F(x^*)\| & = \|[x_n, x^*; F](x_n - x^*)\| \\
& \leq \|[x_n, x^*; F]\|\|x_n - x^*\| \\
& \leq |c|\|x_n - x^*\| \leq |c|r \leq R.
\end{aligned}$$

Concerning, the second substep of the method (9.1), we get

$$\begin{aligned}\|z_n - x^*\| = \|T_4(x_n, y_n) - x^*\| &\leq w_1(\|x_n - x^*\|, \|y_n - x^*\|)\|x_n - x^*\| \\ &\leq g_2(\|x_n - x^*\|)\|x_n - x^*\| \leq \|x_n - x^*\|\end{aligned}$$

So, the iterate $z_n \in U(x^*, r) - \{x^*\}$ and the estimate (9.7) holds.
Next, we show $B_n^{-1} \in \mathcal{L}(Y, X)$ which shall establish the existence of the iterate x_{n+1} from the last substep of the method (9.1). We have in turn that

$$\begin{aligned} & \|M^{-1}([z_n, x_n; F] + [z_n, y_n; F] - [y_n, x_n; F] - M)\| \\ \leq \; & \|M^{-1}([z_n, y_n; F] - M)\| \\ + \; & \|M^{-1}([z_n, x_n; F] - [y_n, x_n; F]\| \\ \leq \; & w_0(\|z_n - x^*\|, \|y_n - x^*\|) \\ + \; & w_2(\|z_n - x^*\|, \|y_n - x^*\|, \|z_n - x^*\|) \\ = \; & q_n < 1, \end{aligned}$$

so $\|B_n^{-1}M\| \leq \dfrac{1}{1 - q_n}$. Then, we can write

$$\begin{aligned} x_{n+1} - x^* &= z_n - x^* - B_n^{-1}F(z_n) \\ &= B_n^{-1}(B_n - [z_n, x^*; F])(z_n - x^*) \end{aligned}$$

leading to

$$\begin{aligned} \|x_{n+1} - x^*\| \leq \; & \frac{1}{1 - q(\|x_n - x^*\|)} \\ & \times \Big(w_2(\|z_n - x^*\|, \|y_n - x^*\|, \|x_n - x^*\|) \\ & + w_3(\|z_n - x^*\|, \|y_n - x^*\|) \Big) \|z_n - x^*\| \\ \leq \; & g_3(\|x_n - x^*\|)\|x_n - x^*\| \leq \|x_n - x^*\|. \end{aligned}$$

Hence, the iterate $x_{n+1} \in U(x^*, r) - \{x^*\}$ and the estimate (9.8) holds. Then, it follows from the estimation

$$\|x_{n+1} - x^*\| \leq d\|x_n - x^*\| < r, \; d = g_3(\|x_n - x^*\|) \in [0, 1)$$

that $x_{n+1} \in U(x^*, r) - \{x^*\}$ *and* $\lim_{n \to \infty} x_n = x^*$.

Next, a uniqueness of the solution region is determined.

PROPOSITION 3. *Suppose:*

1. *There exists a solution $y^* \in S(x^*, p_0)$ of the equation $F(x) = 0$ for some $p_0 > 0$.*
2. *$\|M^{-1}([y, x^*; F] - M)\| \leq V(\|y - x^*\|)$ for some continuous and nondecreasing function $V : M \to \mathbb{R}$.*
3. *There exists $p \geq p_0$ such that*
$$V(p) < 1.$$

Set $\Omega_1 = \Omega \cap U[x^*, p]$. Then, the only solution of the equation $F(x) = 0$ in the region Ω_1 is x^*.

Proof. *Let* $Q = [y^*, x^*; F]$. *It follows from* (1) – (3) *in turn that*

$$\|M^{-1}(Q - M)\| \le V(\|y^* - x^*\|) \le V(p) < 1,$$

so the linear operator $Q^{-1} \in \mathbb{L}(Y, X)$. *Then, from the identity* $y^* - x^* = Q^{-1}(F(y^*) - F(x^*)) = Q^{-1}(0)$, *we conclude that* $y^* = x^*$.

Remark. (i) Under the conditions $(H_1) - (H_4)$ in Proposition (3), we can set $p = r$.
(ii) A possible choice for $M = F'(x^*)$. Notice we did not assume x^* is a simple solution of the equation $F(x) = 0$ which is a usual hypothesis in earlier studies. Moreover, the results can be used to solve non-differentiable equations.

3. Special Cases

Let us pick an interesting choice for the iteration function T_4 [3, 4, 5]. Set $\alpha = 0$, $\beta = 1$ and define

$$T_4(x_n, y_n) = y_n - (3I - [v_n, x_n; F]^{-1}([y_n, x_n; F] + [y_n, v_n; F]))[v_n, x_n; F]^{-1}F(y_n)$$

Then, we can write in turn that

$$\begin{aligned} y_n - x^* &= [v_n, x_n; F]^{-1}F(y_n) + [v_n, x_n; F]^{-1}([v_n, x_n; F] - [y_n, x^*; F])(y_n - x^*) \\ &+ [v_n, x_n; F]^{-1}([v_n, x_n; F] - [y_n, x_n; F]) \\ &\quad +([v_n, x_n; F] - [y_n, x_n; F])[v_n, x_n; F]^{-1}F(y_n) \end{aligned}$$

leading to

$$\begin{aligned} \|y_n - x^*\| &\le \Bigg[\frac{w_4(\|v_n - x^*\|, \|y_n - x^*\|, \|x_n - x^*\|)}{1 - \varphi(\|x_n - x^*\|)} \\ &\quad + \frac{c}{(1 - \varphi(\|x_n - x^*\|))^2} \\ &\quad \Big(w_2(\|v_n - x^*\|, \|y_n - x^*\|, \|x_n - x^*\|) \\ &\quad + w_5(\|v_n - x^*\|, \|y_n - x^*\|, \|x_n - x^*\|)\Big)\Bigg] \|y_n - x^*\| \\ &\le g_2(\|x_n - x^*\|)\|x_n - x^*\| \le \|x_n - x^*\|. \end{aligned}$$

where

$$g_2(t) = \left[\frac{w_4(bt, g_1(t)t, t)}{1 - \varphi(t)} + c\frac{(w_2(bt, g_1(t), t) + w_5(bt, g_1(t), t))}{(1 - \varphi(t))^2}\right] g_1(t),$$

$$\begin{aligned} \|M^{-1}([v_n, x_n; F] - M)\| &\le w_0(\|v_n - x^*\|, \|x_n - x^*\|) \\ &\le w_0(b\|x_n - x^*\|, \|x_n - x^*\|) \\ &= \varphi(\|x_n - x^*\|), \end{aligned}$$

and we also used

$$\|M^{-1}([v_n, x_n; F] - [y_n, x^*; F])\| \le w_4(\|v_n - x^*\|, \|y_n - x^*\|, \|x_n - x^*\|)$$

and

$$\|M^{-1}([v_n,x_n;F]-[y_n,v_n;F])\| \leq w_5(\|v_n-x^*\|,\|y_n-x^*\|,\|x_n-x^*\|)$$

and the functions w_4 and w_5 are as w_2. Further specializations of the function "w" can be found in [1, 2]

4. References

[1] Argyros, C. I., Regmi, S., Argyros, I. K., George, S., *Contemporary Algorithms: Theory and Applications*, Volume II, NOVA Publishers, (2022).

[2] Argyros, I. K., *The theory and applications of iteration methods*, 2nd Edition, CRC Press, Engineering Series, Taylor and Francis Group, (2022).

[3] Liu, Z., Zheng, Q., Zhao, P., *A variant of Steffensen's method of fourth-order convergence and its applications*, Appl. Math. Comput., 216(7), (2010), 1978-1983.

[4] Sharma, J. R., Arora, H., *An efficient derivative-free iterative method for solving systems of nonlinear equations*, Appl. Anal. Discrete Math., 7, (2013), 390-403.

[5] Wang, X., Zhang, T., Qian, W., Teng, M., *Seventh order derivative-free iterative methods for solving nonlinear equations*, Numer. Algor., 70, (2015), 545-558.

Chapter 10

Extended Convergence for a Derivative-Free Fourth-Order Steffensen-Like Method with Parameters

1. Introduction

The method is fourth order, using fifth derivative, $F : \Omega \subset B \to B$, B- Banach space

$$F(x) = 0 \tag{10.1}$$

The method is [14]:

$$\begin{aligned} u_n &= x_n + bF(x_n) \\ y_n &= x_n - [x_n, u_n; F]^{-1}F(x_n) \\ H_n &= [x_n, u_n; F]^{-1}[z_n, y_n; F], , z_n = y_n + cF(y_n), \\ x_{n+1} &= y_n - (\alpha I + H_n(\beta I + \gamma H_n))[x_n, y_n; F]^{-1}F(y_n), \end{aligned} \tag{10.2}$$

where $b, c, \alpha, \beta, \gamma \in \mathbb{R}$ with $\beta = 3 - 2\alpha$, $\gamma = \alpha - 2$, and $[\cdot,\cdot;F] : \Omega \times \Omega \to L(B,B)$ is a divided difference of order one [1]-[4]. Relevant work can be done on the methods in [1]-[13],[15]-18].

2. Local Analysis

In Section 2 the local convergence analysis relies on scalar parameters and functions. Let λ and μ be nonnegative parameters. Set $T = [0, \infty)$.
Suppose equation:

$$w_0(t, \lambda t) - 1 = 0 \tag{10.3}$$

has a least solution $R_0 \in T - \{0\}$ for some function $w_0 : T \times T \to T$ which is nondecreasing and continuous. Set $T_0 = [0, R_0)$.
Suppose equation

$$g_1(t) - 1 = 0 \tag{10.4}$$

has a least solution $r_1 \in T_0 - \{0\}$, for some function $w : T_0 \to T$ continuous and nondecreasing and function $g_1 : T_0 \to T$ is given by

$$g_1(t) = \frac{w_0(t, \lambda t) + w(t)}{1 - w_0(t, \lambda t)}.$$

Suppose equation

$$g_1(t) - 1 = 0 \tag{10.5}$$

has a least solution $r_2 \in T_0 - \{0\}$, for some function $w_1 : T_0 \to T$ continuous and nondecreasing and function $g_2 : T_0 \to T$ is defined by

$$\begin{aligned} g_2(t) &= [\frac{w_0(t,\lambda t) + w(g_1(t)t)}{1 - w_0(t,\lambda t)} + (|\gamma|(\frac{w_0(\mu g_1(t)t, g_1(t)t) + w_0(t,\lambda t)}{1 - w_0(t,\lambda t)})^2 \\ &+ |\beta + 2\gamma|(\frac{w_0(\mu g_1(t)t, g_1(t)t) + w_0(t,\lambda t)}{1 - w_0(t,\lambda t)}))\frac{w_1(g_1(t)t)}{(1 - w_0(t,\lambda t))^2}]g(t) \end{aligned}$$

It shall be shown that

$$r = \min\{r_m\}, m = 1,2 \tag{10.6}$$

is a convergence radius for method (10.2).

The definition of radius r gives

$$0 \le w_0(t,\lambda t) < 1 \tag{10.7}$$

and

$$0 \le g_m(t) < 1 \tag{10.8}$$

hold for all $t \in T_1$.

Consider $S(x_*,\delta)$ to denote a ball in B with radius $\delta > 0$ and center $x_* \in B$. Then, $\bar{S}(x_*,\delta)$ is its closure.

The local convergence analysis of method (10.2) uses the conditions (A) with the "w" functions as defined previously.
Assume:

(A_1) $F : \Omega \to B$ is differentiable at $x = x_*$, which is a simple solutions of equation (10.1) and $[\cdot,\cdot;F] : \Omega \times \Omega \to L(B,B)$ is a standard divided difference of order one.

(A_2) $\|I + b[x,x_*;F]\| \le \lambda$ and

$$\|F'(x_*)^{-1}([x,x+bF(x);F] - F'(x_*))\| \le w_0(\|x - x_*\|, \|x + bF(x) - x_*\|)$$

for each $x \in \Omega$, and some $\lambda \ge 0$
Set $\Omega_0 = \Omega \cap S(x_*,R_0)$.

(A_3) $\|I + c[x,x_*;F]\| \le \mu$,$\|F'(x_*)^{-1}([x,x_*;F] - F'(x_*))\| \le w(\|x - x_*\|)$ and $\|F'(x_*)^{-1}[x,x_*;F]\| \le w_1(\|x - x_*\|)$, for each $x \in \Omega_0$.

(A_4) $\bar{S}(x_*,\rho) \subset \Omega$, where $\rho = \max\{r, \lambda^r, \mu^r\}$
and

(A_5) There exists $r_* \ge r$ satisfying $w(r_*) < 1$.
Set $\Omega_1 = \Omega \cap \bar{S}(x_*,r_*)$.

Next, we develop the local convergence analysis of method (10.2) relying on the conditions (A), and the notation.

Theorem 17. Suppose the conditions (A) hold and pick starter point $x_0 \in S(x_*, r) - \{x_*\}$ Then, the following assertions hold: Iteration $\{x_n\}$ generated by method (10.2) is well defined in $S(x_*, r)$ for each $n = 0, 1, 2, \cdots$, stays in $S(x_*, r)$ and converges to x_* so that

$$\|y_n - x_*\| \leq g_1(\|x_n - x_*\|)\|x_n - x_*\| \leq \|x_n - x_*\| < r \tag{10.9}$$

and

$$\|x_{n+1} - x_*\| \leq g_2(\|x_n - x_*\|)\|x_n - x_*\| \leq \|x_n - x_*\| \tag{10.10}$$

with radius r is defined by (10.6) and the function g_m are as given previously.
Moreover, the only solution of equation (10.1) in the domain Ω_1 given by (A_5) is point x_*.
Proof. Assertions (10.9) and (10.10) are shown using induction on integer k. Let $v \in S(x_*, r) - \{x_*\}$. By (A_1), (A_2), (A_4) and (10.7), we have in turn

$$\begin{aligned} & \|F'(x_*)^{-1}([v, v + bF(v); F] - F'(x_*))\| \\ \leq \ & w_0(\|v - x_*\|, \|v - x_* + bF(v)\|) \\ = \ & w_0(\|v - x_*\|, \|(I + b[v, x_*; F])(v - x_*)\|) \\ \leq \ & w_0(\|v - x_*\|, \lambda\|v - x_*\|) \\ \leq \ & w_0(r, \lambda r) < 1 \end{aligned} \tag{10.11}$$

Estimation (10.11) gives together with a lemma by Banach [1]-[4] for inverses of linear operators that $[v, v + bF(v); F]^{-1} \in L(B, B)$ with

$$\|[v, v + bF(v); F]^{-1} F'(x_*)\| \leq \frac{1}{1 - w_0(\|v - x_*\|, \lambda\|v - x_*\|)} \tag{10.12}$$

By hypothesis $x_0 \in S(x_*, r) - \{x_*\}$. So for $v = x_0$, (10.12) gives $[x_0, x_0 + bF(x_0); F]^{-1} \in L(B, B)$ and y_0 is well defined by the first substep of method (10.2) if $n = 0$, from which we can also have

$$\begin{aligned} y_0 - x_* &= x_0 - x_* - [x_0, u_0; F]^{-1} F(x_0) \\ &= [x_0, u_0; F]^{-1}([x_0, u_0; F] - [x_0, x_*; F])(x_0 - x_*) \\ &= [x_0, u_0; F]^{-1}(([x_0, u_0; F] - F'(x_*)) \\ &\quad + (F'(x_*) - [x_0, x_*; F]))(x_0 - x_*). \end{aligned} \tag{10.13}$$

Then, using (10.6), (10.8) (for $m = 1$), (10.12) (for $v = x_0$), (A_3) and the triangle inequality, we obtain in turn.

$$\begin{aligned} \|y_0 - x_*\| &\leq \frac{(w_0(\|x_0 - x_*\|, \lambda\|x_0 - x_*\|) + w(\|x_0 - x_*\|))\|x_0 - x_*\|}{1 - w_0(\|x_0 - x_*\|, \lambda\|x_0 - x_*\|)} \\ &= g_1(\|x_0 - x_*\|)\|x_0 - x_*\| \leq \|x_0 - x_*\| < r, \end{aligned} \tag{10.14}$$

showing $y_0 \in S(x_*, r)$ and (10.9) if $n = 0$.
Similarly, since x_1 is well defined, by the second substep of method (10.2) if $n = 0$, we can write,

$$\begin{aligned} x_1 - x_* &= y_0 - x_* - [x_0, u_0; F]^{-1} F(y_0) - ((\alpha - 1)I + \beta H_0 + \gamma H_0^2)[x_0, u_0; F]^{-1} F(y_0) \\ &= (y_0 - x_* - [x_0, u_0; F]^{-1} F(y_0)) - (\gamma(H_0 - I)^2 \\ &+ (\beta + 2\gamma)(H_0 - I))[x_0, u_0; F]^{-1} F(y_0) \end{aligned} \tag{10.15}$$

Then, in view of (10.6), (10.8) (for $m = 2$), (10.12) (for $v = x_0$) (H_3), (10.14) and (10.15), we have in turn

$$\begin{aligned}
\|x_1 - x_*\| &\leq [\frac{w_0(\|x_0-x_*\|,\lambda\|x_0-x_*\|)+w(\|y_0-x_*\|)}{1-w_0(\|x_0-x_*\|,\lambda\|x_0-x_*\|)} \\
&+ (|\gamma|(\frac{w_0(\|z_0-x_*\|,\|y_0-x_*\|)+w_0(\|x_0-x_*\|,\lambda\|x_0-x_*\|)}{1-w_0(\|x_0-x_*\|,\lambda\|x_0-x_*\|)})^2 \\
&+ |\beta+2\gamma|\frac{w_0(\|z_0-x_*\|,\|y_0-x_*\|)+w_0(\|x_0-x_*\|,\lambda\|x_0-x_*\|)}{1-w_0(\|x_0-x_*\|,\lambda\|x_0-x_*\|)}) \\
&+ \frac{w_1(\|y_0-x_*\|)}{1-w_0(\|x_0-x_*\|,\lambda\|x_0-x_*\|)}]\|y_0-x_*\| \\
&= g_2(\|x_0-x_*\|)\|x_0-x_*\| \leq \|x_0-x_*\| < r, \qquad (10.16)
\end{aligned}$$

showing $x_1 \in D(x_*, r)$ and (10.10) if $n = 0$. The induction for estimations (10.9) and (10.10) is terminated if we simply exchange x_0, u_0, y_0, z_0, x_1 by $x_i, u_i, y_i, z_i, x_{i+1}$ in the previous calculations. Hence we have

$$\|x_{i+1} - x_*\| \leq q\|x_i - x_*\| < r, \qquad (10.17)$$

with $q = g_2(\|x_0 - x_*\|) \in [0,1)$, which imply $x_{i+1} \in S(x_*, r)$, and $\lim_{n\to\infty} x_i = x_*$. Moreover, let $M = [p, x_*; F]$ for $p \in \Omega_1$ with $F(p) = 0$. Then, (A_2) and (A_5) give

$$\|F'(x_*)^{-1}(M - F'(x_*))\| \leq w(\|p - x_*\|) \leq w(r_*) < 1.$$

Hence, $p = x_*$ follows by the existence of the inverse of M together with the estimation $0 = F(p) - F(x_*) = M(p - x_*)$.

3. References

[1] Amiri, A., Cordero, A., Darvishi, M. T., Torregrosa, J. R., *Preserving the order of convergence: Low-complexity Jacobian-free iterative schemes for solving nonlinear systems*, J. Comput. Appl. Math., 337, (2018), 87-97.

[2] Argyros, I. K., *Computational theory of iterative methods*, Elsevier (Academic Press), New York, (2007).

[3] Argyros, I. K., George, S., *Mathematical modeling for the solution of equations and systems of equations with applications*, Volume-4, Nova Publishers, NY, (2020).

[4] Argyros, I. K., Magreñán, Á. A., *A contemporary study of iterative methods*, Elsevier (Academic Press), New York, (2018).

[5] Argyros, I. K., Magreñán, Á.A., *Iterative methods and their dynamics with applications: A Contemporary Study*, CRC Press, New York, USA, (2017).

[6] Cordero, A., García-Maimó, J., Torregrosa, J. R., Vassileva, M. P., *Solving nonlinear problems by Ostrowski-Chun type parametric families*, Math. Chem., 53(1), (2015), 430-449.

[7] Grau-Sánchez, M., Grau, A., Noguera, M., *On the computational efficiency index and some iterative methods for solving systems of nonlinear equations*, J. Comput. Appl. Math., 236(6), (2011), 1259-1266.

[8] Hueso, J.L., Martínez, E., Torregrosa, J. R., *Third and fourth order iterative methods free from second derivative for nonlinear systems*, Appl. Math. Comput., 211(1), (2009), 190-197.

[9] Jarratt, P., *Some fourth order multipoint iterative methods for solving equations*, Math. Comp., 20(95), (1966), 434-437.

[10] Liu, Z., Quan, Z., Peng, Z., *A variant of Steffensen's method of fourth-order convergence and its applications*, Appl. Math. Comput., 216(7), (2010), 1978-1983.

[11] Magreñán, Á. A., *Different anomalies in a Jarratt family of iterative root-finding methods*, Appl. Math. Comput., 233, (2014), 29-38.

[12] Ortega, J. M., Rheinboldt, W. C., *Iterative methods for the solution of nonlinear equations in several variables*, Academic Press, New York, (1970).

[13] Ostrowski, A. M., *Solutions of equations and systems of equations*, Academic Press, New York, London, (1966).

[14] Sharma, J. R., Arora, H., *Improved Newton-like methods for solving systems of nonlinear equations*, SeMA Journal, 74(2), (2016), 1-7.

[15] Sharma, J. R., Arora, H., *On efficient weighted-Newton methods for solving systems of nonlinear equations*, Appl. Math. Comput., 222, (2013), 497-506.

[16] Sharma, J. R., Arora, H., Petkovi, M. S., *An efficient derivative-free family of fourth-order methods for solving systems of nonlinear equations*, Appl. Math. Comput., 235, (2014), 383-393.

[17] Sharma, J. R., Himani, A., *An efficient derivative-free method for solving systems of nonlinear equations*, Appl. Anal. Discr. Math., 7, (2013), 390-403.

[18] Traub, J. F., *Iterative methods for the solution of equations*, American Mathematical Soc, (1982).

Chapter 11

A Comparison Between Two Derivative-Free Fourth-Order Methods under the Same Conditions for Equations

1. Introduction

$F : \Omega \subset B \rightarrow B,$

$$F(x) = 0 \tag{11.1}$$

The methods are [9]

$$\begin{aligned} u_n &= x_n + bF(x_n) \\ y_n &= x_n - [x_n, u_n; F]^{-1}F(x_n) \\ x_{n+1} &= y_n - [x_n, y_n; F]^{-1}([x_n, y_n; F] - [u_n, y_n; F] \\ &+ [x_n, u_n; F])[x_n, y_n; F]^{-1}F(y_n), \end{aligned} \tag{11.2}$$

and in [12]

$$\begin{aligned} y_n &= x_n - [x_n, u_n; F]^{-1}F(x_n) \\ x_{n+1} &= y_n - (3I - [x_n, u_n; F]^{-1}([x_n, y_n; F] + [u_n, y_n; F])) \\ &\times [x_n, u_n; F]^{-1}F(y_n). \end{aligned} \tag{11.3}$$

The idea of this Chapter can also be used in [1]-[12], [14]-[17] along the same lines.

2. Ball Convergence

Some scalar parameters and functions are developed which are crucial in the ball convergence analysis of the method (11.2). Let $\lambda \geq 0$ and $\mu \geq 0$. Set $M = [0, \infty)$.

Suppose equation:

1. $w_0(t, \lambda t) - 1 = 0$ has a least solution $\rho_0 \in M - \{0\}$ for function $w_0 : M \times M \rightarrow M$ being nondecreasing and continuous. Set $M_0 = [0, \rho_0)$ and consider function $w : M_0 \rightarrow M$ to be continuous and nondecreasing.

2. $g_1(t) - 1 = 0$ has a least solution $r_1 \in M_0 - \{0\}$, where function $g_1 : M_0 \to M$ is defined by
$$g_1(t) = \frac{w_0(t,\lambda t) + w(t)}{1 - w_0(t,\lambda t)}.$$

3. $w_0(t, g_1(t)t) - 1 = 0$ has a least solution $\rho_1 \in M_0 - \{0\}$ Set $\rho = \min\{\rho_0, \rho_1\}$ and $M_1 = [0, \rho)$.

4. $g_2(t) - 1 = 0$ has a least solution $r_2 \in M_1 - \{0\}$, where $g_2(t) : M_1 \to M$ is defined by
$$g_2(t) = \left[\frac{w_0(t,g_1(t)t) + w(g_1(t)t)}{1 - w_0(t,g_1(t)t)} + \frac{(w_0(t,\lambda t) + w_0(\lambda t, g_1(t)t))w_1(g_1(t)t)}{(1 - w_0(t,g_1(t)t))^2}\right] g_1(t)$$

The parameter r defined by
$$r = \min\{r_m\}, m = 1,2, \tag{11.4}$$
is shown next to be a radius of convergence for method (11.2).

Notice that this definition implies
$$0 \le w_0(t,\lambda t) < 1, \tag{11.5}$$
$$0 \le w_0(t,g_1(t)t) < 1 \tag{11.6}$$
and
$$0 \le g_m(t) < 1 \tag{11.7}$$
hold for each $t \in [0, r)$.

We denote by $D(x_*, \delta)$ the open ball in B of center $x_* \in B$, and of radius $\delta > 0$. The notation $\bar{D}(x_*, \delta)$ is used for the closure of ball $D(x_* \delta)$.

The conditions (H) are developed to be used in the ball convergence of both methods with the "w" functions as given previously. Assume:

(H_1) $F : \omega \to B$ is differentiable at x_*, a simple solution of (11.1) and $[\cdot,\cdot;F] : \Omega \times \Omega \to L(B,B)$ is a divided difference of order one.

(H_2) $\|I + b[x, x_* F]\| \le \lambda$ and
$$\|F'(x_*)^{-1}([x, x + bF(x); F] - F'(x_*))\| \le w_0(\|x - x_*\|, \|x + bF(x) - x^*\|)$$
for each $x \in \Omega$, and some $\lambda \ge 0$.
Set $\Omega_0 = \Omega \cap D(x_*, \rho_0)$.
Set $\Omega_0 = \Omega \cap D(x_*, \rho_0)$.

(H_3) $\|F'(x_*)^{-1}([x, x_*; F] - F'(x_*))\| \le w(\|x - x_*\|)$ and
$\|F'(x_*)^{-1}[x, x_*; F]\| \le w_1(\|x - x_*\|)$, for each $x \in \Omega_0$.

(H_4) $\bar{D}(x_*, R) \subset \Omega$, where $R > 0$ is to be determined and

(H_5) There exists $R_* \ge R$ satisfying $w(R_*) < 1$.
Set $\Omega_1 = \Omega \cap \bar{D}(x_*, R_*)$.

Next, we develop the Ball convergence analysis first of method (11.2) utilizing the conditions (H) and the notation.

Theorem 18. Suppose the conditions (H) hold with $R = min\{r, \lambda r\}$ and r defined by (11.4). Then, the following assertions hold provided $x_0 \in D(x_*, r) - \{x_*\}$,

$$\{x_n\} \subset D(x_*, r), \lim_{n\to\infty} = x_*, \tag{11.8}$$

$$\|y_n - x_*\| \le g_1(\|x_n - x_*\|)\|x_n - x_*\| \le \|x_n - x_*\| < r \tag{11.9}$$

and

$$\|x_{n+1} - x_*\| \le g_2(\|x_n - x_*\|)\|x_n - x_*\| \le \|x_n - x_*\|, \tag{11.10}$$

where the functions g_m are given previously. Moreover, the only solution in the domain Ω_1 defined in (H_5) is x_*.

Proof. Assertions (11.8)-(11.10) shall be shown by mathematical induction on k. Let $z \in D(x_*, r) - \{x_*\}$. By (H_1), (H_2), (H_4) and (11.5), we get in turn

$$\begin{aligned} \|F'(x_*)^{-1}([z, z + bF(z); F] - F'(x_*))\| &\le w_0(\|z - x_*\|, \|z - x_* + bF(z)\|) \\ &= w_0(\|z - x_*\|, \|(I + b[z, x_*; F])z - x_*\|) \\ &\le w_0(\|z - x_*\|, \lambda\|z - x_*\|) \\ &\le w_0(r, \lambda r) < 1 \end{aligned} \tag{11.11}$$

In view of (11.11) and the Banach lemma on invertible operators [3, 6] , linear $[z, z + bF(z); F]$ is invertible and

$$\|[z, z + bF(z); F]^{-1}F'(x_*)\| \le \frac{1}{1 - w_0(\|z - x_*\|, \lambda\|z - x_*\|)} \tag{11.12}$$

So, for $z = x_0, y_0$ is well defined by the first substep of method (11.2) from which we can write,

$$\begin{aligned} y_0 - x_* &= x_0 - x_* - [x_0, u_0; F]^{-1}F(x_0) \\ &= [x_0, u_0; F]^{-1}([x_0, u_0; F] - [x_0, x_*; F])(x_0 - x_*) \\ &= [x_0, u_0; F]^{-1}(([x_0, u_0; F] - F'(x_*)) \\ &+ (F'(x_*) - [x_0, x_*; F]))(x_0 - x_*). \end{aligned} \tag{11.13}$$

Then, using (11.4), (11.7) (for m=1), (11.12) (for $z = x_0$ and (H_3), we have that

$$\begin{aligned} \|y_0 - x_*\| &\le \frac{(w_0(\|x_0 - x_*\|, \lambda\|x_0 - x_*\|) + w(\|x_0 - x_*\|))\|x_0 - x_*\|}{1 - w_0(\|x_0 - x_*\|, \lambda\|x_0 - x_*\|)} \\ &= g_1(\|x_0 - x_*\|)\|x_0 - x_*\| \\ &\le \|x_0 - x_*\| < r, \end{aligned} \tag{11.14}$$

showing $y_0 \in D(x_*, r)$ and (11.9) for $n = 0$. Notice also that (11.12) holds for $z = y_0, x_1$ is well defined by the second subset of method (11.2) and we can also write

$$\begin{aligned} x_1 - x_* &= y_0 - x_* - [x_0, y_0; F]^{-1}F(y_0) - [x_0, y_0; F]^{-1}([x_0, u_0; F] \\ &- [u_0, y_0; F])[x_0, y_0; F]^{-1}F(y_0). \end{aligned} \tag{11.15}$$

Next, using (11.4), (11.7) (for m=2), (11.12) (for$z = x_0$), (H_3), (11.14) and (11.15), we obtain in turn

$$\begin{aligned}
\|x_1 - x_*\| &\le \Bigg[\frac{w_0(\|x_0 - x_*\|, \|y_0 - x_*\|) + w(\|y_0 - x_*\|)}{1 - w_0(\|x_0 - x_*\|, \|y_0 - x_*\|)} \\
&\quad + \frac{1}{(1 - w_0(\|x_0 - x_*\|, \|y_0 - x_*\|))^2} \\
&\quad \times (w_0(\|x_0 - x_*\|, \lambda\|x_0 - x_*\|) \\
&\quad + w_0(\lambda\|x_0 - x_*\|, \|y_0 - x_*\|) w_1(\|y_0 - x_*\|))\Bigg] \|y_0 - x_*\| \\
&\le g_2(\|x_0 - x_*\|)\|x_0 - x_*\| \le \|x_0 - x_*\|, \qquad (11.16)
\end{aligned}$$

showing $x_1 \in D(x_*, r)$ and (11.10) for $n = 0$. Switch x_0, y_0, x_1 by x_k, y_k, x_{k+1} in the preceding estimates to complete the induction for (11.8)-(11.10). The estimate

$$\|x_{k+1} - x_*\| \le \mu\|x_k - x_*\| < r, \qquad (11.17)$$

where $\mu = g_2(\|x_0 - x_*\|) \in [0, 1)$ gives $x_{k+1} \in D(x_*, r)$, and $\lim_{n\to\infty} x_k = x_*$. Finally, set $A = [q, x_*; F]$ for $q \in \Omega_1$ with $F(q) = 0$. It then follows by (H_2) and (H_5) that

$$\|F'(x_*)^{-1}(A - F'(x_*))\| \le w(\|q - x_*\|) \le w(R_*) < 1 \qquad (11.18)$$

Hence, $q = x_*$ follows from the existence of A^{-1} and $0 = F(q) - F(x_*) = A(q - x_*)$.

Next we present the ball convergence of method (11.3) in a similar way. But the second function is defined by

$$\begin{aligned}
\bar{g}_2(t) &= [\frac{w_0(t, \lambda t) + w(g_1(t)t)}{1 - w_0(t, \lambda t)} + \frac{1}{(1 - w_0(t, \lambda t))^2}(w_0(t, \lambda t) + w_0(t, g_1(t)t) \\
&+ w_0(\lambda t, g_1(t)t)) w_1(g_1(t)t)] g_1(t). \qquad (11.19)
\end{aligned}$$

This definition is given by

$$\begin{aligned}
x_1 - x_* &= y_0 - x_* - [x_0, u_0; F]^{-1}F(y_0) + (2I - [x_0, u_0; F]^{-1}([x_0, y_0; F] \\
&+ [u_0, y_0; F]))[x_0, u_0; F]^{-1}F(y_0) \\
&= y_0 - x_* - [x_0, u_0; F]^{-1}F(y_0) + [x_0, u_0; F]^{-1}(2([x_0, u_0; F] - F'(x_*)) \\
&+ (F'(x_*) - [x_0, y_0; F]) + (F'(x_*) - [x_0, u_0; F]))[x_0, u_0; F]^{-1}F(y_0).
\end{aligned}$$

So as in (11.16)

$$\begin{aligned}
\|x_1 - x_*\| &\le \Bigg[\frac{w_0(\|x_0 - x_*\|, \|u_0 - x_*\|) + w(\|y_0 - x_*\|)}{1 - w_0(\|x_0 - x_*\|, \lambda\|x_0 - x_*\|)} \\
&\quad + \frac{1}{(1 - w_0(\|x_0 - x_*\|, \lambda\|x_0 - x_*\|))^2} \\
&\quad \times \Bigg(w_0(\|x_0 - x_*\|, \lambda\|x_0 - x_*\|) + w_0(\|x_0 - x_*\|, \|y_0 - x_*\|) \\
&\quad + w_0(\lambda\|x_0 - x_*\|, \|y_0 - x_*\|)\Bigg)\Bigg] w_1(\|y_0 - x_*\|)\|y_0 - x_*\| \\
&\le \bar{g}_2(\|x_0 - x_*\|)\|x_0 - x_*\| \le \|x_0 - x_*\|. \qquad (11.20)
\end{aligned}$$

Set

$$\bar{r} = \min\{r_1, \bar{r}_2\}, \tag{11.21}$$

where $\bar{r}_2$ is the least positive solution of equation

$$\bar{g}_2(t) - 1 = 0 \tag{11.22}$$

in $M_0 - \{0\}$ (if it exists). Hence we arrived the corresponding result for method (11.3).

Theorem 19. Suppose the conditions (H) hold with $R = \min\{\bar{r}, p\bar{r}\}$ and $\bar{r}$ as defined by (11.21). Then, the conclusions of Theorem 2.1 hold for method (11.3) but with r, g_2 replaced by $\bar{r}$ and $\bar{g}_2$ respectively.

3. References

[1] Amiri, A., Cordero, A., Darvishi, M. T., Torregrosa, J. R., *Preserving the order of convergence: Low-complexity Jacobian-free iterative schemes for solving nonlinear systems*, J. Comput. Appl. Math., 337, (2018), 87-97.

[2] Argyros, I. K., *Computational theory of iterative methods*, Elsevier (Academic Press), New York, (2007).

[3] Argyros, I. K., George, S., *Mathematical modeling for the solution of equations and systems of equations with applications*, Volume-4, Nova Publishers, NY, (2020).

[4] Argyros, I. K., Magreñán, Á. A., *A contemporary study of iterative methods*, Elsevier (Academic Press), New York, (2018).

[5] Argyros, I. K., Magreñán, Á. A., *Iterative methods and their dynamics with applications: A Contemporary Study*, CRC Press, New York, USA, (2017).

[6] Grau-Sánchez, M., Grau, A., Noguera, M., *On the computational efficiency index and some iterative methods for solving systems of nonlinear equations*, J. Comput. Appl. Math., 236(6), (2011), 1259-1266.

[7] Hueso, J. L., Martínez, E., Torregrosa, J. R., *Third and fourth order iterative methods free from second derivative for nonlinear systems*, Appl. Math. Comput., 211(1), (2009), 190-197.

[8] Jarratt, P., *Some fourth order multipoint iterative methods for solving equations*, Math. Comp., 20(95), (1966), 434-437.

[9] Liu, Z., Quan, Z., Peng, Z., *A variant of Steffensen's method of fourth-order convergence and its applications*, Appl. Math. Comput., 216(7), (2010), 1978-1983.

[10] Magreñán, Á. A., *Different anomalies in a Jarratt family of iterative root-finding methods*, Appl. Math. Comput., 233, (2014), 29-38.

[11] Ortega, J. M., Rheinboldt, W. C., *Iterative methods for the solution of nonlinear equations in several variables*, Academic Press, New York, (1970).

[12] Ostrowski, A. M., *Solutions of equations and systems of equations*, Academic Press, New York, London, (1966).

[13] Sharma, J. R., Arora, H., Petkovi, M. S., *An efficient derivative-free family of fourth-order methods for solving systems of nonlinear equations*, Appl. Math. Comput., 235, (2014), 383-393.

[14] Sharma, J. R., Arora, H., *Improved Newton-like methods for solving systems of nonlinear equations*, SeMA Journal, 74(2), (2016), 1-7.

[15] Sharma, J. R., Arora, H., *On efficient weighted-Newton methods for solving systems of nonlinear equations*, Appl. Math. Comput., 222, (2013), 497-506.

[16] Sharma, J. R., Himani, A., *An efficient derivative-free method for solving systems of nonlinear equations*, Appl. Anal. Discr. Math., 7, (2013), 390-403.

[17] Traub, J. F., *Iterative methods for the solution of equations*, American Mathematical Soc, (1982).

Chapter 12

Extended Newton-Traub Method of Order Six

1. Introduction

The method is defined for $x_0 \in \Omega$ and each $n = 0, 1, 2, \cdots$ by

$$\begin{aligned} y_n &= x_n - F'(x_n)^{-1}F(x_n) \\ y_n &= x_n - \Big[\frac{3}{2}I - \frac{1}{2}F'(x_n)^{-1}F'(y_n)\Big]F'(x_n)^{-1}F'(x_n) \\ x_{n+1} &= z_n - \Big[\frac{7}{2}I + \Big(-4I + \frac{3}{2}F'(x_n)^{-1}F'(y_n)\Big)F'(x_n)^{-1}F'(y_n)\Big]F'(x_n)^{-1}F'(z_n). \end{aligned}$$

This method uses two operator evaluations, two inverses and only one inversion of $F'(x_n)$ per iteration. The local convergence order is shown to be six in the special case when $B = \mathbb{R}^j$ for j a natural number [1,12,13]. In the same article the construction and a favorable comparison to other methods using similar information is presented. However, the convergence assumptions require the existence of at least the seventh derivative. This restrict the applicability of the method to solving equations, where $F^{(2)}$ exists although the method may converge even if $F^{(2)}, \cdots, F^{(7)}$ do not exist. Other problems with the Taylor series approach have been reported in earlier chapters. In order to extend the applicability of this method we present the local convergence using w-continuity on F'. Only the operators on the method and in the more general setting of a Banach space. Moreover the more interesting case of majorizing sequences is presented not given previously using majorizing sequences.

2. Local Convergence Analysis

The following conditions are used for $T = [0, \infty)$:

(A_1) There exists a solution $x^\star \in \Omega$ of the equation $F(x) = 0$.

(A_2) There exists a continuous and nondecreasing function $w_0 : T \to T$ such that the equation $w_0(t) - 1 = 0$ has a smallest positive solution ρ_0. Set $T_0 = [0, \rho_0)$

(A_3) There exists an operator $L \in \mathcal{L}(B_1, B_2)$ such that $L^{-1} \in \mathcal{L}(B_1, B_2)$ and for each $x \in \Omega$, $\|L^{-1}(F'(x) - L)\| \leq w_0(\|x - x^\star\|)$. Set $\Omega_0 = U(x^\star, \rho_0) \cap \Omega$.

(A_4) There exists a continuous and nondecreasing function $w : T_0 \to T$ such that the equations $h_1(t) - 1 = 0, h_2(t) - 1 = 0$, where the functions h_1, h_2 with domain T_0 given for $\overline{w}(t) = \{ \begin{matrix} w\Big((1+h_1(t))t\Big) \\ w_0(t) + w_0(h_1(t)) \end{matrix}$ by

$$h_1(t) = \frac{\int_0^1 ((1-\theta)t)d\theta}{1-w_0(t)}, h_2(t) = \frac{\int_0^1 w((1-\theta)t)d\theta}{1-w_0(t)} + \frac{1}{2}\frac{\overline{w}(t)\Big(1+\int_0^1 w_0(\theta t)d\theta\Big)}{\Big(1-w_0(t)\Big)^2}$$

have smallest solutions $r_1, r_2 \in T_0 - \{0\}$, respectively.

(A_5) There exists smallest solution $\rho \in T_0 - \{0\}$ of the equation $w_0(h_1(t)t) - 1 = 0$. Set $T_1 = [0, \rho)$

(A_6) The equation $h_3(t) - 1 = 0$ has a smallest solution $r_3 \in T_1 - \{0\}$, where the function h_3 with domain T_1 is defined by

$$\begin{aligned} h_3(t) &= \left[\frac{\int_0^1 w((1-\theta)h_2(t)t)d\theta}{1-w_0(h_2(t)t)} + \frac{1}{2}\frac{\overline{\overline{w}}(t)\Big(1+\int_0^1 w_0(\theta h_2(t)t)d\theta\Big)}{\Big(1-w_0(t)\Big)\Big(1-w_0(h_2(t)t)\Big)}\right. \\ &\left. + \frac{1}{2}\frac{\overline{p}(t)\Big(1+\int_0^1 w_0(\theta h_2(t)t)d\theta\Big)}{\Big(1-w_0(t)\Big)}\right] h_2(t), \end{aligned}$$

where $\overline{\overline{w}}(t) = \{ \begin{matrix} w\Big((1+h_2(t))t\Big) \\ w_0(t) + w_0(h_2(t)) \end{matrix}$ and

$$\overline{p}(t) = \frac{1}{2}\left[3\left(\frac{\overline{w}(t)}{1-w_0(t)}\right)^2 + 14\left(\frac{\overline{w}(t)}{1-w_0(t)}\right)\right].$$

The radius of convergence for the method (12.1) is defined by

$$r = \min\{r_i : i = 1, 2, 3\}.$$

and

(A_7) $U[x^\star, r] \subset D$.

Notice that $L = F'(x^\star)$ is a possible choice. In this case $x^\star$ is a simple solution of the equation $F(x) = 0$. However, L can be any other linear operator satisfying these conditions. Then, the solution $x^\star$ does not have to be simple.

The motivation for the introduction of these real functions is given in a series of the following calculations starting from the first sub step of the method

$$\begin{aligned}
y_n - x^\star &= x_n - x^\star - F'(x_n)^{-1}F(x_n)\\
&= \int_0^1 F'(x_n)^{-1}\Big(F'(x^\star + \theta(x_n - x^\star)) - F'(x_n)\Big)d\theta\\
&\quad (x_n - x^\star)\\
\|y_n - x^\star\| &\leq \frac{\int_0^1 w\Big((1-\theta)\|x_n - x^\star\|\Big)d\theta\|x_n - x^\star\|}{1 - w_0(\|x_n - x^\star\|)}\\
&\leq g(\|x_n - x^\star\|)\|x_n - x^\star\|\\
&\leq \|x_n - x^\star\| < r,\\
z_n - x^\star &= x_n - x^\star - F'(x_n)^{-1}F(x_n)\\
&\quad + \frac{1}{2}F'(x_n)^{-1}\Big(F'(x_n) - F'(y_n)\Big)F'(x_n)^{-1}F(z_n)\\
\|z_n - x^\star\| &\leq \left[\frac{\int_0^1 w((1-\theta)\|x_n - x^\star\|)d\theta}{1 - w_0(\|x_n - x^\star\|)}\right.\\
&\quad \left. + \frac{1}{2}\frac{\overline{w}_n\Big(1 + \int_0^1 w_0(\theta\|x_n - x^\star\|)d\theta\Big)}{\Big(1 - w_0(\|x_n - x^\star\|)\Big)^2}\right]\|x_n - x^\star\|\\
&\leq h_2(\|x_n - x^\star\|)\|x_n - x^\star\| \leq \|x_n - x^\star\|,
\end{aligned}$$

where

$$\begin{aligned}
\|L^{-1}\Big(F'(x_n) - F'(y_n)\Big)\| &\leq w(\|x_n - y_n\|)\\
&\leq w(\|x_n - x^\star\| + \|y_n - x^\star\|) = \overline{w}_n\\
\|L^{-1}\Big(F'(x_n) - F'(y_n)\Big)\| &\leq \|L^{-1}(F'(x_n) - L)\| + \|L^{-1}(F'(y_n) - L)\|\\
&\leq w_0(\|x_n - x^\star\|) + w_0(\|y_n - x^\star\|) = \overline{w}_n,\\
x_{n+1} - x^\star &= z_n - x^\star - F'(z_n)^{-1}F(z_n)\\
&\quad + \Big(F'(z_n)^{-1} - F'(x_n)^{-1}\Big)F(z_n)\\
&\quad - \Big[\frac{5}{2}I - 4F'(x_n)^{-1}F'(y_n) + \frac{3}{2}\Big(F'(x_n)^{-1}F'(y_n)\Big)\Big]^2\\
&\quad \times F'(x_n)^{-1}F(z_n)\\
&= z_n - x^\star - F'(z_n)^{-1}F(z_n) + F'(z_n)^{-1}\Big(F'(x_n) - F'(z_n)\Big)\\
&\quad \times F'(z_n)^{-1}F(z_n)\\
&\quad - \frac{1}{2}\Big[3\Big(F'(x_n)^{-1}(F'(x_n) - F'(y_n))\Big)^2\\
&\quad - 14F'(x_n)^{-1}(F'(x_n) - F'(y_n))\Big]F'(x_n)^{-1}F(z_n)
\end{aligned}$$

$$\begin{aligned}
\|x_{n+1}-x^\star\| &\le \Bigg[\frac{\int_0^1 w((1-\theta)\|z_n-x^\star\|)d\theta}{1-w_0(\|z_n-x^\star\|)} \\
&\quad+\frac{1}{2}\frac{\overline{\overline{w}}(t)\left(1+\int_0^1 w_0(\|z_n-x^\star\|)d\theta\right)}{\left(1-w_0(\|x_n-x^\star\|)\right)\left(1-w_0(\|z_n-x^\star\|)\right)} \\
&\quad+\frac{\overline{p}(t)\left(1+\int_0^1 w_0(\theta\|z_n-x^\star\|)d\theta\right)}{\left(1-w_0(\|x_n-x^\star\|)\right)}\Bigg]\|z_n-x^\star\| \\
&\le h_3(\|x_n-x^\star\|)\|x_n-x^\star\| \\
&\le \|x_n-x^\star\|,
\end{aligned}$$

where $\overline{\overline{w}}(t)=\begin{cases} w(\|x_n-x^\star\|+\|z_n-x^\star\|) \\ w_0(\|x_n-x^\star\|)+w_0(\|z_n-x^\star\|),\end{cases}$

$$\begin{aligned}
&\frac{1}{2}\left[3\|F'(x_n)^{-1}(F'(x_n)-F'(y_n))\|^2+14\|F'(x_n)^{-1}(F'(x_n)-F'(y_n))\|\right] \\
&\le \frac{1}{3}\left[3\left(\frac{\overline{w}_n}{1-w_0(\|x_n-x^\star\|)}\right)+14\left(\frac{\overline{w}_n}{1-w_0(\|x_n-x^\star\|)}\right)\right]=\overline{p}_n
\end{aligned}$$

and since

$$\begin{aligned}
F(z_n) &= F(z_n)-F(x^\star)=\int_0^1 F'(x^\star+\theta(z_n-x^\star))d\theta(z_n-x^\star), \\
\|L^{-1}F(z_n)\| &= \|L^{-1}\int_0^1\left(F'(x^\star+\theta(z_n-x^\star))d\theta-L+L\right)(z_n-x^\star)\| \\
&\le \left(1+\int_0^1 w_0(\theta\|z_n-x^\star\|)d\theta\right)\|z_n-x^\star\|.
\end{aligned}$$

Then, under the conditions $(A_1)-(A_6)$ the iterates $\{x_n\}\subset U(x^\star,r)$ and $\lim_{n\to\infty} x_n=x^\star$.

3. Semi-local Convergence

The Conditions are:

(H_1) There exists a continuous and non decreasing function $v_0: T\to T$ such that the equation $v_0(t)-1=0$ has a smallest positive solution denoted by s.

(H_2) There exists an operator $L\in\mathcal{L}(B_1,B_2)$ such that $L^{-1}\in\mathcal{L}(B_2,B_1)$ and each $x\in\Omega$, $\|L^{-1}(F'(x)-L)\|\le v_0(\|x-x_0\|)$ for some x_0. It follows by (H_1) that $v_0(0)<1$, thus $F'(x_0)^{-1}\in\mathcal{L}(B_2,B_1)$ and the iterate y_0 is well defined by the first sub step of the method (12.1). Set $\|F'(x_0)^{-1}F(x_0)\|\le b_0$ and $\Omega_2=U(x_0,s)\cap\Omega$.

(H_3) There exists a continuous and nondecreasing function v with values in T such that for each $x,y\in\Omega_2$, $\|L^{-1}(F'(y)-F'(x))\|\le w(\|y-x\|)$. Define the scalar sequence $\{a_n\}$ for $a_0=0$ and $b_0\in[0,s)$ such that

$$\overline{v}(n) = \{ \begin{array}{c} v(b_n - a_n) \\ v_0(a_n) + v_0(b_n), \end{array}$$

$$c_n = b_n + \left(1 + \frac{1}{2}\frac{\overline{v}_n}{1 - v_0(a_n)}\right)(b_n - a_n),$$

$$q_n = \left(1 + \int_0^1 v_0(a_n + \theta(c_n - a_n))d\theta\right)(c_n - a_n) + (1 + v_0(a_n))(b_n - a_n),$$

$$a_{n+1} = c_n + \left[1 + 3\left(\frac{v_n}{1 - v_0(a_n)}\right)^2 + 14\left(\frac{v_n}{1 - v_0(a_n)}\right)\right]\frac{q_n}{1 - v_0(a_n)},$$

$$\delta_{n+1} = \int_0^1 v((1-\theta)(a_{n+1} - a_n))d\theta(a_{n+1} - a_n) + (1 + v_0(a_n))(a_{n+1} - b_n),$$

and

$b_{n+1} = a_{n+1} + \dfrac{\delta_{n+1}}{1 - v_0(a_{n+1})}$. There exists $s_0 \in [0, s)$ such that for each $n = 0, 1, 2, \cdots$ $v_0(a_n) < 1$ and $a_n \leq s_0$.

It follows that $0 \leq a_n \leq b_n \leq c_n \leq s^\star \leq s_0$, where $s^\star = \lim_{n\to\infty} a_n$.

and

(H_4) $U[x_0, s^\star] \subset \Omega$.
The motivations for these conditions are:

$$z_n - y_n = F'(x_n)^{-1}F(x_n) - \frac{1}{2}F'(x_n)^{-1}(F'(x_n) - F'(y_n))F'(x_n)^{-1}F(x_n)$$

$$\|z_n - y_n\| \leq \left(1 + \frac{1}{2}\frac{\overline{\overline{v}}}{1 - v_0(a_n)}\right)\|y_n - x_n\| \leq c_n - b_n,$$

where $\overline{\overline{v}}(n) = \{ \begin{array}{c} v(b_n - a_n) \\ v_0(a_n) + v_0(b_n), \end{array}$

$$\begin{aligned} x_{n+1} - z_n &= -[I + 3(F'(x_n)^{-1}(F'(x_n) - F'(y_n)))^2 + 14F'(x_n)^{-1}(F'(x_n) - F'(y_n))] * \\ &* \quad F'(x_n)^{-1}F(z_n), \end{aligned}$$

$$\|x_{n+1} - z\| \leq \left[1 + 3\left(\frac{\overline{\overline{v}}_n}{1 - v_0(a_n)}\right)^2 + 14\left(\frac{\overline{\overline{v}}_n}{1 - v_0(a_n)}\right)\right]\frac{q_n}{1 - v_0(a_n)},$$

since from

$$\begin{aligned}
F(z_n) &= F(z_n) - F(x_n) + F(x_n) \\
&= \int_0^1 F'(x_n + \theta(z_n - x_n))d\theta(z_n - x_n) - F'(x_n)(y_n - x_n), \\
\|L^{-1}F(z_n)\| &\leq \Big(1 + \int_0^1 v_0(\|x_n - x_0\| + \theta\|z_n - x_n\|)d\theta\Big)\|z_n - x_n\| \\
&\quad + (1 + v_0(\|x_n - x_0\|))\|y_n - x_n\| = \overline{q}_n \leq q_n, \\
F(x_{n+1}) &= F(x_{n+1}) - F(x_n) + F(x_n) \\
&= F(x_{n+1}) - F(x_n) - F'(x_n)(x_{n+1} - x_n) \\
&\quad + F'(x_n)(x_{n+1} - y_n) \\
&= \int_0^1 \Big[F'(x_n + \theta(x_{n+1} - x_n))d\theta - F'(x_n)\Big] \\
&\quad \times (x_{n+1} - x_n) + F'(x_n)(x_{n+1} - y_n), \\
\|L^{-1}F(x_{n+1})\| &\leq \int_0^1 v(\theta\|x_{n+1} - x_n\|)d\theta\|x_{n+1} - x_n\| \\
&\quad + \Big(1 + v_0(\|x_n - x_0\|)\Big)\|x_{n+1} - y_n\| \\
&= \overline{\delta}_{n+1} \leq \delta_{n+1} \\
\|y_{n+1} - x_{n+1}\| &\leq \|F'(x_{n+1})^{-1}L\|\|L^{-1}F(x_{n+1})\| \\
&\leq \frac{\overline{\delta}_{n+1}}{1 - v_0(\|x_{n+1} - x_0\|)} \leq \frac{\delta_{n+1}}{1 - v_0(a_{n+1})} = b_{n+1} - a_{n+1}.
\end{aligned}$$

The iterates $\{x_n\}, \{y_n\}, \{z_n\} \in U(x_0, s^\star)$. Indeed, we have the estimates

$$\begin{aligned}
\|y_0 - x_0\| &\leq b_0 - a_0 < s^\star \\
\|z_n - x_0\| &\leq \|z_n - y_n\| + \|y_n - x_0\| \leq c_n - b_n + b_n - a_0 = c_n < s^\star \\
\|x_{n+1} - x_0\| &\leq \|x_{n+1} - z_n\| + \|z_n - x_0\| \leq a_{n+1} - c_n + c_n - a_0 = a_{n+1} < s^\star, \\
\|y_{n+1} - x_0\| &\leq \|y_{n+1} - x_n\| + \|x_{n+1} - x_0\| \leq b_{n+1} - a_{n+1} + a_{n+1} - a_0 = b_{n+1} < s^\star.
\end{aligned}$$

Then, there exists $x^\star \in U[x_0, s^\star]$ solving the equation $F(x) = 0$ and satisfying the error estimate

$$\|x^\star - x_n\| \leq s^\star - a_n.$$

4. References

[1] Cordero, A., Torregrosa, J. R., *Variants of Newtons method for functions of several variables*, Appl. Math. Comput., 183, (2006), 199-208.

[2] Darvishi, M. T., Barati, A., *A third-order Newton-type method to solve systems of nonlinear equations*, Appl. Math. Comput., 187, (2007), 630-635.

[3] Frontini, M., Sormani, E., *Third-order methods from quadrature formulae for solving systems of nonlinear equations*, Appl. Math. Comput., 149, (2004), 771-782.

[4] Grau-Sánchez, M., Grau, A., Noguera, M., *On the computational efficiency index and some iterative methods for solving systems of nonlinear equations*, J. Comput. Appl. Math., 236, (2011), 1259-1266.

[5] Homeier, H. H. H., *A modified Newton method with cubic convergence: the multivariable case*, J. Comput. Appl. Math., 169, (2004), 161-169.

[6] Jarratt, P., *Some fourth-order multipoint methods for solving equations*, Math. Comput., 20, (1966), 434-437.

[7] Magreñán, Á. A., *A new tool to study real dynamics: the convergence plane*, Appl. Math. Comput., 248, (2014), 215-224.

[8] Noor, M.A., Waseem, M., *Some iterative methods for solving a system of nonlinear equations*, Comput. Math. Appl., 57, (2009), 101-106.

[9] Ortega, J. M., Rheinboldt, W. C., *Iterative Solution of Nonlinear Equations in Several Variables*, Academic Press, New York, (1970).

[10] Sharma, J. R., Arora, H., *Efficient Jarratt-like methods for solving systems of nonlinear equations*, Calcolo, 51, (2014), 193-210.

[11] Sharma, J. R., Gupta, P., *An efficient fifth-order method for solving systems of nonlinear equations*, Comput. Math. Appl., 67, (2014), 591-601.

[12] Sharma, J. R., Sharma, S., Bahl, A., *An improved Newton-Traub composition for solving systems of nonlinear equations*, Appl. Math. Comput., 290, (2016), 98-110.

[13] Xiao, X., Yin, H., *A new class of methods with higher order of convergence for solving systems of nonlinear equations*, Appl. Math. Comput., 264, (2015), 300-309.

Chapter 13

Derivative-Free Method With or Without Memory

1. Introduction

The methods without memory, followed by the method with memory, are given by:

$$\begin{aligned} y_n &= x_n - [u_n, x_n; F]^{-1} F(x_n), \\ u_n &= x_n + bF(x_n), \, b \in \mathbb{R} - \{0\} \\ A_n &= [u_n, x_n; F]^{-1} [z_n, y_n; F], \, z_n = y_n + cF(y_n), c \in \mathbb{R}, \\ x_{n+1} &= y_n - (aI + A_n((3-2a)I + (a-2)A_n)[u_n, x_n; F]^{-1} F(y_n), \, a \in \mathbb{R}. \end{aligned} \tag{13.1}$$

If $a = 3$ and $B_n = -[v_{n-1}, x_{n-1}; F]^{-1}$ replaces b for $v_{n-1} = x_{n-1} + B_{n-1} F(x_{n-1}$ then (13.1) reads

$$\begin{aligned} u_n &= x_n - [v_n, x_n; F]^{-1} F(x_n) \\ x_{n+1} &= y_n - (3I - A_n(3I - A_n))[v_n, x_n : F]^{-1} F(y_n). \end{aligned} \tag{13.2}$$

The fourth order is shown in [8] and the increased order $2 + \sqrt{6} \approx 4.45$ in [6] in the special case $B = \mathbb{R}^k$. The Taylor series approach is utilized for these results [1]-[7], [9, 10], in the local convergence analysis. The limitations of this approach are given in earlier Chapters under conditions on the fifth derivative of the operator F. But derivatives are not on the methods. In this Chapter the local as well the semi-local convergence (not given in [6] or[8]) is provided by using only conditions on the operators appearing on the methods and in the more general setting of a Banach space.

2. Local Convergence

Method (13.1)

The majorant function g_i, $i = 1, 2$ are:

$$\begin{aligned} g_1(t) &= \frac{\varphi(f_1(t), t)}{1 - \varphi_0(f_1(t), t)}, \\ g_2(t) &= \frac{1}{1 - \varphi_0(f_1(t), t)} \Big[\varphi(f_1(t), g_1(t)t) + \varphi_1(t, g_1(t)t, f_1(t), f_2(t)) \\ &\quad \times (1 + \varphi_3(g_1(t)t))(1 + |a-2|\varphi_1(t, g_1(t)t, f_1(t), f_2(t))) \Big] g_1(t). \end{aligned}$$

The motivation follows from the estimates

$$\begin{aligned} y_n - x^* &= [u_n, x_n; F]^{-1}([u_n, x_n; F] - [x_n, x^*; F])(x_n - x^*) \\ \|y_n - x^*\| &\leq \frac{\varphi(\|x_n - x^*\|, \|u_n - x^*\|)\|x_n - x^*\|}{1 - \varphi_0(\|u_n - x^*\|, \|x_n - x^*\|)} \\ &\leq g_1(\|x_n - x^*\|)\|x_n - x^*\| \\ &\leq \|x_n - x^*\| < r \end{aligned}$$

(for $\|u - x^*\| \leq f_1(\|x - x^*\|)$). Then, by the second substep one can write

$$\begin{aligned} x_{n+1} - x^* &= y_n - x^* - [u_n, x_n; F]^{-1} F(y_n) \\ &\quad -(A_n - I)[(a-2)(A_n - I) - I][u_n, x_n; F]F(y_n), \\ \|x_{n+1} - x^*\| &\leq \frac{1}{1 - \varphi_0(\|u_n - x^*\|, \|x_n - x^*\|)} \Big[\varphi(\|u_n - x^*\|, \|y_n - x^*\|) \\ &\quad + \lambda_n(1 + |a-2|\lambda_n)(1 + \varphi_3(\|y_n - x^*\|))\Big] \|y_n - x^*\|, \end{aligned}$$

where

$$\|A_n - I\| \leq \varphi_1(\|x_n - x^*\|, \|y_n - x^*\|, \|u_n - x^*\|, \|z_n - x^*\|) = \bar{\lambda}_n \leq \lambda_n$$

so

$$\|x_{n+1} - x^*\| \leq g_2(\|x_n - x^*\|)\|x_n - x^*\| \leq \|x_n - x^*\|.$$

3. Semi-local Convergence

The majorant sequence $\{\alpha_n\}$ is defined for $\alpha_0 = 0, \beta_0 \geq \|[u_0, x_0; F]^{-1} F9x_0)\|$ and each $n = 0, 1, 2, \ldots$ by

$$\begin{aligned} \lambda_n &= \psi(\alpha_n, \beta_n, f_1(\alpha_n))(\beta_n - \alpha_n), \\ \alpha_{n+1} &= \beta_n + [1 + \psi_1(\alpha_n, \beta_n, f_3(\alpha_n), f_4(\beta_n))] \frac{\lambda_n}{1 - \psi_0(f_3(\alpha_n), \alpha_n)}, \\ \delta_{n+1} &= \psi(\alpha_n, \alpha_{n+1}, f_3(\alpha_n))(\alpha_{n+1} - \alpha_n) \\ &\quad + (1 + \psi_0(f_3(\alpha_n), \alpha_n))(\alpha_{n+1} - \beta_n), \\ \beta_{n+1} &= \alpha_{n+1} + \frac{\delta_{n+1}}{1 - \psi_0(f_3(\alpha_n), \alpha_n)}. \end{aligned}$$

The motivation follows from the estimates

$$\begin{aligned} x_{n+1} - y_n &= -[I - (A_n - I) + (a-2)(A_n - I)^2][u_n, x_n; F]^{-1} F(y_n), \\ \|x_{n+1} - y_n\| &\leq [1 + \|A_n - I\| + |a-2|\|A_n - I\|^2] \frac{\bar{\lambda}_n}{1 - \psi_0(f_3(\alpha_n), \alpha_n)} \\ &\leq \alpha_{n+1} - \beta_n, \\ F(x_{n+1}) &= F(x_{n+1}) - F(x_n) - [u_n, x_n; F](x_{n+1} - x_n) \\ &\quad + [u_n, x_n; F](x_{n+1} - y_n), \\ \|L^{-1} F(x_{n+1})\| &\leq \bar{\delta}_{n+1} \leq \delta_{n+1}, \\ \|y_{n+1} - x_{n+1}\| &\leq \|[u_{n+1}, x_{n+1}; F]^{-1} L\| \|L^{-1} F(x_{n+1})\| \\ &\leq \beta_{n+1} - \alpha_{n+1}, \end{aligned}$$

where we also used

$$\begin{aligned}\|L^{-1}F(y_n)\| &= \|L^{-1}(F(y_n)-F(x_n)-[u_n,x_n;F](y_n-x_n)\| \\ &\leq \bar{\lambda}_n \leq \lambda_n.\end{aligned}$$

Remark. Let us provide some explicit choice of the functions f_j, $j=1,2,3,4$.

4. Local Convergence of Method (13.2)

Notice the estimates: We have starting from the local case,

$$\begin{aligned}u-x^* &= x-x^*+bF(x) \\ &= (I+b[x,x^*:F](x-x^*) \\ &= (I+bLL^{-1}([x,x^*;F]_L+L))(x-x^*) \\ &= [(I+bL)+bLL^{-1}([x,x^*;F]-L)](x-x^*),\end{aligned}$$

so for $|b|\leq\xi_0$, $\|I+bL\|\leq\xi_1$, we get $|c|\leq\xi_2$ and $\|I+cL\|\leq\xi_3$,

$$\begin{aligned}\|u-x^*\| &\leq [\|I+bL\|+|b|\|L\|\|L^{-1}([x,x^*;F]-L)\|]\|x-x^*\| \\ &\leq (\xi_1+\xi_0\|L\|\varphi_3(\|x-x^*\|)\|x-x^*\|,\end{aligned}$$

thus we can choose

$$f_1(t)=(\xi_1+\xi_0\|L\|\varphi_3(t))t.$$

Similarly,

$$z-x^*=[(I+cL)+cLL^{-1}([y,x^*;F]-L)](y-x^*),$$

hence we can set

$$f_2(t)=(\xi_3+\xi_2\|L\|\varphi_3(g_1(t)t))g_1(t)t.$$

Moreover, for the semi-local case

$$u-x_0=[(I+bL)+bLL^{-1}([x,x);F]-L)](x-x_0)+bF(x_0),$$

thus we can choose

$$f_3(t)=(\xi_1+\xi_0\|L\|\psi_3(t))t+\xi_0\|F(x_0)\|$$

and similarly

$$f_4(t)=(\xi_3+\xi_2\|L\|\psi_3(t))t+\xi_2\|F(x_0)\|.$$

Method (13.2)

Suppose for each $n=0,1,2,\ldots$

$$\|B_n\|\leq\xi_0,\ \|I+B_nL\|\leq\xi_1,$$

then the functions f_j and the majorant sequences are the same as in method (13.1).

5. References

[1] Argyros, C. I., Regmi, S., Argyros, I. K., George, S., *Contemporary Algorithms: Theory and Applications*, Volume III, NOVA Publishers, (2023).

[2] Argyros, I. K., *The theory and applications of iteration methods, 2nd Edition, CRC Press, Engineering Series*, Taylor and Francis Group, (2022).

[3] Grau-Sánchez, M., Grau, Á., Noguera, M., *Frozen divided difference scheme for solving systems of nonlinear equations*, J. Comput. Appl. Math., 235, (2011), 1739-1743.

[4] Ortega, J. M., Rheinboldt, W. C., *Iterative Solution of Nonlinear Equations in Several Variables*, Academic Press, New York, (1970).

[5] Ostrowski, A. M., *Solution of Equations and Systems of Equations*, Academic Press, New York, (1966).

[6] Petković, M. S., Sharma, J. R., *On some efficient derivative-free iterative methods with memory for solving systems of nonlinear equations*, Numer. Algor., 71, (2016), 457-474.

[7] Sharma, J. R., Arora, H., Petković, M. S., *An efficient derivative-free family of fourth order methods for solving systems of nonlinear equations*, Appl. Math. Comput., 235, (2014), 383-393.

[8] Steffensen, J. F., *Remarks on iteration*, Skand. Aktuar Tidskr., 16, (1933), 64-72.

[9] Traub, J. F., *Iterative Methods for the Solution of Equations*, Prentice-Hall, Englewood Cliffs, (1964).

[10] Wolfram, S., *The Mathematica Book*, 5th edn., Wolfram Media, (2003).

Chapter 14

Increasing the Order of Convergence by Utilizing Memory for Nondifferentiable Equations

1. Introduction

The methods are:

$$x_{n+1} = x_n - [u_n, x_n; F]^{-1} F(x_n),\ u_n = x_n + \alpha F(x_n),\ \alpha \in \mathbb{R}, \tag{14.1}$$

and

$$\begin{aligned} u_n &= x_n - [2x_n - x_{n-1}, x_{n-1}; F]^{-1} F(x_n) \\ x_{n+1} &= x_n - [u_n, x_n : F]^{-1} F(x_n). \end{aligned} \tag{14.2}$$

Method (14.1) is due to Shamanski [6]. If $\alpha = 1$ the method reduces to Steffensen,s method [st]. It is a method without memory of order two. The Taylor series approach requires the existence of at least the third derivative limiting its applicability. In order to increase the convergence order the method (14.2) with memory is developed in [4, 7]. Notice that α is approximated by $-[2x_n, x_{n-1}, x_{n-1}]^{-1}$. Then convergence order is shown to be . But the existence of at least the fourth derivative is required which again limits the applicability of the method. We employ conditions only on the operators on the methods in both the local and the semilocal convergence (not given in [2]). Hence, the applicability of these methods [1, 2, 3, 5] is extended under weaker conditions.

2. Local Convergence

The conditions are:

$$\begin{aligned} \|L^{-1}([u,x;F] - L)\| &\leq \varphi_0(\|u - x^*\|, \|x - x^*\|), \\ \|L^{-1}([u,x;F] - [x,x^*;F])\| &\leq \varphi(\|u - x^*\|, \|x - x^*\|), \\ \|L^{-1}([2y - x, x; F] - L)\| &\leq \varphi_1(\|x - x^*\|, \|y - x^*\|), \\ \|L^{-1}([2y - x, x; F] - [y, x^*; F])\| &\leq \varphi_2(\|x - x^*\|, \|y - x^*\|) \end{aligned}$$

and

$$\|u - x^*\| \leq f(\|x - x^*\|).$$

Possible choices for the function f have been reported in previous chapters.
Method (14.1)
The majorant function is

$$g_1(t) = \frac{\varphi(f(t),t)}{1-\varphi_0(f(t),t)}.$$

The motivation follows from the estimate

$$\begin{aligned}
x_{n+1}-x^* &= x_n - x^* - [u_n,x_n;F]^{-1}F(x_n)\\
&= [u_n,x_n;F]^{-1}([u_n,x_n;F]-[x_n,x^*;F])(x_n-x^*),\\
\|x_{n+1}-x^*\| &\leq \frac{\varphi(\|u_n-x^*\|,\|x_n-x^*\|)\|x_n-x^*\|}{1-\varphi_0(\|u_n-x^*\|,\|x_n-x^*\|)}\\
&\leq \frac{\varphi(f(\|x_n-x^*\|),\|x_n-x^*\|)\|x_n-x^*\|}{1-\varphi_0(\|f(\|x_n-x^*\|),\|x_n-x^*\|)}\\
&\leq g_1(\|x_n-x^*\|)\|x_n-x^*\| \leq \|x_n-x^*\| < r.
\end{aligned}$$

Method (14.2)
The majorant functions $g_i,\ i=1,2$ are :

$$g_2(t) = \frac{\varphi_2(t,t)}{1-\varphi_1(t,t)}$$

and

$$g_3(t) = \frac{\varphi_2(g_2(t),t)}{1-\varphi_1(g_2(t)t,t)}.$$

The motivation follows from the estimates

$$\begin{aligned}
u_n-x^* &= x_n-x^*-[2x_n-x_{n-1},x_{n-1};F]^{-1}F(x_n)\\
&= [2x_n-x_{n-1},x_{n-1};F]^{-1}([2x_n-x_{n-1},x_{n-1};F]-[x_n,x^*;F])(x_n-x^*)\\
\|u_n-x^*\| &\leq \frac{\varphi_2(\|x_{n-1}-x^*\|,\|x_n-x^*\|)\|x_n-x^*\|}{1-\varphi_1(\|x_{n-1}-x^*\|,\|x_n-x^*\|)}\\
&= g_2(\|x_n-x^*\|)
\end{aligned}$$

By replacing the left hand side of the above estimate by the right hand side in the estimate $\|x_{n+1}-x^*\|$ obtained for the method (14.1), we arrive at the definition of the function g_3.

3. Semi-local Convergence

Exchange the role of x^* by x_0 in proofs of the previous section. Moreover, use instead the condition:

$$\begin{aligned}
\|L^{-1}([u,x;F]-L)\| &\leq \psi_0(\|u-x_0\|,\|x-x_0\|),\\
\|L^{-1}([y,x;F]-[u,x;F])\| &\leq \psi(\|x-x_0\|,\|y-x_0\|.\|u-x_0\|),\\
\|L^{-1}([2y-x,x;F]-L)\| &\leq \psi_1(\|x-x_0\|,\|y-x_0\|),\\
\|L^{-1}([2y-x,x;F]-[y,x_0;F])\| &\leq \psi_2(\|x-x_0\|,\|y-x_0\|)
\end{aligned}$$

and

$$\|u-x_0\| \leq g(\|x-x_0\|).$$

Possible choices for the function g have been given in previous chapters.

Method (14.1)

The majorant sequence $\{\alpha_n\}$ is defined for $\alpha_0 = 0, \alpha_1 \geq \|[u_0, x_0; F]^{-1}F(x_0)\|$ and each $n = 0, 1, 2, \ldots$ by

$$\begin{aligned} \delta_{n+1} &= \psi(\alpha_n, \alpha_{n+1}, g(\alpha_n))(\alpha_{n+1} - \alpha_n). \\ \alpha_{n+2} &= \alpha_{n+1} + \frac{\delta_{n+1}}{1 - \psi_0(g(\alpha_{n+1}), \alpha_{n+1})}. \end{aligned}$$

The estimates leading to the definition of the sequence $\{\alpha_n\}$ are:

$$\begin{aligned} F(x_{n+1}) &= F(x_{n+1}) - F(x_n) - [u_n, x_n; F](x_{n+1} - x_n) \\ &= ([x_{n+1}, x_n; F] - [u_n, x_n; F])(x_{n+1} - x_n), \\ \|L^{-1}F(x_{n+1})\| &\leq \psi(\|x_n - x_0\|, \|x_{n+1} - x_0\|, \|u_n - x_0\|)\|x_{n+1} - x_n\| \\ &= \bar{\delta}_{n+1} \leq \delta_{n+1} \\ &= \psi(\alpha_n, \alpha_{n+1}, g(\alpha_n))(\alpha_{n+1} - \alpha_n), \\ \|x_{n+2} - x_{n+1}\| &\leq \|[u_{n+1}, x_{n+1}; f]^{-1}L\|\|L^{-1}F(x_{n+1})\| \\ &\leq \frac{\delta_{n+1}}{1 - \psi_0(g(\alpha_{n+1}), \alpha_{n+1})} = \alpha_{n+2} - \alpha_{n+1}. \end{aligned}$$

Method (14.2)

The majorant sequence $\{\alpha_n\}$ for each $n = -1, 0, 1, 2, \ldots$ and $\alpha_{-1} = 0, \alpha_0 \geq \|x_0 - x_{-1}\|, \alpha_1 \geq \|[u_0, x_0; F]^{-1}F(x_0)\|$,

$$\bar{g}(\alpha_{n+1}) = \frac{\psi_2(\alpha_{n-1}, \alpha_n)(\alpha_n) + \|L^{-1}F(x_0)\|}{1 - \psi_0(\alpha_{n-1}, \alpha_n)} \tag{14.3}$$

and α_{n+2} is defined as in method (14.1) with $\bar{g}(\alpha_{n+1})$ replacing $g(\alpha_{n+1})$.
The motivation for the development of the sequence $\{\alpha_n\}$ is:

$$\begin{aligned} u_n - x_0 &= x_n - x_0 - [2x_n - x_{n-1}, x_{n-1}; F]^{-1}(F(x_n) - F(x_0) + F(x_0)) \\ &= [2x_n - x_{n-1}, x_{n-1}; F]^{-1}([2x_n - x_{n-1}, x_{n-1}; F] - [x_n, x_0; F])(x_n - x_0) \\ &\quad -[2x_n - x_{n-1}, x_{n-1}; F]^{-1}F(x_0), \\ \|u_n - x_0\| &\leq \frac{\psi_2(\|x_{n-1} - x_0\|, \|x_n - x_0\|)\|x_n - x_0\| + \|L^{-1}F(x_0)\|}{1 - \psi_0(\|x_{n-1} - x_0\|, \|x_n - x_0\|)} \\ &\leq \bar{g}(\alpha_{n+1}). \end{aligned}$$

4. References

[1] Chicharoo, F. I., Cordero, A., Garrido, N., *On the improvement of the order of convergence of iterative methods for solving nonlinear systems by means of memory*, Appl. Math. Lett., 104, (2020), 106277.

[2] Cordero, A., Torregrosa, J. R., *Variants of Newtons method using fifth order quadrature formulas*, Appl. Math. Comput., 190, (2007), 686-698.

[3] Narang, M., Bathia, S., Alshomrani, A. S., Kanwar, V., *General efficient class of Steffensen type methods with memory for solving systems of nonlinear equations*, J. Comput. Appl. Math., 352, (2019), 23-39.

[4] Ortega, J. M., Rheinboldt, W. C., *Iterative Solution of Nonlinear Equations in Several Variables*, Academic Press, New York, (1970).

[5] Petković, M. S., Sharma, J. R., *On some efficient derivative-free iterative methods with memory for solving systems of nonlinear equations*, Numer. Algorithms, 71, (2016), 457-474.

[6] Samanskii, V., *On a modification of the Newton method*, Ukrain. Math., 19, (1967), 133-138.

[7] Traub, J. F., *Iterative Methods for the Solution of Equations*, Prentice-Hall, Englewood Cliffs, (1964).

Chapter 15

Two Competing Sixth-Order Methods

1. Introduction

The methods are

$$
\begin{aligned}
y_n &= x_n - F'(x_n)^{-1}F(x_n) \\
z_n &= x_n - \frac{1}{2}(3I - F'(x_n)^{-1}F'(y_n))F'(x_n)^{-1}F(x_n), \qquad (15.1) \\
x_{n+1} &= z_n - (\frac{7}{2}I - F'(x_n)^{-1}F'(y_n)(4I - \frac{3}{2}F'(x_n)^{-1}F'(y_n))F'(x_n)^{-1}F(z_n), \\
y_n &= x_n - \frac{2}{3}F'(x_n)^{-1}F(x_n) \qquad (15.2) \\
z_n &= x_n - \frac{1}{2}(3F'(y_n) - F'(x_n))^{-1}(3F'(y_n) + F'(x_n))F'(x_n)^{-1}F(x_n), \\
x_{n+1} &= z_n - 2(3F'(y_n) - F'(x_n))^{-1}F(z_n).
\end{aligned}
$$

The local sixth order of these methods is shown in [13] under assumptions on at least the seventh derivative and the Taylor series expansion approach. The limitations of this approach are presented in previous chapters. Our methodology uses only assumptions on the operators on the method in the more general setting of a Banach space. Our analysis includes the semi-local convergence not previously given. Relevant work can be found in [1]-[12], [14]

2. Local Convergence

The conditions are:

$$\|L^{-1}(F'(x) - L)\| \leq w_0(\|x - x^*\|)$$

and

$$\|L^{-1}(F'(y) - F'(x))\| \leq w(\|y - x\|).$$

Method (15.1)
The majorant functions $g_i, i = 1,2,3$ are :

$$
\begin{aligned}
g_1(t) &= \frac{\int_0^1 w((1-\theta)t)d\theta}{1-w_0(t)},\\
\bar{w}(t) &= \begin{cases} w(1+g_1(t)t) \\ w_0(t)+w_0(f_1(t)t), \end{cases}\\
g_2(t) &= \frac{\int_0^1 w((1-\theta)t)d\theta}{(1-w_0(t))} + \frac{\bar{w}(t)(1+\int_0^1 w_0(\theta t)d\theta)}{2(1-w_0(t))^2}
\end{aligned}
$$

and

$$
\begin{aligned}
g_3(t) &= \left[\frac{\int_0^1 w((1-\theta)g_2(t)t)d\theta}{1-w_0(g_2(t)t)}\right.\\
&\left. + \frac{\bar{w}(t)\left(2+\frac{3\bar{w}(t)}{1-w_0(t)}\right)(1+\int_0^1 w_0(\theta f_2(t)t)d\theta}{(1-w_0(t))^2}\right] g_2(t).
\end{aligned}
$$

The estimates leading to the functions g_i are:

$$
\begin{aligned}
y_n - x^* &= x_n - x^* - F'(x_n)^{-1}F(x_n),\\
\|y_n - x^*\| &\le \frac{\int_0^1 w((1-\theta)\|x_n - x^*\|d\theta\|x_n - x^*\|}{1-w_0(\|x_n - x^*\|)}\\
&\le g_1(\|x_n - x^*\|)\|x_n - x^*\|\\
&\le \|x_n - x^*\| < r,\\
z_n - x^* &= x_n - x^* - F'(x_n)^{-1}F(x_n)\\
&\quad -\frac{1}{2}(I - F'(x_n)^{-1}F'(y_n))F'(x_n)^{-1}F(x_n),\\
\|z_n - x^*\| &\le \left[\frac{\int_0^1 w((1-\theta)\|x_n - x^*\|)d\theta}{1-w_0(\|x_n - x^*\|)}\right.\\
&\quad \left. + \frac{\bar{w}_n(1+\int_0^1 w_0(\theta\|x_n - x^*\|)d\theta)}{2(1-w_0(\|x_n - x^*\|))^2}\right]\\
&\quad \times\|x_n - x^*\|\\
&\le g_1(\|x_n - x^*\|)\|x_n - x^*\|\\
&\le \|x_n - x^*\|,
\end{aligned}
$$

where

$$\bar{w}_n = \begin{cases} w(\|y_n - x_n\|) \\ OR \\ w_0(\|x_n - x^*\|) + w_0(\|y_n - x^*\|), \end{cases}$$

$$\begin{aligned}
x_{n+1} - x^* &= z_n - x^* - F'(z_n)^{-1}F(z_n) + (F'(z_n)^{-1} - F'(x_n)^{-1})F(z_n) \\
&\quad -\frac{1}{2}(5I - 8F'(x_n)^{-1}F'(y_n) \\
&\quad +3F'(x_n)^{-1}F'(y_n)F'(x_n)^{-1}F'(y_n))F'(x_n)^{-1}F(z_n), \\
&= z_n - x^* - F'(z_n)^{-1}F(z_n) \\
&\quad +F'(z_n)^{-1}(F'(x_n) - F'(z_n))F'(x_n)^{-1}F(z_n) \\
&\quad -\frac{1}{2}F'(x_n)^{-1}(F'(y_n) - F'(x_n)) \\
&\quad [2I + 3F'(x_n)^{-1}(F'(y_n) - F'(x_n))]F'(x_n)^{-1}F(z_n), \\
\|x_{n+1} - x^*\| &\leq \left[\frac{w((1-\theta)\|z_n - x^*\|)d\theta}{1 - w_0(\|z_n - x^*\|)}\right. \\
&\quad \left. + \frac{\bar{w}_n\left(2 + \frac{3\bar{w}_n}{1 - w_0(\|x_n - x^*\|)}\right)(1 + \int_0^1 w_0(\theta\|z_n - x^*\|)d\theta)}{2(1 - w_0(\|x_n - x^*\|)}\right] \\
&\quad \times \|z_n - x^*\| \\
&\leq g_3(\|x_n - x^*\|)\|x_n - x^*\| \leq \|x_n - x^*\|.
\end{aligned}$$

Method(15.2)
The functions $g_i, i = 1,2,3$ are:

$$\begin{aligned}
g_1(t) &= \frac{\int_0^1 w((1-\theta)t)d\theta + \frac{1}{3}(1 + \int_0^1 w_0(\theta t)d\theta)}{1 - w_0(t)}, \\
p(t) &= \frac{1}{2}(3w_0(g_1(t)t) + w_0(t)), \\
\bar{\bar{w}}(t) &= \begin{cases} w(1 + g_2(t)t) \\ w_0(t) + w_0(f_2(t)t), \end{cases} \\
g_2(t) &= \frac{\int_0^1 w((1-\theta)t)d\theta}{1 - w_0(t)} + \frac{3\bar{w}(t)(1 + \int_0^1 w_0(\theta t)d\theta)}{4(1 - p(t))},
\end{aligned}$$

and

$$g_3(t) = \left[\frac{\int_0^1 w(1-\theta)g_2(t)t)d\theta}{1 - w_0(g_2(t)t)} + \frac{(\bar{w}(t) + 2\bar{\bar{w}}(t))(1 + \int_0^1 w_0(\theta g_2(t)t)d\theta)}{2(1 - p(t))(1 - p(t))}\right] g_2(t).$$

The motivation is coming from the estimates

$$\begin{aligned}
y_n - x^* &= x_n - x^* - F'(x_n)^{-1}F(x_n) + \frac{1}{3}F'(x_n)^{-1}F(x_n), \\
\|y_n - x^*\| &\leq \frac{[\int_0^1 w((1-\theta)\|x_n - x^*\|)d\theta + \frac{1}{3}(1 + \int_0^1 w_0(\theta\|x_n - x^*\|)d\theta)]}{1 - w_0(x_n - x^*\|)} \\
&\leq g_1(\|x_n - x^*\|)\|x_n - x^*\| \leq \|x_n - x^*\| < r, \qquad (15.3)
\end{aligned}$$

$$\begin{aligned}
&\|(2L)^{-1}(3F'(y_n)-F'(x_n)-2L)\| \\
&\leq \frac{1}{2}[3\|L^{-1}(F'(y_n)-L)\|+\|L^{-1}(F'(x_n)-L)\|] \\
&\leq \frac{1}{2}(3\varphi_0(\|y_n-x^*\|)+\varphi_0(\|x_n-x^*\|)) \\
&= \bar{p}_n \leq p_n < 1, \\
\|(3F'(y_n)-F'(x_n)^{-1}L\| &\leq \frac{1}{2(1-p_n)},
\end{aligned}$$

$$\begin{aligned}
z_n - x^* &= x_n - x^* - F'(x_n)^{-1}F(x_n) \\
&+[I-\frac{1}{2}(3F'(y_n)-F'(x_n))^{-1}(3F'(y_n) \\
&+F'(x_n))]F'(x_n)^{-1}F(x_n) \\
&= x_n - x^* - F'(x_n)^{-1}F(x_n)+\frac{1}{2}(3F'(y_n)-F'(x_n))^{-1} \\
&\times[2(3F'(y_n)-F'(x_n)-3F'(y_n) \\
&-F'(x_n)]F'(x_n)^{-1}F(x_n), \\
\|z_n-x^*\| &\leq \left[\frac{\int_0^1 w((1-\theta)\|x_n-x^*\|)d\theta}{1-w_0(\|x_n-x^*\|)}\right. \\
&+\left.\frac{3\bar{w}_n(1+\int_0^1 w_0(\theta\|x_n-x^*\|)d\theta)}{4(1-p_n)}\right]\|x_n-x^*\| \\
&\leq g_2(\|x_n-x^*\|)\|x_n-x^*\|, \\
x_{n+1}-x^* &= z_n - x^* - F'(z_n)^{-1}F(z_n)+(F'(z_n)^{-1} \\
&-2(3F'(y_n)-F'(x_n)^{-1})F(z_n) \\
&= z_n - x^* - F'(z_n)^{-1}F(z_n)+F'(z_n)^{-1} \\
&[(F'(y_n)-F'(x_n))+2(F'(y_n)-F'(z_n))] \\
&\times(3F'(y_n)-F'(x_n)^{-1}F(z_n), \\
\|x_{n+1}-x^*\| &\leq \left[\frac{\int_0^1 w((1-\theta)\|z_n-x^*\|)d\theta}{1-w_0(\|z_n-x^*\|)}\right. \\
&+\left.\frac{(\bar{w}_n+\bar{\bar{w}}_n)(1+\int_0^1 w_0(\theta\|z_n-x^*\|)d\theta)}{2(1-p_n)(1-w_0(\|z_n-x^*\|))}\right]\|z_n-x^*\| \\
&\leq g_3(\|x_n-x^*\|)\|x_n-x^*\| \leq \|x_n-x^*\|.
\end{aligned}$$

3. Semi-local Convergence

The conditions are:

$$\|L^{-1}(F'(x)-L)\| \leq v_0(\|x-x_0\|)$$

and

$$\|L^{-1}(F'(y)-F'(x)\| \leq v(\|y-x\|).$$

<u>**Method(15.1)**</u>

The majorizing sequence $\{x_n\}$ is defined for $\alpha_0 = 0, \beta_0 \geq \|F'(x_0)^{-1}F(x_0)\|$ and each $n =$

$0, 1, 2, \ldots$ by

$$\begin{aligned}
\bar{v}_n &= \begin{cases} v(\beta_n - \alpha_n) \\ v_0(\alpha_n) + v_0(\beta_n), \end{cases} \\
\gamma_n &= \beta_n + \frac{\bar{v}_n(\beta_n - \alpha_n)}{2(1 - v_0(\alpha_n))}, \\
\alpha_{n+1} &= \gamma_n + \frac{\zeta_n}{2}\frac{(3\bar{v}_n^2 + 2\bar{v}_n + 2)}{1 - v_0(\alpha_n)} \\
\zeta_n &= (1 + \int_0^1 v_0(\alpha_n + \theta(\gamma_n - \alpha_n))d\theta)(\gamma_n - \alpha_n) \\
&\quad +(1 + v_0(\alpha_n))(\beta_n - \alpha_n), \\
\delta_{n+1} &= \int_0^1 v((1-\theta)(\alpha_{n+1} - \alpha_n))d\theta(\alpha_{n+1} - \alpha) \\
&\quad +(1 + v_0(\alpha_n))(\alpha_{n+1} - \beta_n)
\end{aligned}$$

and

$$\beta_{n+1} = \alpha_{n+1} + \frac{\delta_{n+1}}{1 - v_0(\alpha_{n+1})}.$$

The estimates leading to this sequence are:

$$\begin{aligned}
z_n - y_n &= [I - \frac{1}{2}(3I - F'(x_n)^{-1}F'(y_n))[-(y_n - x_n)] \\
&= \frac{1}{2}F'(x_n)^{-1}(F'(x_n) - F'(y_n))(y_n - x_n), \\
\|z_n - y_n\| &\leq \frac{\bar{v}_n(\beta_n - \alpha_n)}{2(1 - v_0(\alpha_n))} = \gamma_n - \beta_n, \\
F(z_n) &= F(z_n) - F(x_n) - F'(x_n)(y_n - x_n), \\
\|L^{-1}F(z_n)\| &\leq (1 + \int_0^1 v_0(\|x_n - x_0\| + \theta\|z_n - x_n\|)d\theta)\|z_n - x_n\| \\
&\quad +(1 + v_0(\|x_n - x_0\|))\|y_n - x_n\| = \bar{\zeta}_n \\
&\leq \zeta,
\end{aligned}$$

$$\begin{aligned}
x_{n+1} - z_n &= -\frac{1}{2}[3(F'(x_n)^{-1}F'(y_n) - I)^2 - 2(F'[(x_n)^{-1}F'(y_n) - I) \\
&\quad +2I]F'(x_n)^{-1}F(z_n), \\
\|x_{n+1} - z_n\| &\leq \frac{1}{2}\frac{(3\bar{v}_n^2 + 2\bar{v}_n + 2)\zeta_n}{1 - v_0(\alpha_n)} = \alpha_{n+1} - \gamma_n, \\
F(x_{n+1}) &= F(x_{n+1}) - F(x_n) - F'(x_n)(x_{n+1} - x_n) + F'(x_n)(x_{n+1} - y_n), \\
\|L^{-1}F(x_{n+1})\| &\leq \int_0^1 v((1-\theta)(\alpha_{n+1} - \alpha_n))d\theta(\alpha_{n+1} - \alpha_n) \\
&\quad +(1 + v_0(\alpha_n))(\alpha_{n+1} - \beta_n) = \delta_{n+1}, \\
\|y_{n+1} - x_{n+1}\| &\leq \|F'(x_{n+1})^{-1}L\|\|L^{-1}F(x_{n+1})\| \leq \frac{\delta_{n+1}}{1 - v_0(\alpha_{n+1})} \\
&= \beta_{n+1} - \alpha_{n+1}.
\end{aligned}$$

Method(15.2)

The majorant sequence $\{\alpha\}$ is defined for $\alpha_0 = 0, \beta_0 \geq \frac{2}{3}\|F'(x_0)^{-1}F(x_0)\|$ and each $n = 0,1,2,...$ by

$$\begin{aligned}
q_n &= \frac{1}{2}(3v_0(\beta_n)+v_0(\alpha_n)),\\
\gamma_n &= \beta_n+\frac{(3\bar{v}_n+4(1+v_0(\alpha_n))(\beta_n-\alpha_n)}{12(1-q_n)},\\
\lambda_n &= (1+\int_0^1 v_0(\alpha_n+\theta(\gamma_n-\alpha_n))d\theta)(\gamma_n-\alpha_n)\\
&\quad +\frac{3}{2}(1+v_0(\alpha_n))(\beta_n-\alpha_n),\\
\alpha_{n+1} &= \gamma_n+\frac{\lambda_n}{1-q_n},\\
\delta_{n+1} &= (1+\int_0^1 v_0(\alpha_n+\theta(\alpha_{n+!}-\alpha_n))d\theta)(\alpha_{n+1}-\alpha_n)\\
&\quad +\frac{2}{3}(1+v_0(\alpha_n)),\\
\beta_{n+1} &= \alpha_{n+1}+\frac{2}{3}\frac{\delta_{n+1}}{1-v_0(\alpha_{n+1})}.
\end{aligned}$$

The motivational estimates are

$$\begin{aligned}
z_n-y_n &= [\frac{2}{3}I-\frac{1}{2}(3F'(y_n)-F'(x_n))^{-1}(3F'(y_n)+F'(x_n))]\\
&\quad \times(-\frac{3}{2}(y_n-x_n))\\
&= \frac{1}{6}(3F'(y_n)-F'(x_n))^{-1}[3(F'(y_n)-F'(x_n))\\
&\quad -4F'(x_n)](y_n-x_n),\\
\|z_n-y_n\| &\leq \frac{(3\bar{v}_n+4(1+v_0(\alpha_n)))(\beta_n-\alpha_n)}{12(1-q_n)}\\
&= \gamma_n-\beta_n,\\
F(z_n) &= F(z_n)-F(x_n)-\frac{3}{2}F'(x_n)(y_n-x_n),
\end{aligned}$$

$$
\begin{aligned}
\|L^{-1}F(z_n)\| &\leq (1+\int_0^1 v_0(\|x_n-x_0\|+\theta\|z_n-x_n\|)d\theta)\|z_n-x_n\| \\
&\quad +\frac{3}{2}(1+v_0(\|x_n-x_0\|))\|y_n-x_n\| \qquad (15.4)\\
&= \bar{\lambda}_n \leq \lambda_n, \\
\|x_{n+1}-z_n\| &\leq \frac{2\lambda_n}{2(1-q_n)} = \alpha_{n+1}-\gamma_n, \\
F(x_{n+1}) &= F(x_{n+1})-F(x_n)-\frac{3}{2}F'(x_n)^{-1}(y_n-x_n), \\
\|L^{-1}F(x_{n+1})\| &\leq (1+\int_0^1 v_0(\alpha_n+\theta(\alpha_{n+1}-\alpha_n))d\theta)(\alpha_{n+1}-\alpha_n) \\
&\quad +\frac{3}{2}(1+v_0(\alpha_n))(\beta_n-\alpha_n) = \delta_{n+1}, \\
\|y_{n+1}-x_n\| &\leq \frac{2}{3}\|F'(x_{n+1})^{-1}L\|\|L^{-1}F(x_{n+1})\| \\
&\leq \frac{2}{3}\frac{\delta_{n+1}}{1-v_0(\alpha_{n+1})} = \beta_{n+1}-\alpha_{n+1}.
\end{aligned}
$$

4. References

[1] Argyros, I. K., *Convergence and Applications of Newton-Type Iterations*, Springer-Verlag, New York, (2008).

[2] Argyros, I. K., George, S., *Mathematical modeling for the solution of equations and systems of equations with applications*, Volume-IV, Nova Publishers, NY, (2020).

[3] Argyros, I. K., George, S., *On the complexity of extending the convergence region for Traubs method*, Journal of Complexity, 56, (2022), 101423.

[4] Cordero, A., Hueso, J. L., Martnez, E., Torregrosa, J. R., *A modified Newton-Jarratt's composition*, Numerical Algorithms, 55, 1, (2010), 87-99.

[5] Darvishi, M. T., Barati, A., *A fourth-order method from quadrature formulae to solve systems of nonlinear equations*, Applied Mathematics and Computation, 188, 1, (2007), 257-261.

[6] Frontini, M., Sormani, E., *Third-order methods from quadrature formulae for solving systems of nonlinear equations*, Applied Mathematics and Computation, 149, 3, (2004), 771-782.

[7] Grau-Snchez, M., Grau, A., Noguera, M., *Ostrowski-type methods for solving systems of nonlinear equations*, Applied Mathematics and Computation, 218, 6, (2011), 2377-2385.

[8] Gutirrez, J. M., Hernndez, M. A., *A family of Chebyshev-Halley type methods in Banach spaces*, Bulletin of the Australian Mathematical Society, 55, 1, (1997), 113-130.

[9] Homeier, H. H. H., *A modified Newton method with cubic convergence: the multivariate case*, Journal of Computational and Applied Mathematics, 169, 1, (2004), 161-169.

[10] Ortega, J. M., Rheinboldt, W. C., *Iterative Solutions of Nonlinear Equations in Several Variables*, Academic Press, New York, NY, USA, (1970).

[11] Ostrowski, A. M., *Solution of Equations and Systems of Equations*, Academic Press, New York, NY, USA, (1966).

[12] Potra, F. A., Ptk, V., *Nondiscrete Induction and Iterative Processes*, Pitman, Boston, Mass, USA, (1984).

[13] Sharma, J. R., Guha, R. K., Sharma, R., *An efficient fourth order weighted-Newton method for systems of nonlinear equations*, Numerical Algorithms, 62, 2, (2013), 307-323.

[14] Sharma, J. R., Gupta, P., *On Some Efficient Techniques for Solving Systems of Nonlinear Equations*, Advances in Numerical Analysis, (2013).

Chapter 16

Seventh-Order Method Mixing Derivatives and Divided Differences

1. Introduction

The method is defined as

$$\begin{aligned} y_n &= x_n - F(x_n)^{-1}F(x_n), \\ s_n &= I - F'(x_n)^{-1}[x_n, y_n; F], \\ z_n &= y_n - G(s_n)[x_n, y_n; F]^{-1}F(y_n) \\ x_{n+1} &= z_n - H(s_n)[y_n, z_n; F]^{-1}F(z_n), \end{aligned} \tag{16.1}$$

where $G, H : \Omega \times \Omega \longrightarrow L(B)$. The seventh order of convergence also uses the initial conditions $G(0) = I, G'(0) = I, \|G''(0)\| < \infty$, $H(0) = I, H'(0) = O$, $H''(0) = I$. The local order was established in [1] when $B = \mathbb{R}^k$ under the Taylor Series expansion approach and assumptions on the eighth derivative. In this Chapter, the applicability of the method (16.1) is expanded under only conditions on the operators appearing on the method.

The more interesting semi-local convergence (not given in [1]) is also presented in this Chapter. Similar work can be done for the methods in [2]-[13].

2. Local Convergence

The conditions used are:

$$\begin{aligned} \|L^{-1}(F'(x) - L)\| &\leq \omega_0(\|x - x^*\|), \\ \|L^{-1}(F'(y) - F'(x))\| &\leq \omega(\|y - x\|), \\ \|L^{-1}([x, y; F] - L)\| &\leq \omega_1(\|x - x^*\|, \|y - x^*\|), \\ s &= I - F'(x)^{'}[x, y; F], \\ \|L^{-1}([x, y; F] - F'(y)G(s))\| &\leq \omega_2(\|x - x^*\|, \|y - x^*\|, \|s - x^*\|), \\ \|[x, y; F] - F'(z)H(s)\| &\leq \omega_3(\|x - x^*\|, \|y - x^*\|, \|z - z^*\|, \|s - x* \| \end{aligned}$$

and

$$\|s - x^*\| \leq \omega_4(\|x - x^*\|, \|y - x^*\|).$$

The functions ω_j, $j=2,3,4$ can be specialized if G and H are further precise.

The majorant functions $g_i, i=1,2,3$ are

$$\begin{aligned}
g_1(t) &= \frac{\int_0^1 \omega((1-\theta)t)d\theta}{1-\omega_0(t)},\\
g_2(t) &= \left[\frac{\int_0^1 \omega((1-\theta)g_1(t)t)d\theta}{1-\omega_0(g_1(t)t)} + \frac{1}{(1-\omega_0(g_1(t)t))(1-\omega_1(t,g_1(t)t))}\right.\\
&\quad \left.\times(\omega_2(t,g_1(t)t,\omega_4(t,g_1(t)t))(1+\int_0^1 \omega_0(\theta g_1(t)t)d\theta)\right]g_1(t),\\
E(t) &= \omega_3(t,g_1(t)t,g_2(t)t,\omega_4(t,g_1(t)t))\\
&\quad \times(1+\int_0^1 \omega_0(\theta g_2(t)t)d\theta)
\end{aligned}$$

and

$$\begin{aligned}
g_3(t) &= \left[\frac{\int_0^1 \omega((1-\theta)g_2(t)t)d\theta}{1-\omega_0(g_2(t)t)}\right.\\
&\quad \left.+\frac{E(t)}{(1-\omega_0(g_2(t)t))(1-\omega_1(g_1(t)t,g_2(t)t))}\right]g_2(t).
\end{aligned}$$

The estimates motivating the introduction of the functions g_i are:

$$\begin{aligned}
y_n - x^* &= x_n - x^* - F'(x_n)^{-1}F(x_n),\\
\|y_n - x^*\| &\leq \frac{\int_0^1 \omega((1-\theta)\|x_n-x^*\|)d\theta\|x_n-x^*\|}{1-\omega_0(\|x_n-x^*\|)}\\
&\leq g_1(\|x_n-x^*\|)\|x_n-x^*\|\\
&\leq \|x_n-x^*\| < r,
\end{aligned}$$

$$\begin{aligned}
z_n - x^* &= y_n - x^* - F'(y_n)^{-1}F(y_n)\\
&\quad +(F'(y_n)^{-1} - G(s_n)[x_n,y_n;F]^{-1})F(y_n)\\
&= y_n - x^* - F'(y_n)^{-1}F(y_n) + F'(y_n)^{-1}([x_n,y_n;F]-F'(y_n)G(s_n))\\
&\quad [x_n,y_n;F]^{-1}F(y_n),
\end{aligned}$$

$$\begin{aligned}
\|z_n - x^*\| &\leq \left[\frac{\omega((1-\theta)\|y_n-x^*\|)d\theta}{1-\omega_0(\|y_n-x^*\|)}\right.\\
&\quad +\frac{1}{(1-\omega_0(\|y_n-x^*\|))(1-\omega_1(\|x_n-x^*\|,\|y_n-x^*\|))}\\
&\quad \left.\times\omega_2(\|x_n-x^*\|,\|s_n-x^*\|,\|y_n-x^*\|)(1+\int_0^1 \omega_0(\theta\|y_n-x^*\|)d\theta)\right]\\
&\quad \times\|y_n-x^*\|\\
&\leq g_2(\|x_n-x^*\|)\|x_n-x^*\| \leq \|x_n-x^*\|\\
x_{n+1} - x^* &= z_n - x^* - F'(z_n)^{-1}F(z_n)\\
&\quad +(F'(z_n)^{-1}H(s_n)[y_n,z_n;F]^{-1})F(z_n)\\
&= z_n - x^* - F'(z_n)^{-1}F(z_n)\\
&\quad +F'(z_n)^{-1}([y_n,z_n;F]-F'(z_n)H(s_n))[y_n,z_n;F]^{-1}F(z_n)
\end{aligned}$$

$$\begin{aligned}
\|x_{n+1}-x^*\| &\leq \Bigg[\frac{\omega((1-\theta)\|z_n-x^*\|)d\theta}{1-\omega_0(\|z_n-x^*\|)} \\
&\quad+\frac{1}{(1-\omega_0(\|z_n-x^*\|))(1-\omega_1(\|y_n-x^*\|,\|z_n-x^*\|))} \\
&\quad\times\omega_3(\|x_n-x^*\|,\|y_n-x^*\|,\|z_n-x^*\|,\|s_n-x^*\|) \\
&\quad\times(1+\int_0^1\omega_0(\theta\|z_n-x^*\|)d\theta)\Bigg]\|z_n-x^*\| \\
&\leq g_3(\|x_n-x^*\|)\|x_n-x^*\|\leq\|x_n-x^*\|.
\end{aligned}$$

3. Semi-local Convergence

The conditions are:

$$\begin{aligned}
\|L^{-1}(F'(x)-L)\| &\leq v_0(\|x-x_0\|), \\
\|L^{-1}(F'(y)-F'(x))\| &\leq v(\|y-x\|), \\
\|L^{-1}([x,y;F]-L)\| &\leq v_1(\|x-x_0\|,\|y-x_0\|), \\
\|G(s)\| &\leq v_2(\|x-x_0\|,\|y-x_0\|)
\end{aligned}$$

and

$$\|H(s)\|\leq v_3(\|x-x_0\|,\|y-x_0\|).$$

The majorant sequence $\{\alpha_n\}$ is defined for $\alpha_0=0,\beta_0\geq\|F'(x_0)^{-1}F(x_0)\|$ by

$$\begin{aligned}
\gamma_n &= \beta_n+\frac{v_2(\alpha_n,\beta_n)\int_0^1 v(\theta(\beta_n-\alpha_n))d\theta(\beta_n-\alpha_n)}{1-v_1(\alpha_n,\beta_n)}, \\
\lambda_n &= (1+\int_0^1 v_0(\beta_n+\theta(\gamma_n-\beta_n))d\theta)(\gamma_n-\beta_n) \\
&\quad+\int_0^1 v(\theta(\beta_n-\alpha_n))d\theta(\beta_n-\alpha_n), \\
\alpha_{n+1} &= \gamma_n+\frac{\lambda_n}{1-v_1(\beta_n,\gamma_n)} \\
\delta_{n+1} &= \int_0^1 v((1-\theta)(\alpha_{n+1}-\alpha_n))d\theta(\alpha_{n+1}-\alpha_n) \\
&\quad+(1+v_0(\alpha_n))(\alpha_{n+1}-\beta_n)
\end{aligned}$$

and

$$\beta_{n+1}=\alpha_{n+1}+\frac{\delta_{n+1}}{1-v_0(\alpha_{n+1})}.$$

The estimates leading to this sequence are:

$$
\begin{aligned}
F(y_n) &= F(y_n) - F(x_n) - F'(x_n)(y_n - x_n) \\
&= \int_0^1 (F'(x_n + \theta(y_n - x_n))d\theta - F'(x_n))(y_n - x_n), \\
\|L^{-1}F(y_n)\| &\leq \int_0^1 v(\theta\|y_n - x_n))d\theta\|y_n - x_n\| \\
&\leq \int_0^1 v(\theta(\beta_n - \alpha_n))d\theta(\beta_n - \alpha_n), \\
\|z_n - y_n\| &\leq \|G(s_n)\|\|[x_n, y_n; F]^{\text{'}}L\|\|L^{-1}F(y_n)\| \\
&\leq \frac{v_2(\|x_n - x_0\|, \|y_n - x_0\|)}{1 - v_0(\|x_n - x_0\|, \|y_n - x_0\|)} \\
&\quad \times \int_0^1 v((1-\theta)(\beta_n - \alpha_n))d\theta(\beta_n - \alpha_n) \\
&\leq \gamma_n - \beta_n,
\end{aligned}
$$

$$
\begin{aligned}
F(z_n) &= F(z_n) - F(y_n) + F(y_n) \\
\|L^{-1}F(z_n)\| &\leq (1 + \int_0^1 v_0(\beta_n + \theta(\gamma_n - \beta_n))d\theta)(\gamma_n - \beta_n) \\
&\quad + \int_0^1 v((1-\theta)(\beta_n - \alpha_n))d\theta(\beta_n - \alpha_n) = \lambda_n, \\
\|x_{n+1} - z_n\| &\leq \|H(s_n)\|\|[y+_n, z_n; F]^{-1}L\|\|L^{-1}F(z_n)\| \\
&\leq \frac{v_3(\|x_n - x_)\|, \|y_n - x_0\|)\lambda_n}{1 - v_)(\|y_n - x_)\|, \|z_n - x_0\|)} \\
&\leq \alpha_{n+1} - \gamma_n, \\
F(x_{n+1}) &= F(x_{n+1}) - F(x_n) - F'(x_n)(x_{n+1} - x_n) \\
&\quad + F'(x_n)(x_{n+1} - y_n),
\end{aligned}
$$

$$
\begin{aligned}
\|L^{-1}F(x_{n+1})\| &\leq \int_0^1 v((1-\theta)(\alpha_{n+1} - \alpha_n))d\theta(\alpha_{n+1} - \alpha_n) \\
&\quad + (1 + v_0(\alpha_n))(\alpha_{n+1} - \beta_n) \\
&= \delta_{n+1}, \\
\|y_{n+1} - x_{n+1}\| &\leq \|F'(x_{n+1})^{-1}L\|\|L^{-1}F(x_{n+1})\| \\
&\leq \frac{\lambda_{n+1}}{1 - v_0(\alpha_{n+1})} \\
&= \beta_{n+1} - \alpha_{n+1}.
\end{aligned}
$$

4. References

[1] Abad, M. F., Cordero, A., Torregrosa, J. R., *A family of seventh-order schemes for solving nonlinear systems*, Bulletin Mathmatique de la Socit des Sciences Mathmatiques de Roumanie, Nouvelle Srie, 57 (105), 2 (2014), 133-145.

[2] Abad, M. F., Cordero, A., Torregrosa, J. R., *Fourth and Fifth-Order Methods for solving Nonlinear Systems of Equations: An Application to the Global Positioning System*, Abstract and Applied Analysis, Article ID 586708 (2013), 10 pages.

[3] Argyros, C. I., Regmi, S., Argyros, I. K., George, S., *Contemporary Algorithms: Theory and Applications*, Volume III, NOVA Publishers, (2023).

[4] Argyros, I. K., *The theory and applications of iteration methods*, 2nd Edition, CRC Press, Engineering Series, Taylor and Francis Group, (2022).

[5] Argyros, I. K., George, S., *Ball comparison between two optimal eight-order methods under weak conditions*, SeMA Journal, 72, (2015), 1-11.

[6] Cordero, A., Torregrosa, J. R., *Variants of Newton's Method using fifth-order quadrature formulas*, Applied Mathematics and Computation, 190, (2007), 686-698.

[7] Jarratt, P., *Some fourth-order multipoint iterative methods for solving equations*, Math. Comput., 20, (1966), 434-437.

[8] King, R. F., *A family of fourth-order methods for nonlinear equations*, SIAM Journal of Numerical Analysis, 10, (1973), 876-879.

[9] Kung, H. T., Traub, J. F., *Optimal order of one-point and multi-point iterations*, J. Assoc. Comput. Math., 21, (1974), 643-651.

[10] Ortega, J. M., Rheinboldt, W. G., *Iterative Solutions of Nonlinear Equations in Several Variables*, Academic Press, New York, (1970).

[11] Ostrowski, A. M., *Solution of Equations and Systems of Equations*, Prentice-Hall, Englewood Cliffs, NJ, USA, (1966).

[12] Soleymani, F., Khattri, S. K., Karimi, S., *Two new classes of optimal Jarratt-type fourth-order methods*, Applied Mathematics Letters, 25, (2012), 847-853.

[13] Traub, J. F., *Iterative Methods for the Solution of Equations*, Prentice-Hall, Englewood Cliffs, (1964).

Chapter 17

Extending Two Efficient Fifth-Order Methods

1. Introduction

The method is defined by

$$\begin{aligned} y_n &= x_n - \frac{1}{2}F'(x_n)^{-1}F(x_n), \\ z_n &= x_n - F'(y_n)^{-1}F(x_n), \\ x_{n+1} &= z_n - (2F'(y_n)^{-1} - F'(x_n)^{-1})F(z_n). \end{aligned} \tag{17.1}$$

and

$$\begin{aligned} y_n &= x_n - F'(x_n)^{-1}F(x_n), \\ z_n &= x_n - 2(F'(y_n) + F(x_n))^{-1}F(x_n), \\ x_{n+1} &= z_n - F'(y_n)^{-1}F(z_n). \end{aligned} \tag{17.2}$$

The convergence order five is established in [9] under the Taylor series expansion approach when $B_1 = B_2 = \mathbb{R}^k$ and conditions up to $F^{(6)}$ not on the method(17.1). The limitations of this approach have been reported in previous chapters. Our idea extends the applicability of the method (17.1) in the local convergence case using only conditions on the operators on the method and in the more general setting of a Banach space. Moreover, the more interesting semi-local convergence (not given in [9]) is developed utilizing majorizing sequences. The approach can also be used with the same advantages on other methods along the same lines [1]-[8].

2. Local Convergence

Method (17.1)
The majorant functions $g_i, i = 1,2,3$ are :

$$
\begin{aligned}
g_1(t) &= \frac{\int_0^1 w((1-\theta)t)d\theta}{1-w_0(t)} + \frac{1+\int_0^1 w_0(\theta t)d\theta}{2(1-w_0(t))}, \\
\bar{w}(t) &= \begin{cases} w(1+g_1(t)t) \\ OR \\ w_0(t)+w_0(g_1(t)t), \end{cases} \\
g_2(t) &= \frac{\int_0^1 w((1-\theta)t)d\theta}{(1-w_0(t))} + \frac{\bar{w}(t)(1+\int_0^1 w_0(\theta t)d\theta)}{(1-w_0(t))(1-w_0(g_1(t)t))}, \\
\bar{\bar{w}}(t) &= \begin{cases} w(g_2(t)+g_3(t))t) \\ OR \\ w_0(g_2(t)t)+w_0(g_3(t)t), \end{cases}
\end{aligned}
$$

and

$$
\begin{aligned}
f_3(t) = & \left[\frac{\int_0^1 w((1-\theta)g_2(t)t)d\theta}{1-w_0(g_2(t)t)} \right. \\
& + \frac{\bar{w}(t)(1+\int_0^1 w_0(\theta g_2(t)t)d\theta)}{(1-w_0(g_1(t)t)(1-w_0(t))} \\
& \left. + \frac{\bar{\bar{w}}(t)(1+\int_0^1 w_0(\theta g_2(t)t)d\theta)}{(1-w_0(g_1(t)t)(1-w_0(g_2(t)t))} \right] g_2(t).
\end{aligned} \tag{17.3}
$$

The calculations leading to these functions are in turn:

$$
\begin{aligned}
y_n - x^* &= x_n - x^* - F'(x_n)^{-1}F(x_n) + \frac{1}{2}F'(x_n)^{-1}F(x_n),\\
\|y_n - x^*\| &\le \Bigg[\frac{\int_0^1 w(1-\theta)\|x_n - x^*\|)d\theta}{1 - w_0(\|x_n - x^*\|)}\\
&\quad + \frac{\bar{w}_n(1 + \int_0^1 w_0(\theta\|x_n - x^*\|)d\theta)}{(1 - w_0(\|x_n - x^*\|))(1 - w_0(\|y_n - x^*\|))}\Bigg]\|x_n - x^*\|\\
&\le g_1(\|x_n - x^*\|)\|x_n - x^*\| \le \|x_n - x^*\| < r,\\
z_n - x^* &= x_n - x^* - F'(x_n)^{-1}F(x_n)\\
&\quad + F'(x_n)^{-1}(F'(y_n) - F'(x_n))F'(y_n)^{-1}F(x_n),\\
\|z_n - x^*\| &\le \Bigg[\frac{\int_0^1 w((1-\theta)\|x_n - x^*\|)d\theta}{1 - w_0(\|x_n - x^*\|)}\\
&\quad + \frac{\bar{w}_n(1 + \int_0^1 w_0(\theta\|x_n - x^*\|)d\theta)}{(1 - w_0(\|x_n - x^*\|))(1 - w_0(\|y_n - x^*\|))}\Bigg]\|x_n - x^*\|\\
&\le g_2(\|x_n - x^*\|)\|x_n - x^*\|,\\
x_{n+1} - x^* &= z_n - x^* - F'(z_n)^{-1}F(z_n)\\
&\quad - (2F'(y_n)^{-1} - F'(x_n)^{-1} - F'(z_n)^{-1})F(z_n)\\
&= z_n - x^* - F'(z_n)^{-1}F(z_n)\\
&\quad - F'(y_n)^{-1}(F'(x_n) - F'(y_n))F'(x_n)^{-1}F(z_n)\\
&\quad - F'(y_n)^{-1}(F'(z_n) - F'(y_n))F'(z_n)^{-1}F(z_n),\\
\|x_{n+1} - x^*\| &\le \Bigg[\frac{\int_0^1 w((1-\theta)\|z_n - x^*\|)d\theta}{1 - w_0(\|z_n - x^*\|)}\\
&\quad + \frac{\bar{w}_n(1 + \int_0^1 w_0(\theta\|z_n - x^*\|)d\theta)}{(1 - w_0(\|y_n - x^*\|))(1 - w_0(\|x_n - x^*\|)}\\
&\quad + \frac{\bar{\bar{w}}_n(1 + \int_0^1 w_0(\theta\|z_n - x^*\|)d\theta)}{(1 - w_0(\|y_n - x^*\|))(1 - w_0(\|z_n - x^*\|))}\Bigg]\|z_n - x^*\|\\
&\le g_3(\|x_n - x^*\|)\|x_n - x^*\| \le \|x_n - x^*\|.
\end{aligned}
$$

Method (17.2)
Similarly, the functions $g_i,\ i=1,2,3$ are

$$
\begin{aligned}
f_1(t) &= \frac{\int_0^1 w((1-\theta)t)d\theta}{1-w_0(t)},\\
f_2(t) &= \frac{\int_0^1 w((1-\theta)t)d\theta}{1-w_0(t)}+\frac{2\bar{w}(t)(1+\int_0^1 w_0(\theta t))d\theta}{2(1-p(t))}\\
p(t) &= \frac{1}{2}(w_0(t)+w_0(g_1(t)t)),\\
f_3(t) &= \left[\frac{\int_0^1 w((1-\theta)g_2(t)t)d\theta}{1-w_0(g_2(t)t)}+\frac{\bar{\bar{w}}(t)(1+\int_0^1 w_0(\theta g_2(t)t))d\theta}{(1-w_0(g_2(t)t))(1-w_0(g_1(t)t))}\right]g_2(t).\\
\bar{w}(t) &= \begin{cases} w((1+g_1(t))t)\\ OR\\ w_0(t)+w_0(g_1(t)t),\end{cases}\\
\bar{\bar{w}}(t) &= \begin{cases} w(g_1(t)+g_2(t))t)\\ w_0(g_1(t)t)+w_0(g_2(t)t),\end{cases}
\end{aligned}
$$

3. Semi-local Convergence

The majorizing sequence $\{\alpha_n\}$ is defined for $\alpha_0=0, \beta_0\geq\frac{1}{2}\|F'(x_0)^{-1}F(x_0)\|$ and each $n=0,1,2,..$ by

$$
\begin{aligned}
\bar{v}(t) &= \begin{cases} v(\beta_n-\alpha_n)\\ OR\\ v_0(\alpha_n)+v_0(\beta_n),\end{cases}\\
\gamma_n &= \beta_n+\left[\frac{\bar{v}_n}{(1-v_0(\alpha_n))(1-v_0(\beta_n))}+\frac{1}{1-v_0(\beta_n)}\right](1+\int_0^1 v_0(\theta\alpha_n)d\theta)\\
\lambda_n &= (1+\int_0^1 v_0(\alpha_n+\theta(\gamma_n-\alpha_n))d\theta)(\gamma_n-\alpha_n)\\
&\quad +(1+v_0(\alpha_n))(\beta_n-\alpha_n),\\
x_{n+1} &= \gamma_n+\left[\frac{\bar{v}_n}{(1-v_0(\alpha_n))(1-v_0(\beta_n))}+\frac{1}{1-v_0(\beta_n)}\right]\lambda_n,\\
\delta_{n+1} &= (1+\int_0^1 v_0(\alpha_n+\theta(\alpha_{n+1}-\alpha_n))d\theta)(\alpha_{n+1}-\alpha_n)\\
&\quad +2(1+v_0(\alpha_n))(\beta_n-\alpha_n),
\end{aligned}
$$

and

$$
\beta_{n+1} = \alpha_{n+1}+\frac{1}{2}\frac{\delta_{n+1}}{1-v_0(\alpha_{n+1})}.
$$

The motivation for the development of the sequence $\{\alpha_n\}$ is justified in turn by the calculations:

$$\begin{aligned}
z_n - y_n &= \frac{1}{2}F'(x_n)^{-1}F(x_n) - F'(y_n)^{-1}F(x_n)\\
&= \frac{1}{2}(F'(x_n)^{-1} - F'(y_n)^{-1})F(x_n) - \frac{1}{2}F'(y_n)^{-1}F(x_n),\\
\|z_n - y_n\| &\leq \frac{1}{2}\left[\frac{\bar{v}_n}{(1-v_0(\alpha_n))(1-v_0(\beta_n)} + \frac{1}{(1-v_0(\beta_n)}\right]\\
&\quad (1+\int_0^1 v_0(\theta\alpha_n)d\theta)(\beta_n - \alpha_n)\\
F(z_n) &= F(z_n) - F(x_n) + F(z_n)\\
&= \int_0^1 F'(x_n + \theta(z_n - x_n))d\theta(z_n - x_n)\\
&\quad +2F'(x_n)(y_n - x_n),\\
\|L^{-1}F(z_n)\| &\leq \lambda_n,\\
x_{n+1} - z_n &= -F'(y_n)^{-1}(F'(x_n) - F'(y_n))F'(x_n)^{-1}F(z_n)\\
&\quad -F'(y_n)^{-1}F(z_n),\\
\|x_{n+1} - z_n\| &\leq \left[\frac{\bar{v}_n}{(1-v_0(\alpha_n))(1-v_0(\beta_n))} + \frac{1}{1-v_0(\beta_n)}\right]\lambda_n\\
&= \alpha_{n+1} - \gamma_n,\\
F(x_{n+1}) &= F(x_{n+1}) - F(x_n) - 2F'(x_n)(y_n - x_n),\\
\|L^{-1}F(x_{n+1})\| &\leq (1+\int_0^1 v_0(\alpha_n + \theta(\alpha_{n+1} - \alpha_n))d\theta)(\alpha_{n+1} - \alpha_n)\\
&\quad +2(1+v_0(\alpha_n))(\beta_n - \alpha_n) = \delta_{n+1},\\
\|y_{n+1} - x_{n+1}\| &\leq \frac{1}{2}\|F'(x_{n+1})^{-1}L\|\|L^{-1}F(x_{n+1})\|\\
&\leq \frac{\delta_{n+1}}{2(1-v_0(\alpha_{n+1}))} = \beta_{n+1} - \alpha_{n+1}.
\end{aligned}$$

Method (17.2) The majorizing sequence $\{\alpha_n\}$ is defined for $\alpha_0 = 0$, $\beta_0 \geq \|F'(x_0)^{-1}F(x_0)\|$ and each $n = 0, 1, 2, \ldots$ by

$$\begin{aligned}
\bar{v} &= \begin{cases} v(\beta_n - \alpha_n) \\ OR \\ v_0(\alpha_n) + v_0(\beta_n), \end{cases}\\
q_n &= \frac{1}{2}(v_0(\alpha_n) + v_0(\beta_n)),\\
\gamma_n &= \beta_n + \frac{\bar{v}_n(\beta_n - \alpha_n)}{2(1-q_n)},\\
\mu_n &= (1+\int_0^1 v_0(\alpha_n + \theta(\gamma_n - \alpha_n))d\theta)(\gamma_n - \alpha_n)\\
&\quad +(1+v_0(\alpha_n))(\beta_n - \alpha_n),\\
x_{n+1} &= \gamma_n + \frac{\mu_n}{1-v_0(\beta_n)},
\end{aligned} \tag{17.4}$$

$$\begin{aligned}\zeta_{n+1} &= \int_0^1 v((1-\theta)(\alpha_{n+1}-\alpha_n))d\theta(\alpha_{n+1}-\alpha_n)\\ &+(1+v_0(\alpha_n))(\alpha_{n+1}-\beta_n),\\ \beta_{n+1} &= \alpha_{n+1}+\frac{\zeta_{n+1}}{1-v_0(\alpha_{n+1})}.\end{aligned}$$

4. References

[1] Argyros, I. K., *Convergence and Applications of Newton-Type Iterations*, Springer-Verlag, New York, (2008).

[2] Argyros, I. K., George, S., *Mathematical modeling for the solution of equations and systems of equations with applications*, Volume-IV, Nova Publishers, NY, (2020).

[3] Argyros, I. K., George, S., *On the complexity of extending the convergence region for Traubs method*, Journal of Complexity, 56, 101423.

[4] Cordero, A., Torregrosa, J. R., *Variants of Newtons method using fifth-order quadrature formulas*, Appl. Math. Comput., 190, (2007), 686-698.

[5] Frontini, M., Sormani, E., *Third-order methods from quadrature formulae for solving systems of nonlinear equations*, Appl. Math. Comput., 149, (2004), 771-782.

[6] Jarratt, P., *Some fourth-order multipoint iterative methods for solving equations*, Math. Comput., 20, (1966), 434-437.

[7] Noor, M. A., Waseem, M., *Some iterative methods for solving a system of nonlinear equations*, Comput. Math. Appl., 57, (2009), 101-106.

[8] Ostrowski, A. M., *Solution of Equations and Systems of Equations*, Academic Press, New York, (1966).

[9] Sharma, J. R., Gupta, P., *An efficient fifth-order method for solving systems of nonlinear equations*, Comput. Math. Appl., 67, (2014), 591-601.

Chapter 18

Extended Ostrowski-Type Method

1. Introduction

The Ostrowski-type method is defined for $x_0 \in \Omega$ and each $n = 0, 1, 2, \ldots$ by

$$\begin{aligned} y_n &= x_n - F(x_n)^{-1}F(x_n), \\ A_n &= 2[y_n, x_n; F] - F'(x_n), \\ z_n &= y_n - A_n^{-1}F(y_n) \\ x_{n+1} &= z_n - A_n^{-1}F(z_n), \end{aligned} \tag{18.1}$$

where $[.,.;F] : \Omega \times \Omega \longrightarrow L(B_1, B_2)$ is a divided difference of order one and F' is the Fréchet derivative of the operator F [1]-[4]. The sixth convergence order is established in [8] under the Taylor expansion series approach when $B_1 = B_2 = \mathbb{R}^k$. The existence of the $F^{(7)}$ is assumed. The limitations of this approach have been explained in previous chapters. The new local convergence uses conditions only on the operators on the method in the more general Banach space setting. Moreover, the more interesting semi-local convergence is provided (not given in [8]) under $\omega-$continuity and utilizing majorizing sequences. The approach can also be used for the same purposes on other methods [1]-[7], [9]-[15].

2. Local Convergence

The conditions used are:

$$\begin{aligned} \|L^{-1}([x,y;F] - L)\| &\leq \varphi_0(\|x - x^*\|, \|y - x^*\|), \\ \|L^{-1}([x,y;F] - F'(y))\| &\leq \varphi(\|x - x^*\|, \|y - x^*\|, \\ \|L^{-1}([x,y;F] - [x,x^*,F])\| &\leq \varphi_1(\|x - x^*\|, \|y - x^*\|, \\ \|L^{-1}(F'(x) - F'(y))\| &\leq \varphi_2(\|x - y\|), \\ \|L^{-1}(F'(x) - L)\| &\leq \varphi_3(\|x - x^*\|) \end{aligned}$$

and

$$\|L^{-1}([x,y;F] - F'(z))\| \leq \varphi_4(\|x - x^*\|, \|y - x^*\|, \|z - x^*\|).$$

The majorant functions g_i, $i = 1,2,3$ are:

$$\begin{aligned}
g_1(t) &= \frac{\int_0^1 \varphi_2((1-\theta)t)d\theta}{1-\varphi_3(t)},\\
q(t) &= \varphi_0(t, g_1(t)t) + \varphi(t, g_1(t)t),\\
g_2(t) &= \left[\frac{[\varphi(t, g_1(t)t) + \varphi_1(t, g_1(t)t)]g_1(t)}{1-q(t)}\right.\\
g_3(t) &= \left[\frac{\int_0^1 \varphi_2((1-\theta)g_2(t)t)d\theta}{1-\varphi_3(g_2(t)t)}\right.\\
&\quad + \frac{\varphi(t, g_1(t)t) + \varphi_4(t, g_1(t)t, g_2(t)t)}{(1-\varphi_3(g_2(t)t))(1-q(t))}\\
&\quad \left.\times(1+\varphi_3(g_2(t)t))\right] g_2(t)
\end{aligned}$$

The estimates leading to the introduction of the functions g_i are justified by

$$\begin{aligned}
y_n - x^* &= x_n - x^* - F'(x_n)^{-1}F(x_n),\\
\|y_n - x^*\| &\leq \frac{\int_0^1 \varphi_2((1-\theta)\|x_n - x^*\|)d\theta\|x_n - x^*\|}{1-\varphi_3(\|x_n - x^*\|)}\\
&\leq g_1(\|x_n - x^*\|)\|x_n - x^*\|\\
&\leq \|x_n - x^*\| < r,\\
\|L^{-1}(A_n - L)\| &\leq \varphi_0(\|x_n - x^*\|, \|y_n - x^*\|) + \varphi(\|x_n - x^*\|, \|y_n - x^*\|)\\
&= \bar{q}_n \leq q_n < 1,\\
\|A_n^{-1}L\| &\leq \frac{1}{1-q_n},
\end{aligned}$$

$$\begin{aligned}
z_n - x^* &= y_n - x^* - A_n^{-1}F(y_n)\\
&= A_n^{-1}[([y_n, x_n; F] - F'(x_n)) + ([y_n, x_N; F] - [y_n, x^*; F])](y_n - x^*)
\end{aligned}$$

$$\begin{aligned}
\|z_n - x^*\| &\leq \left[\frac{\varphi(\|x_n - x^*\|, \|y_n - x^*\|) + \varphi_1(\|x_n - x^*\|, \|y_n - x^*\|)}{1-q_n}\right]\|y_n - x^*\|\\
&\leq g_2(\|x_n - x^*\|)\|x_n - x^*\| \leq \|x_n - x^*\|\\
x_{n+1} - x^* &= z_n - x^* - F'(z_n)^{-1}F(z_n)\\
&\quad + F'(z_n)^{-1}(A_n - F'(z_n))A_n^{-1}F(z_n),
\end{aligned}$$

$$\begin{aligned}
\|x_{n+1} - x^*\| &\leq \left[\frac{\int_0^1 \varphi_2((1-\theta)\|z_n - x^*\|)d\theta}{1-\varphi_3(\|z_n - x^*\|)}\right.\\
&\quad + \frac{\varphi_2(\|x_n - x^*\|, \|y_n - x^*\|) + \varphi_4(\|x_n - x^*\|, \|y_n - x^*\|, \|z_n - x^*\|)}{(1-\varphi_3(\|z_n - x^*\|))(1-q_n)}\\
&\quad \left.\times(1+\varphi_3(\|z_n - x^*\|))\right]\|z_n - x^*\|\\
&\leq g_3(\|x_n - x^*\|)\|x_n - x^*\| \leq \|x_n - x^*\|.
\end{aligned}$$

3. Semi-local Convergence

The conditions are:

$$\begin{aligned}
\|L^{-1}([x,y;F]-L)\| &\leq \psi_0(\|x-x_0\|,\|y-x_0\|),\\
|L^{-1}([x,y;F]-F'(x))\| &\leq \psi_1(\|x-x_0\|,\|y-x_0\|),\\
\|L^{-1}(F'(y)-F'(x))\| &\leq \psi_2(\|y-x\|),
\end{aligned}$$

and

$$\|L^{-1}(F'(x)-L)\| \leq \psi_3(\|x-x_0\|).$$

The majorant sequence $\{\alpha_n\}$ is defined for $\alpha_0 = 0, \beta_0 \geq \|F'(x_0)^{-1}F(x_0)\|$ and for each $n = 0,1,2,\ldots$ by

$$\begin{aligned}
p_n &= \psi(\alpha_n,\beta_n)+\psi_1(\alpha_n,\beta_n),\\
\gamma_n &= \beta_n+\frac{\int_0^1 \psi_2(\theta(\beta_n-\alpha_n))d\theta(\beta_n-\alpha_n)}{1-p_n},\\
\lambda_n &= 1+\int_0^1 \psi_3(\beta_n+\theta(\gamma_n-\beta_n))d\theta(\gamma_n-\beta_n)+\int_0^1 \psi_2(\theta(\beta_n-\alpha_n))d\theta(\beta_n-\alpha_n)\\
\alpha_{n+1} &= \gamma_n+\frac{\lambda_n}{1-p_n}\\
\delta_{n+1} &= \int_0^1 \psi_2(\theta(\alpha_{n+1}-\alpha_n))d\theta(\alpha_{n+1}-\alpha_n)\\
&\quad +(1+\psi_3(\alpha_n))(\alpha_{n+1}-\beta_n)
\end{aligned}$$

and

$$\beta_{n+1} = \alpha_{n+1}+\frac{\delta_{n+1}}{1-\psi_3(\alpha_{n+1})}.$$

The motivation for the introduction of the majorizing sequence $\{\alpha_n\}$ is justified by the calculations:

$$\begin{aligned}
F(y_n) &= F(y_n)-F(x_n)-F'(x_n)(y_n-x_n)\\
z_n-y_n &= -A_n^{-1}F(y_n),\\
\|z_n-y_n\| &\leq \frac{\int_0^1 \psi_2(\theta\|y_n-x_n\|)d\theta\|y_n-x_n\|}{1-p_n}\\
&\leq \gamma_n-\beta_n,
\end{aligned}$$

since $\psi(\|x_n-x_0\|,\|y_n-x_0\|)+\psi_1(\|x_n-x_0\|,\|y_n-x_0\|) = \bar{p}_n \leq p_n < 1$,

$$A_n^{-1}L\| \leq \frac{1}{1-p_n},$$

$$\begin{aligned}
x_{n+1}-z_n &= -A_n^{-1}(F(z_n)-F(y_n)+F(y_n)),\\
\|x_{n+1}-z_n\| &\leq \frac{\lambda_n}{1-p_n}\\
&\leq \alpha_{n+1}-\gamma_n,\\
F(x_{n+1}) &= F(x_{n+1})-F(x_n)-F'(x_n)(x_{n+1}-x_n)\\
&\quad +F'(x_n)(x_{n+1}-y_n),
\end{aligned}$$

$$\begin{aligned}
\|L^{-1}F(x_{n+1})\| &\leq \int_0^1 \psi_2(\theta(\|x_{n+1}-x_n \\
)d\theta\|x_{n+1}-x_n\| \\
&+(1+\psi_3(\|x_n-x_0\|))\|x_{n+1}-y_n\| \\
&= \bar{\delta}_{n+1} \leq \delta_{n+1}, \\
\|y_{n+1}-x_{n+1}\| &\leq \|F'(x_{n+1})^{-1}L\|\|L^{-1}F(x_{n+1})\| \\
&\leq \frac{\delta_{n+1}}{1-\psi_3(\alpha_{n+1})} \\
&= \beta_{n+1}-\alpha_{n+1}.
\end{aligned}$$

4. References

[1] Argyros, C. I., Regmi, S., Argyros, I. K., George, S., *Contemporary Algorithms: Theory and Applications*, Volume III, NOVA Publishers, (2023).

[2] Argyros, I. K., *Advances in the Efficiency of Computational Methods and Applications*, World Scientific Publishing Oxon, United Kingdom (2000).

[3] Argyros, I. K., *The theory and applications of iteration methods*, 2nd Edition, CRC Press, Engineering Series, Taylor and Francis Group, (2022).

[4] Argyros, I. K., George, S., *Ball comparison between two optimal eight-order methods under weak conditions*, SeMA Journal, 72, (2015), 1-11.

[5] Cordero, A., Hueso, J. L., Martnez, E., Torregrosa, J. R., *Steffensen type methods for solving nonlinear equations*, J. Comput. Appl. Math., (2010), 10.1016/j.cam.2010.08.043.

[6] Grau, M., Daz-Barrero, J. L., *An improvement to Ostrowski root-finding method*, Appl. Math. Comput., 173, (2006), 450-456.

[7] Grau-Snchez, M., Grau, A., Noguera, M., *On the computational efficiency index and some iterative methods for solving systems of nonlinear equations*, J. Comput. Appl. Math., (2011), 10.1016/j.cam.2011.08.008.

[8] Grau-Snchez, M., Noguera, M., Amat, S., *On the approximation of derivatives using divided difference operators preserving the local convergence order of iterative methods*, Journal of Computational and Applied Mathematics, 237, 1, (2013), 363-372.

[9] Grau-Snchez, M., Noguera, M., Gutirrez, J. M., *On some computational orders of convergence*, Appl. Math. Lett., 23, (2010), 472-478.

[10] Kou, J., Li, Y., Wang, X., *Some variants of Ostrowskis method with seventh-order convergence*, J. Comput. Appl. Math., 209, (2007), 153-159.

[11] Kou, J., Wang, X., *Some improvements of Ostrowskis method*, Appl. Math. Lett., 23, (2010), 92-96.

[12] Ortega, J. M., Rheinboldt, W. G., *Iterative solutions of nonlinear equations in several variables*, Academic Press, New York, (1970).

[13] Potra, F. A., Ptk, V., *Nondiscrete Induction and Iterative Processes*, Pitman Publishing, Boston, (1984).

[14] Shakno, S. M., *On an iterative algorithm with superquadratic convergence for solving nonlinear operator equations*, J. Comput. Appl. Math., 231, (2009), 222-235.

[15] Traub, J. F., *Iterative Methods for the Solution of Equations*, Prentice-Hall, Englewood Cliffs, (1964).

Chapter 19

Frozen Derivative Iterative Method of High Order

1. Introduction

The iterative method is defined for $x_0 \in \Omega$ by

$$
\begin{aligned}
F'(x_0)u_j &= F(x_{j-1}),\ j=1,2,3 \\
x_j &= x_{j-1}-u_j, \\
F'(x_0)u_4 &= F(x_3),\ j=5,6,7,8, \\
F'(x_0)u_j &= F'(x_3)u_{j-1}, \\
x_4 &= x_3-\frac{21}{4}u_4+11u_5-\frac{23}{2}u_6+6u_7-\frac{5}{4}u_8, \\
F'(x_0)u_9 &= F(x_{i-1}),\ i=5,...,k \\
F'(x_0)u_j &= F'(x_3)u_{j-1},\ j=10,...,13, \\
x_i &= x_{i-1}-\frac{11}{2}u_9+12u_{10}-13u_{11}+7u_{12}-\frac{3}{2}u_{13}, \\
x_0 &= x_k,
\end{aligned}
\tag{19.1}
$$

where $k \geq 3$ is a fixed natural number. Notice: Number of steps k_j derivative evaluation 2; LU-factorization; matrix-vector multiplication $4(k-3)$; vector-vector multiplication $5(k-3)$ and number of lower and upper triangular systems $5k-12$. The local convergence analysis was carried out in [1] when $B_1 = B_2 = \mathbb{R}^m$, where m is a natural number under the Taylor series expansion approach and assumptions reaching the $F^{(9)}$ not on the method limiting its applicability to solving equations whose operator F is at least that many times differentiable. But the method may converge otherwise. That is why in order to expand the applicability of the method we provide a new local convergence using information only for F, F' appearing on it under generalized continuity assumptions. Other benefits of our approach include computable upper error bounds on the distances $\|x_n - x^*\|$ and uniqueness of the solution not given in [1].

Our approach can extend the applicability of other methods along the same lines [2]-[9]. The semi-local convergence can also be given (not provided in [1]) by exchanging the role of the solution x^* by the initial point x_0 and using majorizing sequence along the lines of the previous chapters. Finally, the analysis is carried out on the more general setting of a Banach space.

2. Local Convergence

The conditions are developed for $M = [0,+\infty)$ as follows
Assume:

(A_1) There exists a continuous and nondecreasing function $\varphi_0 : M \longrightarrow M$ such that the equation $\varphi_0(t) - 1 = 0$ has a smallest positive solution denoted by ρ. Let $M_0 = [0,\rho)$.
It is convenient to set $h_2(t) = 1 - \varphi_0(t)$ and adopt the notation $h_2 = h_2(t)$.

(A_2) There exists a continuous and nondecreasing function $\varphi : M_0 \longrightarrow M$.
Define the real function on the interval M_0 in turn by

$$\begin{aligned}
h &= h(t) = \int_0^1 \varphi((1+\theta)t)d\theta, \\
h_1 &= h_1(t) = 1 + \int_0^1 \varphi_0(\theta t)d\theta, \\
h_3 &= h_3(t) = 1 + \varphi_0(t), \\
h_0 &= h_0(t) = \frac{h_3}{h_2} \\
g_1 &= g_1(t) = 1 + \frac{h_1}{h_2}\left(\frac{21}{4} + h_0 + \frac{23}{2}h_0^2 + 6h_0^3 + \frac{5}{4}h_0^4\right), \\
g_2 &= g_2(t) = 1 + \frac{h_1}{h_2}\left(\frac{11}{2} + 12h_0 + 13h_0^2 + 7h_0^3 + \frac{3}{2}h_0^4\right),
\end{aligned}$$

and $h_4 = h_4(t) = g_1 h$
$g = g(t) = g_2 g_4$.

(A_3) The equations $h_4(t) - 1 = 0$ and $g(t) - 1 = 0$ have smallest solutions $r_1, r_2 \in M_0 - \{0\}$, respectively.
Let

$$r = \min\{r_1, r_2\}.$$

The real functions are related to the operators on the method.

(A_4) There exists a simple solution $x^* \in \Omega$ of the equation $F(x) = 0$ such that for each $x \in \Omega$

$$\|F'(x^*)^{-1}(F'(x) - F'(x^*))\| \leq \varphi_0(\|x - x^*\|).$$

Let $\Omega_0 = \Omega \cap U(x^*, \rho)$.

(A_5)

$$\|F'(x^*)^{-1}(F'(y) - F'(x))\| \leq \varphi(\|y - x\|)$$

for each $x, y \in \Omega_0$.
and

(A_6)

$$U[x^*, r] \subset \Omega.$$

The motivation for the introduction of the real functions is given by a series of estimates:

$$\begin{aligned} \|u_4\| &= \|F'(x_0)^{-1}F(x_3)\| \\ &\leq \frac{(1+\int_0^1 \varphi_0(\theta\|x_3-x^*\|)d\theta)\|x_3-x^*\|}{1-\varphi_0(\|x_3-x^*\|)}, \end{aligned} \tag{19.2}$$

$$\begin{aligned} &\|F'(x_0)^{-1}F'(x_3)F'(x_0)^{-1}F(x_3)\| \\ &\leq \frac{(1+\varphi_0(\|x_3-x^*\|))(1+\int_0^1 \varphi_0(\theta\|x_3-x^*\|)d\theta)\|x_3-x^*\|}{(1-\varphi_0(\|x_0-x^*\|))^2}, \\ &\|F'(x_0)^{-1}F'(x_3)F'(x_0)^{-1}F'(x_3)F'(x_0)^{-1}F(x_3)\| \\ &\leq \frac{(1+\varphi_0(\|x_3-x^*\|))^2(1+\int_0^1 \varphi_0(\theta\|x_3-x^*\|)d\theta)\|x_3-x^*\|}{(1-\varphi_0(\|x_0-x^*\|))^3}, \\ &\|F'(x_0)^{-1}F'(x_3)F'(x_0)^{-1}F'(x_3)F'(x_0)^{-1}F'(x_3)F'(x_0)^{-1}F(x_3)\| \\ &\leq \frac{(1+\varphi_0(\|x_3-x^*\|))^3(1+\int_0^1 \varphi_0(\theta\|x_3-x^*\|)d\theta)\|x_3-x^*\|}{(1-\varphi_0(\|x_0-x^*\|))^4}, \\ &\|F'(x_0)^{-1}F'(x_3)F'(x_0)^{-1}F'(x_3)F'(x_0)^{-1}F'(x_3)F'(x_0)^{-1}F'(x_3)\| \\ &\leq \frac{(1+\varphi_0(\|x_3-x^*\|))^4(1+\int_0^1 \varphi_0(\theta\|x_3-x^*\|)d\theta)\|x_3-x^*\|}{(1-\varphi_0(\|x_0-x^*\|))^5}, \end{aligned}$$

where we also used

$$\begin{aligned} \|F'(x_0)^{-1}F'(x^*)\| &\leq \frac{1}{1-\varphi_0(\|x_0-x^*\|)}, \\ F(x_3) &= F(x_3)-F(x^*) = \int_0^1 F'(x^*+\theta(x_3-x^*))d\theta(x_3-x^*) \end{aligned}$$

$$\begin{aligned} &\|F'(x^*)^{-1}F(x_3)\| \\ &= \|F'(x^*)^{-1}[\int_0^1 F'(x^*+\theta(x_3-x^*))d\theta - F'(x^*)+F'(x^*)]\| \\ &\leq (1+\int_0^1 \varphi_0(\theta\|x_3-x^*\|)d\theta)\|x_3-x^*\|. \end{aligned}$$

Then, we have by the first four substeps the definition of the functions $''h''$ and the summing up of the proceeding estimates for $\|x_0 - x^*\| \leq t$:

$$\begin{aligned}
\|x_4 - x^*\| &\leq \|x_3 - x^*\| + \frac{21}{4}\|F'(x_0)^{-1}F'(x_3)\| + \ldots + \frac{5}{4}\|F'(x_0)^{-1}F'(x_3)u_7\| \\
&\leq \|x_3 - x^*\| + \frac{21}{4}\|F'(x_0)^{-1}F(x_3)\| \\
&\quad + 11\|F'(x_0)^{-1}F'(x_3)F'(x_0)^{-1}F(x_3)\| \\
&\quad + \frac{23}{2}\|F'(x_0)^{-1}F'(x_3)F'(x_0)^{-1}F'(x_3)F'(x_0)^{-1}F(x_3)\| \\
&\quad + 6\|F'(x_0)^{-1}F'(x_3)F'(x_0)^{-1}F'(x_3) \\
&\quad \times F'(x_0)^{-1}F'(x_3)F'(x_0)^{-1}F(x_3)\| \\
&\quad + \frac{5}{4}\|F'(x_0)^{-1}F'(x_3)F'(x_0)^{-1}F'(x_3)F'(x_0)^{-1}F'(x_3) \\
&\quad \times F'(x_0)^{-1}F'(x_3)F'(x_0)^{-1}F(x_3)\| \\
&\leq g_4(\|x_0 - x^*\|)\|x_3 - x^*\| \\
&\leq g_1(\|x_0 - x^*\|)h(\|x_0 - x^*\|)\|x_0 - x^*\| \\
&= h_4(\|x_0 - x^*\|)\|x_0 - x^*\| \leq \|x_0 - x^*\| < r, \\
u_9 &= F'(x_0)^{-1}F(x_{i-1}),
\end{aligned}$$

and similarly

$$\begin{aligned}
\|x_i - x^*\| &\leq \|x_{i-1} - x^*\| + \frac{11}{2}\|u_9\| + 12\|u_{10}\| \\
&\quad + 13\|u_1\| + 7\|u_{12}\| + \frac{3}{2}\|u_{13}\| \\
&\leq \left[1 + \frac{11}{2}\frac{h_1}{h_2} + 12\frac{h_3 h_1}{h_2^2} \right. \\
&\quad \left. + 13\frac{h_3^2 h_1}{h_2^2} + 7\frac{h_3^3 h_1}{h_2^4} + \frac{3}{2}\frac{h_3^4 h_1}{h_2^5}\right] \|x_{i-1} - x^*\|
\end{aligned}$$

and for $c_0 = g_2(\|x_0 - x^*\|)$,

$$\begin{aligned}
\|x_5 - x^*\| &\leq c_0\|x_4 - x^*\|, \\
\|x_6 - x^*\| &\leq c_0\|x_5 - x^*\| \leq c_0^2\|x_4 - x^*\|, \\
&\ \ \vdots \\
\|x_k - x^*\| &\leq c_0^{k-4}\|x_4 - x^*\|, \ k = 5, \ldots \\
&\leq c_0^{k-1}g_1(\|x_0 - x^*\|)\|x_0 - x^*\| \\
&= c\|x_0 - x^*\|, \ c = c_0^{k-1}g_1(\|x_0 - x^*\|) \in [0, 1), \\
&\ \ \vdots \\
\|x_{n+k} - x^*\| &\leq c^n\|x_0 - x^*\| < r
\end{aligned}$$

Thus $\lim_{n\to+\infty} x_n = x^*$ and $\{x_n\} \subset U(x^*, r)$.
Hence we arrived at:

Theorem 20. Suppose that the conditions $(A_1) - (A_5)$ hold. Then, $\lim_{n\to+\infty} x_n = x^*$ provided that $x_0 \in U(x^*, r) - \{x^*\}$.

3. References

[1] Ahmad, F., Rehman, S. U., Ullah, M. Z., Aljahdali, H. M., Ahmad, S., Alshomrani, A. S., Carrasco, J. A., Ahmad, S., Sivasankaran, S., *A frozen Jacobian multistep iterative method for solving nonlinear initial value problems (IVPs) and boundary value problems (BVPs)*, Complexity, (2017), Article ID 9407656.

[2] Argyros, I. K., *Convergence and Applications of Newton-Type Iterations*, Springer-Verlag, New York, (2008).

[3] Argyros, I. K., George, S., *Mathematical modeling for the solution of equations and systems of equations with applications*, Volume IV, Nova Publishers, NY, (2020).

[4] Argyros, I. K., George, S., *On the complexity of extending the convergence region for Traubs method*, Journal of Complexity, 56, (2022), Article ID 101423.

[5] Arroyo, V., Cordero, A., Torregrosa, J. R., *Approximation of preliminary orbits of artificial satellites: Addressing the efficiency challenge*, Mathematical and Computer Modelling, 54, 7-8, (2011), 18021807.

[6] Cordero, A., Kansal, M., Kanwar, V., Torregrosa, J. R., *A stable class of improved second-derivative free Chebyshev-Halley type methods with optimal eighth-order convergence*, Numerical Algorithms, 72, 4, (2016), 937958.

[7] Ortega, J. M., Rheinboldt, W. C., *Iterative solution of nonlinear equations in several variables*, Academic Press, New York, NY, USA, (1970).

[8] Traub, J. F., *Iterative methods for the solution of equations, Prentice-Hall Series in Automatic Computation*, Prentice-Hall, Englewood Cliffs, NJ, USA, (1964).

[9] Ullah, M. Z., Serra-Capizzano, S., Ahmad, F., *An efficient multistep iterative method for solving systems of nonlinear equations related to ordinary differential equations (ODEs)*, Applied Mathematics and Computation, 250, (2015), 249259.

Chapter 20

Generalized Jarratt-Type Method

1. Introduction

The method is defined for $x_0 \in \Omega$ and each $n = 0.1.2....$ by

$$\begin{aligned} y_n &= x_n - \frac{2}{3}F'(x_n)^{-1}F(x_n) \\ x_{n+1} &= A(s)F'(x_n)^{-1}F(x_n), s = F'(x_n)^{-1}F'(y_n). \end{aligned} \tag{20.1}$$

The convergence order 4 is estimated in [5], when $B_1 = B_2 = \mathbb{R}^k$, provided that $F', F'', \ldots, F^{(5)}$ exists, A''' is bounded, $A(1) = 1, A'(1) = -\frac{3}{4}$ and $A''(1) = \frac{9}{4}$. The local convergence is established under the Taylor series expansion approach without computable error bounds or uniqueness of the solution information. We extend the applicability of the method in the local case using generalized conditions on F'. Moreover, in the semi-local case (not given in [5]) majorant sequences are employed. Our analysis includes computational error estimates of $\|x_n - x^*\|$ and uniqueness of the solution information. Our approach can be applied to other methods using inverses of linear operators along the same lines [1]-[4].

2. Local Convergence

The conditions are:

(C_1) There exists an invertible operator L, a solution $x^* \in \Omega$ of the equation $F(x) = 0$ and a continuous and nondecreasing function $\varphi_0 : M = [0,\infty) \longrightarrow [0,\infty)$ such that for each $x \in \Omega$

$$\|L^{-1}(F'(x) - L)\| \leq \varphi_0(\|x - x^*\|).$$

(C_2) The equation $\varphi_0(t) - 1 = 0$ has a smallest solution denoted by $\rho \in M - \{0\}$. Set $M_0 = [0,\rho)$.

(C_3) There exists continuous and nondecreasing functions $\varphi : M_0 \longrightarrow M$, $\varphi_1 : M_0 \times M_0 \longrightarrow M$ such that for each $x, y \in \Omega_0 = \Omega \cap U(x^*, \rho)$

$$\|L^{-1}(F'(y) - F'(x))\| \leq \varphi(\|y - x\|)$$

and

$$\|I - A(s)\| \leq \varphi_1(\|x - x^*\|, \|y - x^*\|), s = F'(x)^{-1}F'(y)$$

(C_4) The equations $g_1(t)-1=0$ and $g_2(t)-1=0$ have positive solutions $r_1, r_2 \in M_0$, where

$$g_1(t) = \frac{\int_0^1 \varphi((1-\theta)t)d\theta + \frac{1}{3}(1+\int_0^1 \varphi_0(\theta t)d\theta)}{1-\varphi_0(t)}$$

and

$$g_2(t) = \frac{\int_0^1 \varphi((1-\theta)t)d\theta + \varphi_1(t, g_1(t)t)(1+\int_0^1 \varphi_0(\theta t)d\theta)}{1-\varphi_0(t)}.$$

Set

$$r = \min\{r_1, r_2\}.$$

and

(C_5)

$$U[x^*, r] \subset \Omega.$$

The motivation for the introduction of these functions g_1, g_2 follows by the estimates:

$$\begin{aligned} y_n - x^* &= x_n - x^* - \frac{2}{3}F'(x_n)^{-1}F(x_n) \\ &= x_n - x^* - F'(x_n)^{-1}F(x_n) + \frac{1}{3}F'(x_n)^{-1}F(x_n), \end{aligned}$$

$$\begin{aligned} \|y_n - x^*\| &\leq \frac{1}{1-\varphi_0(\|x_n - x^*\|)}\left[\int_0^1 \varphi((1-\theta)\|x_n - x^*\|)d\theta \right. \\ &\left. + \frac{1}{3}(1+\int_0^1 \varphi_0(\theta\|x_n - x^*\|)d\theta)\right]\|x_n - x^*\| \\ &\leq g_1(\|x_n - x^*\|)\|x_n - x^*\| \leq \|x_n - x^*\| < r, \end{aligned}$$

$$\begin{aligned} x_{n+1} - x^* &= x_n - x^* - F'(x_n)^{-1}F(x_n) \\ &\quad + (I - A(s))F'(x_n)^{-1}F9x_n), \\ \|x_{n+1} - x^*\| &\leq \frac{1}{1-\varphi_0(\|x_n - x^*\|)}\left[\int_0^1 \varphi((1-\theta)\|x_n - x^*\|)d\theta \right. \\ &\left. + \varphi_1(\|x_n - x^*\|, \|y_n - x^*\|)(1+\int_0^1 \varphi_0(\theta\|x_n - x^*\|)d\theta)\right]\|x_n - x^*\| \\ &\leq g_2(\|x_n - x^*\|)\|x_n - x^*\| \leq \|x_n - x^*\|. \end{aligned}$$

The uniqueness of the solution in formation is as given in previous Chapters. Further specialization of the operator A determines the function φ_1.

3. Semi-Local Convergence

The role of $x^*, \varphi_0, \varphi, \varphi_1$ is exchanged by x_0, ψ_0, ψ and ψ_1, respectively where the latter functions are defined in what follows.

(H_1) There exists an invertible operator $L, x_0 \in \Omega$ and a continuous and nondecreasing function $\psi_0 : M \longrightarrow [0,\infty)$ such that for each $x \in \Omega$

$$\|L^{-1}(F'(x) - L)\| \leq \psi_0(\|x - x_0\|).$$

(H_2) The equation $\psi_0(t) - 1 = 0$ has a smallest solution denoted by $\rho_1 \in M$. Let $\Omega_1 = \Omega \cap U(x_0, \rho_1)$ and $M_1 = [0, \rho_1)$. Notice that in particular $\|L^{-1}(F'(x_0) - L)\| \leq \psi_0(0) < 1$. Thus $F'(x_0)^{-1}$ exists and we can set

$$\|F'(x_0)^{-1}F(x_0)\| \leq \beta_0.$$

(H_3) There exists continuous and nondecreasing functions $\psi : M_1 \longrightarrow M$, $\psi_2 : M_1 \times M_1 \longrightarrow M$ such that for each $x, y \in \Omega_1$

$$\|L^{-1}(F'(y) - F'(x))\| \leq \psi(\|y - x\|)$$

and

$$\|I - A(s)\| \leq \psi_1(\|x - x_0\|, \|y - x_0\|).$$

Define the scalar sequence $\{x_n\}$ for $\alpha_0 = 0$, $\beta_0 \in [0, \rho_1)$ and each $n = 0, 1, 2, \ldots$ by

$$\begin{aligned}
\alpha_{n+1} &= \beta_n + \frac{1}{2}\psi_1(\alpha_n, \beta_n)(\beta_n - \alpha_n), \\
\gamma_{n+1} &= (1 + \int_0^1 \psi(\alpha_n + \theta(\alpha_{n+1} - \alpha_n))d\theta)(\alpha_{n+1} - \alpha_n) \\
&\quad + \frac{3}{2}(1 + \psi_0(\alpha_n))(\beta_n - \alpha_n)
\end{aligned}$$

(H_4) There exists $p_0 \in [\beta_0, \rho_1)$ such that for each $n = 0, 1, 2, \ldots$

$$\psi_0(\alpha_n) < 1 \text{ and} alpha_0 \leq p_0.$$

Clearly, it follows that

$$0 \leq \alpha_n \leq \beta_n \leq \alpha_{n+1} < p_0$$

and there exists $p \in [\beta_0, p_0]$ such that

$$\lim_{n \longrightarrow \infty} \alpha_n = p.$$

The limit point p is the least upper bound of the sequence $\{\alpha_n\}$.
and

(H_5)

$$U[x_0, p] \subset \Omega.$$

The motivation for the introduction of the majorant sequence $\{\alpha_n\}$ is justified from the estimates:

$$\begin{aligned}
x_{n+1} - y_n &= (\frac{2}{3}I - A(s))F'(x_n)^{-1}F(x_n) \\
&= \frac{1}{3}(2I - 3A(s))(-\frac{3}{2}(y_n - x_n)) \\
&= \frac{1}{2}(3A(s) - 2I)(y_n - x_n), \\
\|x_{n+1} - y_n\| &\le \frac{1}{2}\psi_1(\|x_n - x_0\|, \|y_n - x_0\|)\|y_n - x_n\| \\
&\le \frac{1}{2}\psi_1(\alpha_n, \beta_n)(\beta_n - \alpha_n) \\
&= \alpha_{n+1} - \beta_n, \\
\|x_{n+1} - x_0\| &\le \|x_{n+1} - y_n\| + \|y_n - x_0\| \\
&\le \alpha_{n+1} - \beta_n + \beta_n - \alpha_0 \\
&= \alpha_{n+1} < p,
\end{aligned}$$

$$\begin{aligned}
F(x_{n+1}) &= F(x_{n+1}) - F(x_n) - \frac{3}{2}F'(x_n)(y_n - x_n) \\
&= \int_0^1 F'(x_n + \theta(x_{n+1} - x_n))d\theta(x_{n+1} - x_n) \\
&\quad + \frac{3}{2}(F'9x_n) - L + L)(y_n - x_n), \\
\|L^{-1}F(x_{n+1})\| &\le (1 + \int_0^1 \psi(\|x_n - x_0\| + \theta\|x_{n+1} - x_n\|)d\theta)\|x_{n+1} - x_n\| \\
&\quad + \frac{3}{2}(1 + \psi_0(\|x_n - x_0\|))\|y_n - x_n\| \\
&= \bar{\gamma}_{n+1} \\
&\le (1 + \int_0^1 \psi(\alpha_n + \theta(\alpha_{n+1} - \alpha_n))d\theta(\alpha_{n+1} - \alpha_n) \\
&\quad + \frac{3}{2}(1 + \psi_0(\alpha_n))(\beta_n - \alpha_n) \\
&= \gamma_{n+1},
\end{aligned}$$

$$\begin{aligned}
\|y_{n+1} - x_{n+1}\| &\le \frac{2}{3}\|F'(x_{n+1})^{-1}L\|\|L^{-1}F(x_{n+1})\| \\
&\le \frac{2}{3}\frac{\bar{\gamma}_{n+1}}{1 - \psi_0(\|x_{n+1} - x_0\|)} \\
&\le \frac{2\gamma_{n+1}}{3(1 - \psi_0(\alpha_{n+1}))} \\
&= \beta_{n+1} - \alpha_{n+1}, \\
\|y_{n+1} - x_0\| &\le \|y_{n+1} - x_{n+1}\| + \|x_{n+1} - x_0\| \\
&\le \beta_{n+1} - \alpha_{n+1} + \alpha_{n+1} - \alpha_0 \\
&= \beta_{n+1} < p.
\end{aligned}$$

Thus, there exists $x^* \in U[x_0, p]$ such that $F(x^*) = 0$ and $\{x_n\} \subset U(x_0, p)$. Uniqueness of the solution results are already given in previous Chapters.

4. References

[1] Argyros, I. K., *Convergence and Applications of Newton-Type Iterations*, Springer-Verlag, New York, (2008).

[2] Argyros, I. K., George, S., *Mathematical modeling for solving equations and systems of equations with applications*, Volume IV, Nova Publishers, NY, (2020).

[3] Argyros, I. K., George, S., *On the complexity of extending the convergence region for Traubs method*, Journal of Complexity, 56, (2022), Article 101423.

[4] Chicharro, F. I., Cordero, A., Garrido, N., Torregrosa, J. R., *On selecting the optimal members of the Kim family and enhancing their convergence*, Mathematical Methods in the Applied Sciences, 43, (2020), 8051-8066.

[5] Padilla, J. J., Chicharro, A., Cordero, F. I., Torregrosa, J. R., *A parametric family of root-finding iterative methods: Fractals of the basin of attraction*, Fractals and Fractional, 6, (2022), 572.

Chapter 21

An Efficient Multistep Method of High Order

1. Introduction

The method is defined for $x_0 \in \Omega$ and each $n = 0.1.2....$ by

$$\begin{aligned} y_m^1 &= x_m - F'(x_m)^{-1}F(x_m) \\ y_m^2 &= x_m - 2(F'(x_m) + F'(y_m^1))^{-1}F(x_m) \\ &\dots \\ y_m^i &= y_m^{i-1} - AF(y_m^{i-1}), i = 3,4,\dots,k, \\ y_{n+1} &= y_m^k, \end{aligned} \tag{21.1}$$

where k is a fixed natural number and $A = (3F'(y_m^1) - F'(x_m))^{-1}(F'(x_m) + F'(y_m^1))F'(x_m)^{-1}$.

The method is shown to be of local convergence order $3(i-1)$ in [5] using the Taylor series expansion approach and assumption reaching the seventh derivative of the operator F. The problems with this approach have already been reported in earlier chapters. The local convergence analysis in this Chapter uses only the derivative appearing on the method (i.e. F'). Hence, the applicability of the method is extended. The same idea can be used on other methods [1]-[4],[6]-[11].

2. Convergence

The local convergence functions $g_1, g_2, \ldots, g_k$ are:

$$\begin{aligned}
g_1(t) &= \frac{\int_0^1 \omega((1-\theta)t)d\theta}{1-\omega_0(t)}, \\
\bar{\omega}(t) &= \begin{cases} \omega(2t) \\ OR, 2\omega_0(t) \end{cases} \\
q(t) &= \omega_0(t), p(t) = 2\omega_0(t), \\
g_2(t) &= g_1(t) + \frac{\bar{\omega}(t)(1+\int_0^1 \omega_0(\theta t)d\theta}{2(1-q(t))(1-\omega_0(t))}, \\
&\ldots \\
g_i(t) &= \Bigg[\frac{\int_0^1 \omega((1-\theta)g_{i-1}(t)t)d\theta}{1-\omega_0(g_{i-1}(t)t)} \\
&\quad + \frac{\bar{\omega}(t)(1+\int_0^1 \omega_0(\theta g_{i-1}(t)t)d\theta)}{(1-\omega_0(t))(1-\omega_0(g_{i-1}(t)t))} \\
&\quad + \frac{\bar{\omega}(t)(1+\int_0^1 \omega_0(\theta g_{i-1}(t)t)d\theta)}{(1-\omega_0(t))(1-p(t))}\Bigg] g_{i-1}(t), \; i = 3, 4, \ldots k, \\
g(t) &= g_1(t) + \frac{\bar{\omega}(t)(1+\int_0^1 \omega_0(\theta t)d\theta)}{2(1-\omega_0(t))(1-q(t))}, \\
h(t) &= g_1(t) + \frac{\bar{\omega}(t)(1+\int_0^1 \omega_0(\theta t)d\theta)}{(1-\omega_0(t))^2} \\
&\quad + \frac{\bar{\omega}(t)(1+\int_0^1 \omega_0(\theta t)d\theta)}{(1-\omega_0(t))(1-p(t))},
\end{aligned}$$

and

$$h_0(t) = g(t)h(t).$$

The radius defined by

$$r = \min\{r_1, r_2, r_i, r_0\}$$

provided that the equations $g_1(t)-1=0, g_2(t)-1=0, g_i(t)-1=0$ and $h_0(t)-1=0$ have smallest positive solutions r_1, r_2, r_i, r_0, respectively in the interval $(0,s)$, where s is the smallest positive solution of the equation $\omega_0(t)-1=0$. The motivation for the development of these functions follows from the calculations in turn:

$$\begin{aligned}
y_0^1 - x^* &= x_0 - x^* - F'(x_0)^{-1}F(x_0), \\
\|y_0^1 - x^*\| &\leq \frac{\int_0^1 \omega((1-\theta)\|x_0-x^*\|)d\theta\|x_0-x^*\|}{1-\omega_0(\|x_0-x^*\|)} \\
&\leq g_1(\|x_0-x^*\|)\|x_0-x^*\| \leq \|x_0-x^*\| < r,
\end{aligned}$$

$$\begin{aligned}
y_0^2 - x^* &= x_0 - x^* - F'(x_0)^{-1}F(x_0) \\
&\quad + (F'(x_0)^{-1} - 2(F'(x_0)+F'(y_0^1))^{-1})F(x_0) \\
&= x_0 - x^* - F'(x_0)^{-1}F(x_0) + (F'(x_0)+F'(y_0^1))^{-1} \\
&\quad \times (F'(y_0^1) - F'(x_0))F'(x_0)^{-1}F(x_0),
\end{aligned}$$

$$\begin{aligned}
\|y_0^2 - x^*\| &\leq \left[\frac{\int_0^1 \omega((1-\theta)\|x_0 - x^*\|)d\theta}{1-\omega_0(\|x_0 - x^*\|)}\right. \\
&\left. + \frac{\bar{\omega}_0(1+\int_0^1 \omega_0(\theta\|x_0 - x^*\|)d\theta)}{2(1-q_0)(1-\omega_0(\|x_0 - x^*\|))}\right]\|x_0 - x^*\| \\
&\leq g_2(\|x_0 - x^*\|)\|x_0 - x^*\| \leq \|x_0 - x^*\|,
\end{aligned}$$

$$\begin{aligned}
y_0^i - x^* &= y_0^{i-1} - x^* - F'(y_0^{i-1})^{-1}F(y_0^{i-1}) \\
&\quad +(F'(y_0^{i-1}) - A)F(y_0^{i-1}) \\
&= y_0^{i-1} - x^* - F'(y_0^{i-1})^{-1}F(y_0^{i-1}) \\
&\quad +F'(y_0^{i-1})^{-1}(F'(x_0) - F'(y_0^{i-1}))F'(x_0)^{-1}F(y_0^{i-1}) \\
&\quad +2(3F'(y_0^1) - F'(x_0))^{-1} \\
&\quad \times(F'(y_0^1) - F'(x_0))F'(x_0)^{-1}F(y_0^{i-1}),
\end{aligned}$$

$$\begin{aligned}
\|y_0^i - x^*\| &\leq \left[\frac{\int_0^1 \omega((1-\theta)\|y_0^{i-1} - x^*\|)d\theta}{1-\omega_0(\|y_0^{i-1} - x^*\|)}\right. \\
&\quad + \frac{\bar{\omega}_0(1+\int_0^1 \omega_0(\theta\|y_0^{i-1} - x^*\|)d\theta}{(1-\omega_0(\|x_0 - x^*\|))(1-\omega_0(\|y_0^{i-1} - x^*\|))} \\
&\quad \left. + \frac{\bar{\omega}_0(1+\int_0^1 \omega_0(\theta\|y_0^{i-1} - x^*\|)d\theta)}{(1-p_0)(1-\omega_0(\|x_0 - x^*\|))}\right]\|y_0^{-1} - x^*\| \\
&\leq g_i(\|x_0 - x^*\|)\|x_0 - x^*\| \leq \|x_0 - x^*\|,
\end{aligned}$$

where we also used

$$\|F'(x_0)^{-1}L\| \leq \frac{1}{1-w_0(\|x_0 - x^*\|)},$$

$$\|(F'(x_0) + F'(y_0^1))^{-1}L\| \leq \frac{1}{2(1-q_0)},$$

$$\|F'(y_0^1)^{-1}L\| \leq \frac{1}{1-\omega_0(\|y_0^1 - x^*\|)},$$

and

$$\|(3F'(y_0^1) - F'(x_0))^{-1}L\| \leq \frac{1}{2(1-p_0)}.$$

It follows that

$$\begin{aligned}
\|x_1 - x^*\| &= \|y_k^1 - x^*\| \\
&\leq g(\|x_0 - x^*\|)h(\|x_0 - x^*\|)\|x_0 - x^*\| \\
&= h_0(\|x_0 - x^*\|)\|x_0 - x^*\| \\
&\leq \|x_0 - x^*\|.
\end{aligned}$$

Thus,

$$\|x_{m+1} - x^*\| \leq c\|x_m - x^*\| \leq r,$$

where $c = h_0(\|x_0 - x^*\|) \in [0,1)$. It follows that $x_{m+1} \in U(x^*, r)$ and $\lim_{m \longrightarrow \infty} x_m = x^*$.

3. References

[1] Argyros, I. K., *Convergence and Applications of Newton-Type Iterations*, Springer-Verlag, New York, (2008).

[2] Argyros, I. K., George, S., *Mathematical modeling for the solution of equations and systems of equations with applications*, Volume IV, Nova Publishers, NY, (2020).

[3] Argyros, I. K., George, S., *On the complexity of extending the convergence region for Traubs method*, Journal of Complexity, 56, (2022), Article 101423.

[4] Argyros, I. K., Hilout, S., *Numerical Methods in Nonlinear Analysis*, World Scientific Publishing Co., New Jersey, (2013).

[5] Argyros, I. K., Magrenn, A. A., *On the convergence of an optimal fourth-order family of methods and its dynamics*, Applied Mathematics and Computation, 252, (2015), 336-346.

[6] Babajee, D. K. R., Dauhoo, M. Z., Darvishi, M. T., Karami, A., Baratib, A., *Analysis of two Chebyshev-like third-order methods free from second derivatives for solving systems of nonlinear equations*, Journal of Computational and Applied Mathematics, 233, (2010), 2002-2012.

[7] Behl, R., Bhalla, S., Magrenn, A. A., Kumar, S., *An efficient high-order iterative scheme for large nonlinear systems with dynamics*, Journal of Computational and Applied Mathematics, 404, (2022), Article 113249.

[8] Cordero, A., Hueso, J. L., Martinez, E., Torregrosa, J. R., *A modified Newtons Jarratts composition*, Numerical Algorithms, 55, (2010), 87-99.

[9] Darvishi, M. T., Barati, A., *A third-order Newton-type method to solve systems of nonlinear equations*, Applied Mathematics and Computation, 187, (2007), 630-635.

[10] Hueso, J. L., Martinez, E., Teruel, C., *Convergence, efficiency and dynamics of new fourth and sixth order families of iterative methods for nonlinear systems*, Journal of Computational and Applied Mathematics, 275, (2015), 412-420.

[11] Jarratt, P., *Some fourth-order multipoint iterative methods for solving equations*, BIT, 9(2), (1969), 119-124.

Chapter 22

Jarratt-Type Family of Methods

1. Introduction

The family is defined for $a \neq 0, 1$, $x_0 \in \Omega$ and each $n = 0.1.2....$ by

$$\begin{aligned} y_n &= x_n - \frac{2}{3} F'(x_n)^{-1} F(x_n) \\ A &= b_1 F'(x_n) + b F'(y_n), \\ x_{n+1} &= x_n - b_3 F'(x_n)^{-1} F(x_n) - b_2 F'(y_n)^{-1} F(x_n) - A^{-1} F(x_n) \end{aligned} \tag{22.1}$$

provided that $b_1 = \frac{b(1-a)}{a}$, $b_2 = \frac{3}{4}\left(1 - \frac{1}{2(a-1)}\right)$ and $b_3 = \frac{1}{4}\left(1 + \frac{3}{2a}\right)$. The Taylor series expansion approach determined the local fourth order convergence assuming the existence of $F^{(5)}$ not on the method [4]. Our convergence analysis uses only the operators on the method. The more interesting semi-local convergence is also provided. Relevant work can be found in [1,2,3], [5]-[9].

2. Local Convergence

The majorant functions $g_i, i = 1, 2$ are:

$$\begin{aligned} g_1(t) &= \frac{\int_0^1 \omega((1-\theta)t)d\theta + \frac{1}{3}(1 + \int_0^1 \omega_0(\theta t)d\theta)}{1 - \omega_0(t)}, \\ q(t) &= \frac{1}{|b + b_1|}(|b_1|\omega_0(t) + |b|\omega_0(g_1(t)t)), \\ g_2(t) &= \frac{\int_0^1 \omega((1-\theta)t)d\theta}{1 - \omega_0(t)} + \left(\frac{|1 - b_3|}{1 - \omega_0(t)}\right. \\ &\quad \left. + \frac{|b_2|}{1 - \omega_0(g_1(t)t)} + \frac{1}{|b_1 + b_2|(1 - q(t))}\right)(1 + \int_0^1 \omega_0(\theta t)d\theta). \end{aligned}$$

The radius of convergence is

$$r = \min\{r_1, r_2\},$$

where r_i are the smallest positive solutions of the equation $g_i(t) - = 0$(if they exist). The motivation for the development of the functions g_i follows in turn by the calculations:

$$\begin{aligned} y_n - x^* &= x_n - x^* - F'(x_n)^{-1}F(x_n) + \frac{1}{3}F'(x_n)^{-1}F(x_n), \\ \|y_n - x^*\| &\leq \frac{1}{1-\omega_0(\|x_n - x^*\|)}\left[\int_0^1 \omega((1-\theta)\|x_n - x^*\|)d\theta \right. \\ &\left. +\frac{1}{3}(1+\int_0^1 \omega_0(\theta\|x_n - x^*\|)d\theta)\right]\|x_n - x^*\| \\ &\leq g_1(\|x_n - x^*\|)\|x_n - x^*\| \\ &\leq \|x_n - x^*\| < r, \end{aligned}$$

$$\begin{aligned} x_{n+1} - x^* &= x_n - x^* - F'(x_n)^{-1}F(x_n) + (1-b_3)F'(x_n)^{-1}F(x_n) \\ &\quad -b_2F'(y_n)^{-1}F(x_n) - A^{-1}F(x_n), \\ \|x_{n+1} - x^*\| &\leq \left[\frac{\int_0^1 \omega((1-\theta)\|x_n - x^*\|)d\theta}{1-\omega_0(\|x_n - x^*\|)} + \left(\frac{|1-b_3|}{1-\omega_0(\|x_n - x^*\|)}\right.\right. \\ &\quad \left. +\frac{|b_2|}{1-\omega_0(\|y_n - x^*\|)} + \frac{1}{|b_1+b_2|(1-q_n)}\right) \\ &\quad \left.\times(1+\int_0^1 \omega_0(\theta\|x_n - x^*\|)d\theta)\right]\|x_n - x^*\| \\ &\leq g_2(\|x_n - x^*\|)\|x_n - x^*\| \leq \|x_n - x^*\|, \end{aligned}$$

where we also used

$$\|L^{-1}(F'(x) - L)\| \leq w_0(\|x - x^*\|) < 1,$$

$$\|F'(x)^{-1}L\| \leq \frac{1}{1 - w_0(\|x - x^*\|)},$$

$$\begin{aligned} &\|((b+b_1)L)^{-1}(b_1(F'(x_n) - L) + b(F'(y_n) - L))\| \\ \leq\ & \frac{1}{|b+b_1|}(|b_1|\omega_0(\|x_n - x^*\|) + |b|\omega_0(\|y_n - x^*\|)) \\ =\ & q_n < 1, \end{aligned}$$

$$\|A^{-1}L\| \leq \frac{1}{|b+b_1|(1-q_n)}.$$

3. Semi-Local Convergence

Let $c = \frac{3}{2}b_3 - 1$. Define the sequence $\{\alpha_n\}$ for $\alpha_0 = 0$, $\beta_0 \geq [0, \rho)$, where ρ is the smallest positive solution of the equation $\omega_0(t) - 1 = 0$(if it exists) and each $n = 0, 12, \dots$ by

$$
\begin{aligned}
p_n &= \frac{1}{|b+b_1|}(|b_1|\omega_0(\alpha_n) + |b|\omega_0(\beta_n)),\\
\alpha_{n+1} &= \beta_n + \left[|c| + \frac{|b_2|(1+\frac{3}{2}v_0(\alpha_n))}{1-v_0(b_n)}\right.\\
&\quad \left. + \frac{1+\frac{3}{2}\omega_0(\alpha_n)}{|b+b_1|(1-p_n)}\right](\beta_n - \alpha_n)\\
\gamma_{n+1} &= (1+\int_0^1 \omega_0(\alpha_n + \theta(\alpha_{n+1}-\alpha_n))d\theta)(\alpha_{n+1}-\alpha_n)\\
&\quad + \frac{3}{2}(1+\omega_0(\alpha_n))(\beta_n - \alpha_n)\\
\beta_{n+1} &= \alpha_{n+1} + \frac{2}{3}\frac{\gamma_{n+1}}{1-\omega_0(\alpha_{n+1})}.
\end{aligned}
$$

The motivation for this sequence is coming from the estimates

$$
\begin{aligned}
\|x_{n+_1} - y_n\| &\leq \left[|c| + \frac{|b_2|(1+\frac{3}{2}v_0(\|x_n - x_0\|))}{1-v_0(\|y_n - x_0\|)}\right.\\
&\quad \left. + \frac{1+\frac{3}{2}\omega_0(\|x_n - x_0\|)}{|b+b_1|(1-p_n)}\right]\|y_n - x_n\|\\
&\leq \left[|c| + \frac{|b_2|(1+\frac{3}{2}v_0(\alpha_n))}{1-v_0(b_n)}\right.\\
&\quad \left. + \frac{1+\frac{3}{2}v_0(\alpha_n)}{|b+b_1|(1-p_n)}\right](\beta_n - \alpha_n)\\
&= \alpha_{n+1} - \beta_n,\\
\|x_{n+1} - x_0\| &\leq \|x_{n+1} - y_n\| + \|y_n - x_0\|\\
&\leq \alpha_{n+1} - \beta_n + \beta_n - \alpha_0 = \alpha_{n+1},
\end{aligned}
$$

$$
\begin{aligned}
F(x_{n+1}) &= F(x_{n+1}) - F(x_n) - \frac{3}{2}F'(x_n)(y_n - x_n),\\
\|L^{-1}F(x_{n+1})\| &\leq (1+\int_0^1 v_0(\|x_n - x_0\| + \theta\|x_{n+1} - x_n\|)d\theta)\|x_{n+1} - x_n\|\\
&\quad + \frac{3}{2}(1+v_0(\|x_n - x_0\|))\|y_n - x_n\|\\
&\leq (1+\int_0^1 v_0(\alpha_n + \theta(\alpha_{n+1} - \alpha_n)d\theta)(\alpha_{n+1} - \alpha_n)\\
&\quad + \frac{3}{2}(1+v_0(\alpha_n))(\beta_n - \alpha_n)\\
&= \gamma_{n+1},
\end{aligned}
$$

$$\begin{aligned}
\|y_{n+1}-x_{n+1}\| &\leq \|F'(x_{n+1})^{-1}L\|\|L^{-1}F(x_{n+1})\| \\
&\leq \frac{\gamma_{n+1}}{1-v_0(\|x_{n+1}-x_0\|)} \\
&\leq \frac{\gamma_{n+1}}{1-v_0(\alpha_{n+1})} \\
&= \beta_{n+1}-\alpha_{n+1}, \\
\|y_{n+1}-x_0\| &\leq \|y_{n+1}-x_{n+1}\|+\|x_{n+1}-x_0\| \\
&\leq \beta_{n+1}-\alpha_{n+1}+\alpha_{n+1}-\alpha_0=\beta_{n+1},
\end{aligned}$$

where we also used

$$\begin{aligned}
&\|((b+b_1)L)^{-1}(A-(b+b_1)L)\| \\
\leq\ & \frac{1}{|b+b_1|}(|b_1|v_0(\|x_n-x_0\|)+|b|v_0(\|y_n-x_0\|)) \\
\leq\ & \frac{1}{|b+b_1|}(|b_1|v_0(\alpha_n)+|b|v_0(\beta_n)) \\
=\ & p_n<1, \\
\|A^{-1}L\| \leq\ & \frac{1}{1-p_n}.
\end{aligned}$$

A convergence criterion for the sequence is

$$v_0(\alpha_n)<1,\ p_n<1 \text{ and } \alpha_n\leq\bar{\rho}<\rho$$

for some $\bar{\rho}\in[0,\rho)$. Then, it follows

$$0\leq\alpha_n\leq\beta_n\leq\alpha_{n+1}\leq\bar{\rho}$$

and there exists $\alpha^*\in[0,\bar{\rho}]$ such that $\lim_{n\longrightarrow\infty}\alpha_n=\alpha^*$. Moreover, $\{x_n\}\in U[x_0,\alpha^*]$ such that $F(x^*)=0$.

4. References

[1] Argyros, I. K., *Convergence and Applications of Newton-Type Iterations*, Springer-Verlag, New York, 2008.

[2] Argyros, I. K., George, S., *Mathematical modeling for the solution of equations and systems of equations with applications*, Volume IV, Nova Publishers, NY, (2020).

[3] Argyros, I. K., George, S., *On the complexity of extending the convergence region for Traubs method*, Journal of Complexity, 56, (2022), Article 101423.

[4] Cordero, A., Garca-Maim, J., Torregrosa, J. R., Vassileva, M. P., Vindel, P., *Chaos in Kings iterative family*, Applied Mathematics Letters, 26, 8, (2013), 842-848.

[5] Cordero, A., Segura, E., Torregrosa, J. R., *Behind Jarratt's steps. Is Jarratt's scheme the best version of itself?*, Discrete Dynamics in Nature and Society, 2023, Article ID 8840525, https://doi.org/10.1155/2023/8840525.

[6] Chun, C., *Some fourth-order iterative methods for solving nonlinear equations*, Applied Mathematics and Computation, 195, 2, (2008), 454-459.

[7] Jarratt, P., *Some fourth-order multipoint iterative methods for solving equations*, BIT, 9 (2), (1969), 119-124.

[8] Kung, H. T., Traub, J. F., *Optimal order of one-point and multipoint iteration*, Journal of the ACM, 21, 4, (1974), 643-651.

[9] Ostrowski, A. M., *Solutions of Equations and System of Equations*, Academic Press, New York, NY, USA, (1960).

Chapter 23

Two Eighth-Order Methods

1. Introduction

The methods are defined for $x_0 \in \Omega$ and each $n = 0, 1, 2, \ldots$ by

$$\begin{aligned}
y_n &= x_n - \frac{1}{2}F(x_n)^{-1}F(x_n), \\
t_n &= x_n - F(x_n)^{-1}F(x_n), \\
A_n &= (F'(x_n) - 3F'(t_n))^{-1}, \\
u_n &= \frac{1}{3}(4y_n - x_n) + A_nF(x_n), \\
z_n &= u_n + 2A_nF(u_n) \\
x_{n+1} &= z_n + 2A_nF(z_n),
\end{aligned} \tag{23.1}$$

and

$$\begin{aligned}
y_n &= x_n - F(x_n)^{-1}F(x_n), \\
B_n &= F'(x_n)^{-1}(F'(x_n) - F'(y_n)), \\
z_n &= y_n - [I + (I + \frac{5}{4}B_n)B_n]F'(x_n)^{-1}F(y_n), \\
x_{n+1} &= z_n - [I + (I + \frac{3}{2}B_n)B_n]F'(x_n)^{-1}F(z_n),
\end{aligned} \tag{23.2}$$

The Taylor expansion series approach and assumptions reaching the ninth derivative of operator F are used in [7] and [13], respectively to show the local eighth order of convergence fro these methods. Other limitations of this approach are reported previous Chapters. Our local convergence approach uses the operators on the methods. Hence, the applicability of both methods is extended. The more interesting semi-local convergence not given previously is also presented in this Chapter.

2. Local Convergence

Method (23.1)

The majorant functions are

$$\begin{aligned} g_1(t) &= \frac{\int_0^1 \omega((1-\theta)t)d\theta + \frac{1}{2}(1+\int_0^1 \omega_0(\theta t)d\theta)}{1-\omega_0(t)}, \\ \bar{g}_1(t) &= \frac{\int_0^1 \omega((1-\theta)t)d\theta}{1-\omega_0(t)}, \\ q(t) &= \frac{1}{2}(\omega_0(t)+3\omega_0(\bar{g}_1(t)t)), \\ \bar{g}_2(t) &= \frac{1}{3}(1+4g_1(t)) + \frac{1+\int_0^1 \omega_0(\theta t)d\theta}{2(1-q(t))}, \end{aligned}$$

$$\begin{aligned} \bar{\omega}(t) &= \begin{cases} \omega((1+\bar{g}_1(t))t), \\ \omega_0(t)+\omega_0(\bar{g}_1(t)t), \end{cases} \\ \bar{\bar{\omega}}(t) &= \begin{cases} \omega((\bar{g}_1(t)+\bar{g}_2(t))t), \\ \omega_0(\bar{g}_1(t)t)+\omega_0(\bar{g}_2(t)t), \end{cases} \end{aligned}$$

$$\begin{aligned} g_2(t) &= \left[\frac{\int_0^1 \omega((1-\theta)\bar{g}_2(t)t)d\theta}{1-\omega_0(\bar{g}_2(t)t)} + \frac{1}{2(1-q(t))}\right. \\ &\quad \left.\times(\bar{\omega}(t)+2\bar{\bar{\omega}}(t))(1+\omega_0(\theta\bar{g}_2(t)t)d\theta)\right]\bar{g}_2(t), \\ \bar{\bar{\bar{\omega}}}(t) &= \begin{cases} \omega((\bar{g}_1(t)+g_2(t))t), \\ \omega_0(\bar{g}_1(t)t)+\omega_0(g_2(t)t) \end{cases} \end{aligned}$$

and

$$\begin{aligned} g_3(t) &= \left[\frac{\int_0^1 \omega((1-\theta)g_2(t)t)d\theta}{1-\omega_0(g_2(t)t)} + \frac{1}{2(1-q(t))(1-\omega_0(g_2(t)t))}\right. \\ &\quad \left.\times(\bar{\omega}(t)+2\bar{\bar{\bar{\omega}}}(t))(1+\omega_0(\theta g_2(t)t)d\theta)\right]g_2(t). \end{aligned}$$

We suppose the equations $g_i(t)-1=0, i=1,2,3$ have smallest positive solutions r_i. set

$$r = \min\{r_i\}$$

and

$$R = \max\{r, \bar{g}_2(r)r\}.$$

The condition

$$U(x^*, R) \subset \Omega$$

also must replace the usual condition

$$U(x^*, r) \subset \Omega.$$

Method (23.2)

The motivation follows from the calculations:

$$
\begin{aligned}
y_n - x^* &= x_n - x^* - F'(x_n)^{-1}F(x_n) + \frac{1}{2}F'(x_n)^{-1}F(x_n), \\
\|y_n - x^*\| &\leq \frac{1}{1-\omega_0(\|x_n - x^*\|)}\left[\int_0^1 \omega((1-\theta)\|x_n - x^*\|))d\theta\right. \\
&\quad \left.+\frac{1}{2}(1+\int_0^1 \omega_0(\theta\|x_n - x^*\|)d\theta)\right]\|x_n - x^*\| \\
&\leq g_1(\|x_n - x^*\|)\|x_n - x^*\| \\
&\leq \|x_n - x^*\| < r, \\
t_n - x^* &= x_n - x^* - F'(x_n)^{-1}F(x_n), \\
\|t_n - x^*\| &\leq \frac{\int_0^1 \omega((1-\theta)\|x_n - x^*\|)d\theta}{1-\omega_0(\|x_n - x^*\|)} \\
&\leq \bar{g}_1(\|x_n - x^*\|)\|x_n - x^*\| \leq \|x_n - x^*\|,
\end{aligned}
$$

$$
\begin{aligned}
u_n x^* &= \frac{1}{3}(4y_n - x_n) - x^* + A_n F(x_n), \\
\|(2L)^{-1}(A_n - 2L)\| &\leq \frac{1}{2}(\omega_0(\|x_n - x^*\|) + 3\omega_0(\|t_n - x^*\|)\| \\
&\leq q_n < 1, \\
\|A_n^{-1}L\| &\leq \frac{1}{2(1-q_n)}, \\
\|u_n - x^*\| &= \|\frac{1}{3}(4y_n - x_n) - x^* + A_n F(x_n)\| \\
&\leq \left[\frac{1}{3}(1+4g_1(\|x_n - x^*\|)))\right. \\
&\quad \left.+\frac{1+\int_)^1 \omega_0(\theta\|x_n - x^*\|)d\theta}{2(1-q_n)}\right]\|x_n - x^*\| \\
&\leq g_2(\|x_n - x^*\|)\|x_n - x^*\| \\
&\leq g_2(r)r \leq R,
\end{aligned}
$$

$$\begin{aligned}
z_n - x^* &= u_n - x^* - F'(u_n)^{-1}F(u_n) \\
&\quad +(F'(u_n)^{-1} + 2A_n)F(u_n) \\
&= u_n - x^* - F'(u_n)^{-1}F(u_n) \\
&\quad +F'(u_n)^{-1}(A_n^{-1} + 2F'(u_n))A_nF(u_n) \\
&= u_n - x^* - F'(u_n)^{-1}F(u_n) \\
&\quad +F'(u_n)^{-1}[(F'(x_n) - F'(t_n))] + 2(F'(u_n) - F'(t_n))A_nF(u_n), \\
\|z_n - x^*\| &\leq \left[\frac{\int_0^1 \omega((1-\theta)\|u_n - x^*\|)d\theta}{1 - \omega_0(\|u_n - x^*\|)}\right. \\
&\quad \left. + \frac{(\bar{\omega}_n + 2\bar{\bar{\omega}}_n)(1 + \int_0^1 \omega_0(\theta\|u_n - x^*\|)d\theta)}{2(1 - q_n)}\right]\|u_n - x^*\| \\
&\leq g_2(\|x_n - x^*\|)\|x_n - x^*\| \leq \|x_n - x^*\|,
\end{aligned}$$

$$\begin{aligned}
x_{n+1} - x^* &= z_n - x^* - F'(z_n)^{-1}F(y_n) \\
&\quad +F'(z_n)^{-1}[(F'(x_n) - F'(t_n)) + 2(F'(z_n) - F'(t_n))]A_nF(z_n),
\end{aligned}$$

and

$$\begin{aligned}
\|x_{n+1} - x^*\| &\leq \left[\frac{\int_0^1 \omega((1-\theta)\|z_n - x^*\|)d\theta}{1 - \omega_0(\|z_n - x^*\|)}\right. \\
&\quad \left. + \frac{(\bar{\omega}_n + 2\bar{\bar{\omega}}_n)(1 + \int_0^1 \omega_0(\theta\|z_n - x^*\|))d\theta}{2(1 - \omega_0(\|z_n - x^*\|))(1 - q_n)}\right]\|z_n - x^*\| \\
&\leq g_3(\|x_n - x^*\|)\|x_n - x^*\| \leq \|x_n - x^*\|.
\end{aligned}$$

Method (23.2) The majorizing functions are motivated by the estimates:

$$\begin{aligned}
y_n - x^* &= x_n - x^* - F'(x_n)^{-1}F(x_n), \\
\|y_n - x^*\| &\leq \frac{\int_0^1 \omega((1-\theta)\|x_n - x^*\|))d\theta\|x_n - x^*\|}{1 - \omega_0(\|x_n - x^*\|)} \\
&\leq g_1(\|x_n - x^*\|)\|x_n - x^*\| \\
&\leq \|x_n - x^*\|,
\end{aligned}$$

$$\begin{aligned}
z_n - x^* &= y_n - x^* - F'(y_n)^{-1}F(y_n) \\
&\quad +(F'(y_n)^{-1} - F'(x_n)^{-1})F(y_n) \\
&\quad (I + \frac{5}{4}B_n)B_nF'(x_n)^{-1}F(y_n) \\
\|z_n - x^*\| &\leq \left[\frac{\int_0^1 \omega((1-\theta)\|y_n - x^*\|)d\theta}{1 - \omega_0(\|y_n - x^*\|)}\right. \\
&\quad + \frac{\bar{\omega}_n(1 + \int_0^1 \omega_0(\theta\|y_n - x^*\|)d\theta)}{(1 - \omega_0(\|x_n - x^*\|))(1 - \omega_0(\|y_n - x^*\|))} \\
&\quad \left. + \frac{\bar{\omega}_n}{(1 - \omega_0(\|x_n - x^*\|))^2}\right]\|u_n - x^*\| \\
&\leq g_2(\|x_n - x^*\|)\|x_n - x^*\| \leq \|x_n - x^*\|,
\end{aligned}$$

$$\begin{aligned} x_{n+1}-x^* &= z_n-x^*-F'(z_n)^{-1}F(z_n) \\ &\quad +(F'(z_n)^{-1}-F'(x_n)^{-1})F(z_n) \\ &\quad -(I+\frac{3}{2}B_n)B_nF'(x_n)^{-1}F(z_n), \end{aligned}$$

and

$$\begin{aligned} \|x_{n+1}-x^*\| &\leq \Bigg[\frac{\int_0^1 \omega((1-\theta)\|z_n-x^*\|)d\theta}{1-\omega_0(\|z_n-x^*\|)} \\ &\quad +\frac{\bar{\bar{\omega}}_n(1+\int_0^1 \omega_0(\theta\|z_n-x^*\|))d\theta}{(1-\omega_0(\|z_n-x^*\|))(1-\omega_)(\|x_n-x^*\|))} \\ &\quad +\frac{\bar{\omega}_n}{(1-\omega_0(\|x_n-x^*\|))^2}(1+\frac{5}{4}\frac{\bar{\omega}_n}{(1-\omega_0(\|x_n-x^*\|))^2}) \\ &\quad \times(1+\int_0^1 \omega_0(\theta\|z_n-x^*\|)d\theta\Bigg]\|z_n-x^*\| \\ &\leq g_3(\|x_n-x^*\|)\|x_n-x^*\| \leq \|x_n-x^*\|, \end{aligned}$$

provided that

$$g_1(t) = \frac{\int_0^1 \omega((1-\theta)t)d\theta}{1-\omega_0(t)},$$

$$\begin{aligned} \bar{\omega}(t) &= \begin{cases} \omega((1+\bar{g}_1(t))t), \\ \omega_0(t)+\omega_0(\bar{g}_1(t)t), \end{cases} \\ \bar{g}_2(t) &= \Bigg[\frac{\int_0^1 \omega((1-\theta)t)d\theta}{(1-\omega_0(t))}+\frac{\bar{\omega}(t)(1+\int_0^1 \omega_0(\theta t)d\theta)}{(1-\omega_0(t))(1-\omega_0(g_1(t)t))}+\frac{\bar{\omega}(t)}{(1-\omega_0(t))^2}(1+\frac{5}{4}\frac{\bar{\omega}(t)}{1-\omega_0(t)}) \\ &\quad \times(1+\int_0^1 \omega_0(\theta g_1(t)t)d\theta)\Bigg]g_1(t) \end{aligned}$$

and

$$\begin{aligned} g_3(t) &= \Bigg[\frac{\int_0^1 \omega((1-\theta)g_2(t)t)d\theta}{1-\omega_0(g_2(t)t)}+\frac{\bar{\bar{\omega}}(t)(1+\int_0^1 \omega_0(\theta g_2(t)t)d\theta)}{(1-\omega_0(t))(1-\omega_0(g_2(t)t))}+\frac{\bar{\omega}(t)}{(1-\omega_0(t))^2}(1+\frac{5}{4}\frac{\bar{\omega}(t)}{1-\omega_0(t)}) \\ &\quad \times(1+\omega_0(\theta g_2(t)t)d\theta)\Bigg]g_2(t), \end{aligned}$$

where

$$\bar{\bar{\omega}}(t) = \begin{cases} \omega(1+\bar{g}_2(t))t), \\ \omega_0(t)+\omega_0(g_2(t)t). \end{cases}$$

3. Semi-local Convergence

The role of x^*, ω is exchanged by x_0, v functions, respectively.

Method (23.1) The majorizing sequence $\{\alpha_n\}$ is defined for $\alpha_0 = 0, \beta_0 \geq \frac{1}{2}\|F'(x_0)^{-1}F(x_0)\|$

and $\bar{\alpha}_0 \geq \|F'(x_0)^{-1}F(x_0)\|$ by

$$\begin{aligned}
p_n &= \frac{1}{2}(v_0(\alpha_n)+3v_0(\bar{\alpha}_n)),\\
\bar{\beta}_n &= \beta_n+\frac{1}{3}(1+\frac{1+v_0(\alpha_n)}{1-p_n})(\beta_n-\alpha_n),\\
\lambda_n &= (1+\int_0^1 v_0(\alpha_n+\theta(\bar{\beta}_n-\alpha_n))d\theta)(\bar{\beta}_n-\alpha_n)+2(1+v_0(\alpha_n))(\beta_n-\alpha_n),\\
\gamma_n &= \bar{\beta}_n+\frac{\lambda_n}{1-p_n},\\
\mu_n &= (1+\int_0^1 v_0(\alpha_n+\theta(\gamma_n-\alpha_n))d\theta)(\gamma_n-\alpha_n)+2(1+v_0(\alpha_n))(\beta_n-\alpha_n),\\
\alpha_{n+1} &= \gamma_n+\frac{\mu_n}{1-p_n}\\
\delta_{n+1} &= (1+\int_0^1 v_0(\alpha_n+\theta(\alpha_{n+1}-\alpha_n))d\theta(\alpha_{n+1}-\alpha_n)\\
&\quad +2(1+v_0(\alpha_n))(\beta_n-\alpha_n),\\
\beta_{n+1} &= \alpha_{n+1}+\frac{\delta_{n+1}}{2(1-v_0(\alpha_{n+1})},\\
\bar{\delta}_{n+1} &= \int_0^1 v_0(\theta(\alpha_{n+1}-\alpha_n)d\theta(\alpha_{n+1}-\alpha_n)+(1+v_0(\alpha_n))(\alpha_{n+1}-\bar{\alpha}_n)
\end{aligned}$$

and

$$\bar{\alpha}_{n+1}=\alpha_{n+1}+\frac{\bar{\delta}_{n+1}}{1-v_0(\alpha_{n+1})}.$$

The motivation for the introduction of the majorizing sequence $\{\alpha_n\}$ is:

$$\begin{aligned}
u_n-y_n &= \frac{1}{3}(y_n-x_n)+2A_nF'(x_n)(y_n-x_n),\\
\|(2L)^{-1}(A_n-L)\| &\leq p_n<1,\\
p_n &= \frac{1}{2}(v_0(\alpha_n)+3v_0(\bar{\alpha}_n)),\\
\|u_n-y_n\| &\leq \frac{1}{3}(1+\frac{1+v_0(\alpha_n)}{1-p_n})(\beta_n-\alpha_n)\\
&\leq \bar{\beta}_n-\beta_n\\
\ell_n &= (1+\int_0^1 v_0(\alpha_n+\theta(\bar{\beta}_n-\alpha_n))d\theta(\bar{\beta}_n-\alpha_n)\\
&\quad +2(1+v_0(\alpha_n))(\beta_n-\alpha_n),\\
\|z_n-u_n\| &\leq \frac{\ell_n}{1-p_n}=\gamma_n-\bar{\beta}_n,\\
\bar{\ell}_n &= (1+\int_0^1 v_0(\alpha_n+\theta(\gamma_n-\alpha_n))d\theta(\gamma_n-\alpha_n)\\
&\quad +2(1+v_0(\alpha_n))(\beta_n-\alpha_n),
\end{aligned}$$

$$\begin{aligned}
\|x_{n+1}-z_n\| &\leq \frac{\bar{\ell}_n}{1-p_n}\\
&\leq \alpha_{n+1}-\gamma_n,\\
F(x_{n+1}) &= F(x_{n+1})-F(x_n)-2F'(x_n)(y_n-x_n),\\
\|LF(x_{n+1})\| &\leq (1+\int_0^1 v_0(\alpha_n+\theta(\alpha_{n+1}-\alpha_n))d\theta)(\alpha_{n+1}-\alpha_n)\\
&\quad +2(1+v_0(\alpha_n))(\beta_n-\alpha_n)=\delta_{n+1},\\
\|y_{n+1}-x_{n+1}\| &\leq \frac{\delta_{n+1}}{1-v_0(\alpha_{n+1})}\\
&= \beta_{n+1}-\alpha_{n+1},\\
F(x_{n+1}) &= F(x_{n+1})-F(x_n)-F'(x_n)(t_n-x_n),\\
&= F(x_{n+1})-F(x_n)-F'(x_n)(x_{n+1}-x_n)+F'(x_{n+1}-t_n),
\end{aligned}$$

$$\begin{aligned}
\|L^{-1}F(x_{n+1})\| &\leq \int_0^1 v_0(\theta(\alpha_{n+1}-\alpha_n)d\theta(\alpha_{n+1}-\alpha_n)\\
&\quad +(1+v_0(\alpha_n))(\alpha_{n+1}-\bar{\alpha}_n)\\
&= \bar{\delta}_{n+1}
\end{aligned}$$

and

$$\begin{aligned}
\|t_{n+1}-x_{n+1}\| &\leq \|F'(x_{n+1})^{-1}L\|\|L^{-1}F(x_{n+1})\|\\
&\leq \frac{\bar{\delta}_{n+1}}{1-v_0(\alpha_{n+1})}\\
&= \bar{\alpha}_{n+1}-\alpha_{n+1}.
\end{aligned}$$

The rest follows as in the previous Chapters.

Method (23.2) The majorant sequence is defined for $\alpha_0=0$, $\beta_0\geq\|F'(x_0)^{-1}F(x_0)\|$ and each $n=0,1,2,\ldots$ by

$$\bar{v}(t)=\begin{cases} v(\beta_n-\alpha_n),\\ v_0(\alpha_n)+v_0(\beta_n),\end{cases}$$

$$\begin{aligned}
\gamma_n &= \beta_n+\frac{1}{4}\left[4+4\frac{\bar{v}_n}{1-v_0(\alpha_n)}+5\left(\frac{\bar{v}_n}{1-v_0(\alpha_n)}\right)^2\right]\times\frac{\int_0^1 v((1-\theta)(\beta_n-\alpha_n))d\theta(\beta_n-\alpha_n)}{1-v_0(\alpha_n)},\\
\lambda_n &= \int_0^1 v((1-\theta)(\beta_n-\alpha_n))d\theta(\beta_n-\alpha_n)+(1+\int_0^1 v_0(\beta_n+\theta(\gamma_n-\beta_n))d\theta)(\gamma_n-\beta_n),\\
\mu_n &= \frac{1}{4}\left[4+4\frac{\bar{v}_n}{1-v_0(\alpha_n)}+5\left(\frac{\bar{v}_n}{1-v_0(\alpha_n)}\right)^2\right],\\
\alpha_{n+1} &= \gamma_n+\frac{\lambda_n\mu_n}{1-v_0(\alpha_n)},\\
\delta_{n+1} &= \int_0^1 v((1-\theta)(\alpha_{n+1}-\alpha_n))d\theta(\alpha_{n+1}-\alpha_n))\\
&\quad +(1+v_0(\alpha_n))(\alpha_{n+1}-\beta_n))
\end{aligned}$$

and

$$\beta_{n+1}=\alpha_{n+1}+\frac{\delta_{n+1}}{1-v_0(\alpha_{n+1})}.$$

The motivation follows from the estimates:

$$
\begin{aligned}
& \|F'(x_n)^{-1}(F'(x_n)-F'(y_n))\| \\
\leq\; & \|F'(x_n)^{-1}L\|\|L^{-1}(F'(x_n)-F'(y_n))\| \\
\leq\; & \frac{\bar{v}_n}{1-v_0(\alpha_n)},
\end{aligned}
$$

$$
\begin{aligned}
z_n-y_n &= -[I+(I+\frac{3}{2}B_n)B_n]F'(x_n)^{-1}F(y_n),\\
\|z_n-y_n\| &\leq \frac{1}{4}\left[4+4\frac{\bar{v}_n}{1-v_0(\alpha_n)}+5\left(\frac{\bar{v}_n}{1-v_0(\alpha_n)}\right)^2\right]\\
&\quad\times\frac{\int_0^1 v((1-\theta)(\beta_n-\alpha_n))d\theta(\beta_n-\alpha_n)}{1-v_0(\alpha_n)},\\
&\leq \gamma_n-\beta_n,\\
x_{n+1}-z_n &= -[I+(I+\frac{3}{2}B_n)B_n]F'(x_n)^{-1}F(z_n),\\
\|x_{n+1}-z_n\| &\leq \frac{1}{4(1-v_0(\alpha_n))}\left[4+4\frac{\bar{v}_n}{1-v_0(\alpha_n)}+5\left(\frac{\bar{v}_n}{1-v_0(\alpha_n)}\right)^2\right]\lambda_n\\
&= \alpha_{n+1}-\gamma_n,
\end{aligned}
$$

$$
\begin{aligned}
F(x_{n+1}) &= F(x_{n+1})-F(x_n)-F'(x_n)(x_{n+1}-x_n)+F'(x_{n+1}-y_n),\\
\|L^{-1}F(x_{n+1})\| &\leq \delta_{n+1},\\
\|y_{n+1}-x_{n+1}\| &\leq \frac{\delta_{n+1}}{1-v_0(\|x_{n+1}-x_0\|)}\\
&\leq \frac{\delta_{n+1}}{1-v_0(\alpha_{n+1})}\\
&= \beta_{n+1}-\alpha_{n+1}.
\end{aligned}
$$

The rest can be found in previous Chapters.

4. References

[1] Argyros, C. I., Regmi, S., Argyros, I. K., George, S., *Contemporary Algorithms: Theory and Applications*, Volume III, NOVA Publishers, (2023).

[2] Argyros, I. K., *Advances in the Efficiency of Computational Methods and Applications*, World Scientific Publishing, Oxon, United Kingdom, (2000).

[3] Argyros, I. K., *The theory and applications of iteration methods*, 2nd Edition, CRC Press, Engineering Series, Taylor and Francis Group, (2022).

[4] Argyros, I. K., George, S., *Ball comparison between two optimal eight-order methods under weak conditions*, SeMA Journal, 72, (2015), 1-11.

[5] Behl, R., Cordero, A., Torregrosa, J. R., *High order family of multivariate iterative methods: Convergence and stability*, J. Comput. Appl. Math., (2020), 113053.

[6] Behl, R., Salimi, M., Ferrara, M., Sharifi, S., Alharbi, S. K., *Some Real-Life Applications of a Newly Constructed Derivative Free Iterative Scheme*, Symmetry, 11, (2019), 239.

[7] Cordero, A., Gómez, E., Torregrosa, J. R., *Efficient High-order iterative methods for solving nonlinear systems and their application on Heat Conduction Problems*, Complexity, (2017), Article ID 6457532.

[8] Darvishi, M. T., Barati, A., *A fourth-order method from quadrature formulae to solve systems of nonlinear equations*, Appl. Math. Comput., 188, (2007), 257-261.

[9] Grau-Sánchez, M., Peris, M. P., Gutiérrez, J. M., *Accelerated iterative methods for finding solutions of system of nonlinear equations*, J. Comput. Appl. Math., 190, (2007), 1815-1823.

[10] Ortega, J. M., Rheinbolt, W. C., *Iterative Solution of Nonlinear Equations in Several Variables*, Academic Press, New York, NY, USA, (1970).

[11] Sharma, J. R., Arora, H., *A novel derivative-free algorithm with seventh order convergence for solving systems of nonlinear equations*, Numer. Algorithms, 4, (2014), 917-933.

[12] Sharma, J. R., Arora, H., *Efficient Jarratt-like methods for solving systems of nonlinear equations*, Calcolo, 51, (2014), 193210.

[13] Wang, X., *Fixed-point iterative method with eighth order constructed by undetermined parameter technique for solving nonlinear systems*, Symmetry, 13(5), (2021), 863, https://doi.org/10.3390/sym13050863.

Chapter 24

Efficient Sixth-Order Methods with Derivatives

1. Introduction

The methods are:

$$\begin{aligned}
y_n &= x_n - \frac{2}{3}F(x_n)^{-1}F(x_n),\\
A_n &= 3F'(y_n) - F'(x_n),\\
z_n &= x_n - \frac{1}{2}A_n^{-1}(3F'(y_n)+F'(x_n))F'(x_n)^{-1}F(x_n) \qquad (24.1)\\
x_{n+1} &= z_n - (\frac{1}{2}A_n^{-1}(3F'(y_n)+F'(x_n)))^2F'(x_n)^{-1}F(z_n),
\end{aligned}$$

and

$$\begin{aligned}
y_n &= x_n - \frac{2}{3}F(x_n)^{-1}F(x_n),\\
z_n &= x_n - [\frac{5}{8}I + \frac{3}{8}(F'(y_n)^{-1}F'(x_n))^2]F'(x_n)^{-1}F(x_n),\\
x_{n+1} &= z_n - [-\frac{9}{4}I + \frac{15}{8}F'(y_n)^{-1}F'(x_n)\\
&\quad + \frac{11}{8}F'(x_n)^{-1}F'(y_n)]F'(y_n)^{-1}F(z_n). \qquad (24.2)
\end{aligned}$$

The Taylor expansion series approach is used in [7] and [10], respectively to show the local sixth order of convergence when $B_1 = B_2 = \mathbb{R}^k$ and under assumptions reaching the seventh derivative. The new local convergence analysis relies on F' which only appears on the methods. The more interesting semi-local convergence analysis not given before is also presented in this chapter. The same benefits can be obtained if the present methodology is applied on other methods [1-6,8,9].

2. Local Convergence

Method (24.1)
The majorant functionsg_i, $i = 1,2,3$ are:

$$\begin{aligned}
g_1(t) &= \frac{\int_0^1 \omega((1-\theta)t)d\theta + \frac{1}{3}(1+\int_0^1 \omega_0(\theta t)d\theta)}{1-\omega_0(t)},\\
\bar{\omega}(t) &= \begin{cases} \omega((1+g_1(t))t), \\ \omega_0(t)+\omega_0(g_1(t)t), \end{cases}\\
q(t) &= \frac{1}{2}(\omega_0(t)+3\omega_0(g_1(t)t)),\\
a(t) &= \frac{3}{4}\frac{\bar{\omega}(t)}{1-q(t)},\\
g_2(t) &= \frac{\int_0^1 \omega((1-\theta)t)d\theta}{1-\omega_0(t)}\\
&+\frac{a(t)(a(t)(1+\int_0^1 \omega_0(\theta t)d\theta}{1-\omega_0(t)},\\
\bar{\bar{\omega}}(t) &= \begin{cases} \omega((1+g_2(t))t), \\ \omega_0(t)+\omega_0(g_2(t)t), \end{cases}
\end{aligned}$$

and

$$g_3(t) =$$
$$= \left[\frac{\int_0^1 \omega((1-\theta)g_2(t)t)d\theta}{1-\omega_0(g_2(t)t)} + \left(\frac{\bar{\bar{\omega}}(t)}{(1-\omega_0(t))(1-\omega_0(g_2(t)t))}+b(t)\right)\frac{1+\int_0^1 \omega_0(\theta g_2(t)t)d\theta}{1-\omega_0(t)}\right]*$$
$$*g_2(t).$$

The motivation for the introduction of the functions g_i follows in turn:

$$\begin{aligned}
y_n - x^* &= x_n - x^* - F'(x_n)^{-1}F(x_n) + \frac{1}{3}F'(x_n)^{-1}F(x_n),\\
\|y_n - x^*\| &\leq \frac{1}{1-\omega_0(\|x_n-x^*\|)}\left[\int_0^1 \omega((1-\theta)\|x_n-x^*\|)d\theta\right.\\
&\left.+\frac{1}{3}(1+\int_0^1 \omega_0(\theta\|x_n-x^*\|)d\theta)\right]\|x_n-x^*\|\\
&\leq g_1(\|x_n-x^*\|)\|x_n-x^*\|\\
&\leq \|x_n-x^*\| < r,\\
z_n - x^* &= x_n - x^* - F'(x_n)^{-1}F(x_n)\\
&+\frac{1}{2}A_n^{-1}(2A_n-(3F'(y_n)+F'(x_n)))F'(x_n)^{-1}F(x_n)\\
&= x_n - x^* - F'(x_n)^{-1}F(x_n)\\
&+\frac{3}{2}A_n^{-1}(F'(y_n)-F'(x_n))F'(x_n)^{-1}F(x_n)
\end{aligned}$$

$$\begin{aligned}
\|z_n - x^*\| &\leq \left[\frac{\int_0^1 \omega((1-\theta)\|x_n - x^*\|)d\theta}{1 - \omega_0(\|x_n - x^*\|)} \right. \\
&\quad \left. + \frac{3\bar{\omega}_n(1 + \int_0^1 \omega_0(\theta\|x_n - x^*\|)d\theta)}{4(1-q_n)(1 - \omega_0(\|x_n - x^*\|))} \right] \|x_n - x^*\| \\
&\leq g_2(\|x_n - x^*\|)\|x_n - x^*\| \leq \|x_n - x^*\|,
\end{aligned}$$

$$\begin{aligned}
x_{n+1} - x^* &= z_n - x^* - F'(z_n)^{-1}F(z_n) \\
&\quad + (F'(z_n)^{-1} - F'(x_n)^{-1})F(z_n) \\
&\quad + [I - (\frac{1}{2}A_n^{-1}(3F'(y_n) + F'(x_n))^2]F'(x_n)^{-1}F(z_n),
\end{aligned}$$

$$\begin{aligned}
\|x_{n+1} - x^*\| &\leq \left[\frac{\int_0^1 \omega((1-\theta)\|z_n - x^*\|)d\theta}{1 - \omega_0(\|z_n - x^*\|)} \right. \\
&\quad + \left(\frac{\bar{\bar{\omega}}_n}{(1 - \omega_0(\|z_n - x^*\|))(1 - \omega_0(\|x_n - x^*\|))} + b_n \right) \\
&\quad \left. \times \frac{1 + \int_0^1 \omega_0(\theta\|z_n - x^*\|)d\theta}{1 - \omega_0(\|x_n - x^*\|)} \right] \|z_n - x^*\| \\
&\leq g_3(\|x_n - x^*\|)\|x_n - x^*\| \leq \|x_n - x^*\|,
\end{aligned}$$

where we also used

$$\|(2L)^{-1}(A_n - 2L)\| \leq \frac{1}{2}(\|L^{-1}(F'(x_n) - L)\| + 3\|L^{-1}(F'(y_n) - L)\| = q_n < 1,$$

$$\|A_n^{-1}L\| \leq \frac{1}{2(1-q_n)},$$

$$\begin{aligned}
& I - [\frac{1}{2}A_n^{-1}(3F'(y_n) + F'(x_n))]^2 \\
= \; & [I - \frac{1}{2}A_n^{-1}(3F'(y_n) + F'(x_n))]\{[I - \frac{1}{2}A_n^{-1}(3F'(y_n) + F'(x_n))] \\
& + A_n^{-1}(3F'(y_n) + F'(x_n))\}, \\
& \|I - [\frac{1}{2}A_n^{-1}(3F'(y_n) + F'(x_n))]^2\| \\
\leq \; & a_n \left(a_n + \frac{4 + \omega_0(\|x_n - x^*\|) + 3\omega_0(\|y_n - x^*\|)}{2(1-q_n)} \right) = b_n.
\end{aligned}$$

Method (24.2)

The functions $g_1, i = 1,2,3$ are:

$$g_1(t) = \frac{\int_0^1 \omega((1-\theta)t)d\theta + \frac{1}{3}(1 + \int_0^1 \omega_0(\theta t)d\theta)}{1 - \omega_0(t)},$$

$$\begin{aligned} g_2(t) &= \frac{\int_0^1 \omega((1-\theta)t)d\theta}{1-\omega_0(t)} \\ &+\frac{3}{8}\left(\frac{\bar{\omega}(t)^2}{(1-\omega_0(t))(1-\omega_0(g_1(t)t))}\right. \\ &\left.+\frac{2\bar{\omega}(t)}{1-\omega_0(g_1(t)t)}\right)(1+\int_0^1 \omega_0(\theta g_1(t)t)d\theta) \end{aligned}$$

and

$$\begin{aligned} g_3(t) &= \left[\frac{\int_0^1 \omega((1-\theta)t)d\theta}{1-\omega_0(t)}\right. \\ &+\left(\frac{\bar{\bar{\omega}}(t)}{(1-\omega_0(g_1(t)t))(1-\omega_0(g_2(t)t))}\right. \\ &+\frac{1}{8}\left(\frac{15\bar{\omega}(t)}{1-\omega_0(g_1(t)t)}+\frac{11\bar{\omega}(t)}{1-\omega_0(t)}\right) \\ &\left.\left.\times\frac{1}{1-\omega_0(g_1(t)t)}\right)\left(1+\int_0^1 \omega_0(\theta g_2(t)t)d\theta\right)\right] g_2(t), \end{aligned}$$

where

$$\bar{\bar{\omega}}(t)=\begin{cases} \omega((\bar{g}_1(t)+g_2(t))t), \\ \omega_0(g_1(t)t)+\omega_0(g_2(t)t), \end{cases}$$

and $\bar{\omega}(t)$ is as in method (24.1). The estimates leading to the development of these function are (for g_1 see method (24.1)):

$$\begin{aligned} z_n-x^* &= y_n-x^*-F'(x_n)^{-1}F(x_n) \\ &+\frac{3}{8}(I-(F'(y_n)^{-1}F'(x_n))^2)F'(x_n)^{-1}F(x_n) \\ &= y_n-x^*-F'(x_n)^{-1}F(x_n) \\ &+\frac{3}{8}\{(I-F'(y_n)^{-1}F'(x_n))(I-F'(y_n)^{-1}F'(x_n))F'(x_n)^{-1} \\ &+2(I-F'(y_n)^{-1}F'(x_n))F'(y_n)^{-1}\}F(x_n), \\ \|z_n-x^*\| &\leq \left[\frac{\int_0^1 \omega((1-\theta)\|x_n-x^*\|)d\theta}{1-\omega_0(\|x_n-x^*\|)}\right. \\ &+\frac{3}{8}\left(\frac{2\bar{\omega}_n^2}{(1-\omega_0(\|x_n-x^*\|))(1-\omega_0(\|y_n-x^*\|))}+\frac{2\bar{\omega}_n}{1-\omega_0(\|y_n-x^*\|)}\right) \\ &\left.\times(1+\int_0^1 \omega_0(\theta\|y_n-x^*\|)d\theta)\right]\|x_n-x^*\| \\ &\leq g_2(\|x_n-x^*\|)\|x_n-x^*\| \leq \|x_n-x^*\|, \end{aligned}$$

$$\begin{aligned} x_{n+1}-x^* &= z_n-x^*-F'(z_n)^{-1}F(z_n) \\ &+(F'(z_n)^{-1}-F'(x_n)^{-1})F(z_n) \\ &+\frac{1}{8}[15(I-F'(y_n)^{-1}F'(x_n)+11(I-F'(x_n)^{-1}F'(y_n))]F'(y_n)^{-1}F(z_n), \end{aligned}$$

$$\begin{aligned}
\|x_{n+1}-x^*\| &\leq \Bigg[\frac{\int_0^1 \omega((1-\theta)\|z_n-x^*\|)d\theta}{1-\omega_0(\|z_n-x^*\|)} \\
&+\Bigg(\frac{\bar{\bar{\omega}}_n}{(1-\omega_0(\|z_n-x^*\|))(1-\omega_0(\|y_n-x^*\|))} \\
&+\frac{1}{8}\left(\frac{15\bar{\omega}_n}{1-\omega_0(\|y_n-x^*\|)}+\frac{11\bar{\omega}_n}{1-\omega_0(\|x_n-x^*\|)}\right)\frac{1}{1-\omega_0(\|y_n-x^*\|)}\Bigg) \\
&\times(1+\int_0^1 \omega_0(\theta\|z_n-x^*\|)d\theta\Bigg]\|z_n-x^*\| \\
&\leq g_3(\|x_n-x^*\|)\|x_n-x^*\| \leq \|x_n-x^*\|.
\end{aligned}$$

3. Semi-local Convergence

Method (24.1)

The majorizing sequence $\{\alpha_n\}$ is defined for $\alpha_0 = 0, \beta_0 \geq \frac{2}{3}\|F'(x_0)^{-1}F(x_0)\|$ and each $n = 0,1,2,\ldots$ by

$$\begin{aligned}
\gamma_n &= \beta_n + \frac{v_0(\alpha_n)+3\bar{v}_n+4}{1-p_n}(\beta_n-\alpha_n), \\
c_n &= (1+\int_0^1 v_0(\alpha_n+\theta(\gamma_n-\alpha_n))d\theta)(\gamma_n-\alpha_n)+\frac{3}{2}(1+v_0(\alpha_n))(\beta_n-\alpha_n), \\
\alpha_{n+1} &= \gamma_n + \left(\frac{v_0(\alpha_n)+3v_0(\beta_n)+4}{4(1-p_n)}\right)^2\frac{c_n}{1-v_0(\alpha_n)}, \\
\delta_{n+1} &= (1+\int_0^1 v_0(\alpha_n+\theta(\alpha_{n+1}-\alpha_n))d\theta(\alpha_{n+1}-\alpha_n) \\
&+\frac{3}{2}(1+v_0(\alpha_n))(\beta_n-\alpha_n),
\end{aligned}$$

and

$$\beta_{n+1} = \alpha_{n+1} + \frac{2\delta_{n+1}}{1-v_0(\alpha_{n+1})}.$$

The motivation follows in turn by the estimates:

$$\begin{aligned}
z_n - y_n &= \left[\frac{2}{3}I-\frac{1}{2}A_n^{-1}(3F'(y_n)+F'(x_n))\right](-\frac{3}{2}(y_n-x_n)) \\
\|z_n-y_n\| &\leq \frac{(4+3v_0(\beta_n)+v_0(\alpha_n))(\beta_n-\alpha_n)}{1-p_n} = \gamma_n-\beta_n, \\
F(z_n) &= F(z_n)-F(x_n)+F(x_n) \\
&= F(z_n)-F(x_n)-\frac{3}{2}F'(x_n)(y_n-x_n), \\
\|L^{-1}F(z_n)\| &\leq (1+\int_0^1 v_0(\alpha_n+\theta(\gamma_n-\alpha_n))d\theta)(\gamma_n-\alpha_n) \\
&+\frac{3}{2}(1+v_0(\alpha_n))(\beta_n-\alpha_n) = c_n,
\end{aligned}$$

$$\begin{aligned}
\|x_{n+1}-z_n\| &\leq \left(\frac{3\bar{v}_n+v_0(\alpha_n)+4}{4(1-p_n)}\right)^2 \frac{c_n}{1-v_0(\alpha_n)} \\
&\leq \alpha_{n+1}-\gamma_n, \\
F(x_{n+1}) &= F(x_{n+1})-F(x_n)-\frac{3}{2}F'(x_n)(y_n-x_n), \\
\|LF(x_{n+1})\| &\leq (1+\int_0^1 v_0(\alpha_n+\theta(\alpha_{n+1}-\alpha_n))d\theta)(\alpha_{n+1}-\alpha_n) \\
&\quad +\frac{3}{2}(1+v_0(\alpha_n))(\beta_n-\alpha_n)=\delta_{n+1}, \\
\|y_{n+1}-x_{n+1}\| &\leq \frac{2\delta_{n+1}}{3(1-v_0(\alpha_{n+1}))} \\
&= \beta_{n+1}-\alpha_{n+1}.
\end{aligned}$$

Method (24.2)

The majorant sequence $\{\alpha_n\}$ for $\alpha_0=0$, β_0 as in method (24.1) and each $n=0,1,2,\ldots$ is

$$\bar{v}(t)=\begin{cases} v(\beta_n-\alpha_n), \\ v_0(\alpha_n)+v_0(\beta_n), \end{cases}$$

$$\begin{aligned}
\gamma_n &= \beta_n+\frac{1}{16}\left[\frac{3\bar{v}_n}{1-v_0(\alpha_n)}\left(\frac{\bar{v}_n}{1-v_0(\beta_n)}\right.\right. \\
&\quad \left.\left.+\frac{2(1+v_0(\alpha_n))}{1-v_0(\beta_n)}\right)+2\right] \\
&\quad (\beta_n-\alpha_n)), \\
\alpha_{n+1} &= \gamma_n+\left[1+\frac{15\bar{v}_n}{8(1-v_0(\beta_n))}+\frac{11\bar{v}_n}{1-v_0(\alpha_n)}\right]\frac{c_n}{1-v_0(\alpha_n)},
\end{aligned}$$

and

$$\beta_{n+1}=\alpha_{n+1}+\frac{2\delta_{n+1}}{3(1-v_0(\alpha_{n+1}))},$$

where c_n and δ_{n+1} are defined in method (24.1) and

$$\bar{v}_n=\begin{cases} v(\beta_n-\alpha_n) \\ v_0(\alpha_n)+v_0(\beta_n). \end{cases}$$

The motivation follows in turn by the estimates:

$$\begin{aligned}
z_n - y_n &= [\frac{2}{3}I - \frac{5}{8}I - \frac{3}{8}(F'(y_n)^{-1}F'(x_n))^2](-\frac{3}{2}(y_n - x_n)) \\
&= -\frac{1}{16}[3(I - F'(y_n)^{-1}F'(x_n))(I - F'(y_n)^{-1}F'9x_n) \\
&\quad +2F'(y_n)^{-1}F'(x_n) - 2I](y_n - x_n), \\
\|z_n - y_n\| &\leq \frac{1}{16}\left[\frac{3\bar{v}_n}{1 - v_0(\beta_n)}\left(\frac{\bar{v}_n}{1 - v_0(\beta_n)}\right.\right. \\
&\quad \left.\left. + \frac{2(1 + v_0(\alpha_n))}{1 - v_0(\alpha_n)}\right) + 2\right](\beta_n - \alpha_n) \\
&\leq \gamma_n - \beta_n.
\end{aligned}$$

The computation for β_{n+1}, identical to that in method (24.1), is omitted.

4. References

[1] Argyros, C. I., Regmi, S., Argyros, I. K., George, S., *Contemporary Algorithms: Theory and Applications*, Volume III, NOVA Publishers, (2023).

[2] Argyros, I. K., *Advances in the Efficiency of Computational Methods and Applications*, World Scientific Publishing, Oxon, United Kingdom, (2000).

[3] Argyros, I. K., *The theory and applications of iteration methods*, 2nd Edition, CRC Press, Engineering Series, Taylor and Francis Group, (2022).

[4] Argyros, I. K., George, S., *Ball comparison between two optimal eight-order methods under weak conditions*, SeMA Journal, 72, (2015), 1-11.

[5] Cordero, A., Jordan, C., Sanabria-Codesal, E., Torregrosa, J. R., *Highly efficient iterative algorithms for solving nonlinear systems with arbitrary order of convergence $p+3, p \geq 5$*, J. Comput. Appl. Math., 330, (2018), 748-758.

[6] Grau-Snchez, M., Grau, A., Noguera, M., *Ostrowski-type methods for solving systems of nonlinear equations*, Appl. Math. Comput., 218, (2011), 2377-2385.

[7] Hueso, J., Martnez, E., Teruel, C., *Convergence, efficiency and dynamics of new fourth and sixth order families of iterative methods for nonlinear systems*, J. Comput. Appl. Math., 275, (2015), 412-420.

[8] Ortega, J. M., Rheinbolt, W. C., *Iterative Solution of Nonlinear Equations in Several Variables*, Academic Press, New York, NY, USA, (1970).

[9] Sharma, J. R., Arora, H., *On efficient weighted Newton methods for solving systems of nonlinear equations*, Appl. Math. Comput., 222, (2013), 497-506.

[10] Soleymani, F., Lofti, T., Bakhtiari, P., *A multistep class of iterative methods for nonlinear systems*, Optim. Lett., 8, (2014), 1001-1015.

Chapter 25

Methods of Order Six with Derivatives and Divided Differences

1. Introduction

The methods are:

$$
\begin{aligned}
y_n &= x_n - F(x_n)^{-1}F(x_n),\\
z_n &= y_n - (3I - 2F'(x_n)^{-1}[x_n, y_n; F])F'(x_n)^{-1}F(y_n),\\
x_{n+1} &= z_n - (3I - 2F'(x_n)^{-1}[x_n, y_n]F'(x_n)^{-1}F(z_n),
\end{aligned}
\tag{25.1}
$$

and

$$
\begin{aligned}
y_n &= x_n - F(x_n)^{-1}F(x_n),\\
z_n &= y_n - (2[x_n, y_n; F]^{-1} - F'(x_n)^{-1})F(y_n),\\
x_{n+1} &= z_n - (2[x_n, y_n; F]^{-1} - F'(x_n)^{-1})F(z_n).
\end{aligned}
\tag{25.2}
$$

The methods are shown to be of local sixth order of convergence in [5] under the Taylor series expansion approach when $B_1 = B_2 = \mathbb{R}^k$ and provided that the seventh derivative exists. The new local approach depends only on the operators on the methods. Hence, the applicability of both methods is extended in a Banach space setting. Moreover, the more interesting semi-local convergence not studied in [5] is presented in this Chapter. The idea in this Chapter is applicable with the same benefits on other methods [1]-[4], [6]-[10].

2. Local Convergence

Method (25.1)
The majorant functionsg_i, $i = 1, 2, 3$ are:

$$
g_1(t) = \frac{\int_0^1 \omega((1-\theta)t)d\theta}{1-\omega_0(t)},
$$

$$\bar{\omega}(t) = \begin{cases} \omega((1+g_1(t))t), \\ \omega_0(t)+\omega_0(g_1(t)t), \end{cases}$$

$$\begin{aligned} g_2(t) &= \Bigg[\frac{\int_0^1 \omega((1-\theta)g_1(t)t)d\theta}{1-\omega_0(g_1(t)t)} \\ &+\left(\frac{\bar{\omega}(t)}{(1-\omega_0(t))(1-\omega_0(g_1(t)t))}+2\frac{\varphi(t,g_1(t)t)}{(1-\omega_0(t))^2}\right) \\ &\times(1+\int_0^1 \omega_0(\theta g_1(t)t)d\theta\Bigg] g_1(t), \end{aligned}$$

$$\bar{\bar{\omega}}(t) = \begin{cases} \omega((1+g_2(t))t), \\ \omega_0(t)+\omega_0(g_2(t)t), \end{cases}$$

and

$$\begin{aligned} g_3(t) &= \Bigg[\frac{\int_0^1 \omega((1-\theta)g_2(t)t)d\theta}{1-\omega_0(g_2(t)t)} \\ &+\left(\frac{\bar{\bar{\omega}}(t)}{(1-\omega_0(t))(1-\omega_0(g_2(t)t))}+2\frac{\varphi(t,g_1(t)t)}{(1-\omega_0(t))^2}\right) \\ &\times\frac{1+\int_0^1 \omega_0(\theta g_2(t)t)d\theta}{1-\omega_0(t)}\Bigg] g_2(t). \end{aligned}$$

The motivation follows from the estimates

$$\begin{aligned} y_n-x^* &= x_n-x^*-F'(x_n)^{-1}F(x_n), \\ \|y_n-x^*\| &\leq \frac{1}{1-\omega_0(\|x_n-x^*\|)}\left[\int_0^1 \omega((1-\theta)\|x_n-x^*\|))d\theta\right]\|x_n-x^*\| \\ &\leq g_1(\|x_n-x^*\|)\|x_n-x^*\| \\ &\leq \|x_n-x^*\| < r, \\ z_n-x^* &= x_n-x^*-F'(y_n)^{-1}F(y_n)+(F'(y_n)^{-1}-F'(x_n)^{-1})F(y_n) \\ &\quad +2F'(x_n)^{-1}(F'(x_n)-[x_n,y_n;F])F'(x_n)^{-1}F(y_n), \end{aligned}$$

$$\begin{aligned} \|z_n-x^*\| &\leq \Bigg[\frac{\int_0^1 \omega((1-\theta)\|y_n-x^*\|)d\theta}{1-\omega_0(\|y_n-x^*\|)} \\ &+\Bigg(\frac{\bar{\omega}_n}{(1-\omega_0(\|y_n-x^*\|))(1-\omega_0(\|x_n-x^*\|))} \\ &+2\frac{\varphi(\|x_n-x^*\|,\|y_n-x^*\|)}{(1-\omega_0(\|x_n-x^*\|))^2}\Bigg) \\ &\times(1+\int_0^1 \omega_0(\theta\|y_n-x^*\|)d\theta)\Bigg]\|y_n-x^*\| \\ &\leq g_2(\|x_n-x^*\|)\|x_n-x^*\| \leq \|x_n-x^*\|, \end{aligned}$$

$$
\begin{aligned}
x_{n+1}-x^* &= z_n - x^* - F'(z_n)^{-1}F(z_n) \\
&\quad +(F'(z_n)^{-1}-F'(x_n)^{-1})F(z_n) \\
&\quad -2F'(x_n)^{-1}(F'(x_n)-[x_n,y_n;F])F'(x_n)^{-1}F(z_n),
\end{aligned}
$$

$$
\begin{aligned}
\|x_{n+1}-x^*\| &\leq \Bigg[\frac{\int_0^1 \omega((1-\theta)\|z_n-x^*\|)d\theta}{1-\omega_0(\|z_n-x^*\|)} \\
&\quad +\Bigg(\frac{\bar{\bar{\omega}}_n}{(1-\omega_0(\|z_n-x^*\|))(1-\omega_0(\|x_n-x^*\|))} \\
&\quad +2\frac{\varphi(\|x_n-x^*\|,\|y_n-x^*\|)}{(1-\omega_0(\|x_n-x^*\|))^2}\Bigg) \\
&\quad \times(1+\int_0^1 \omega_0(\theta\|z_n-x^*\|)d\theta)\Bigg]\|z_n-x^*\| \\
&\leq g_3(\|x_n-x^*\|)\|x_n-x^*\| \leq \|x_n-x^*\|.
\end{aligned}
$$

Method (25.2)

The functions $g_1, i = 1,2,3$ are:

$$
g_1(t) = \frac{\int_0^1 \omega((1-\theta)t)d\theta}{1-\omega_0(t)},
$$

$$
\begin{aligned}
g_2(t) &= \Bigg[\frac{\int_0^1 \omega((1-\theta)g_1(t)t)d\theta}{1-\omega_0(g_1(t)t)} \\
&\quad +\Bigg(\frac{\varphi(t,g_1(t)t)}{(1-\omega_0(t))(1-\varphi_0(t,g_1(t)t))} \\
&\quad +\frac{\varphi_1(t,g_1(t)t)}{(1-\omega_0(g_1(t)t))(1-\varphi_0(t,g_1(t)t))}\Bigg) \\
&\quad \times(1+\int_0^1 \omega_0(\theta g_1(t)t)d\theta)\Bigg]g_1(t)
\end{aligned}
$$

and

$$
\begin{aligned}
g_3(t) &= \Bigg[\frac{\int_0^1 \omega((1-\theta)g_2(t)t)d\theta}{1-\omega_0(g_2(t)t)} \\
&\quad +\Bigg(\frac{\varphi(t,g_1(t)t)}{(1-\omega_0(t))(1-\varphi_0(t,g_1(t)t))} \\
&\quad +\frac{\varphi_2(t,g_1(t)t,g_2(t)t)}{(1-\omega_0(g_2(t)t))(1-\varphi_0(t,g_1(t)t))}\Bigg) \\
&\quad \times\left(1+\int_0^1 \omega_0(\theta g_2(t)t)d\theta\right)\Bigg]g_2(t).
\end{aligned}
$$

The motivation follows in turn from the estimates:

$$\begin{aligned}
y_n - x^* &= x_n - x^* - F'(x_n)^{-1}F(x_n) \\
\|y_n - x^*\| &\leq \left[\frac{\int_0^1 \omega((1-\theta)\|x_n - x^*\|)d\theta}{1 - \omega_0(\|x_n - x^*\|)}\|x_n - x^*\|\right. \\
&\leq g_1(\|x_n - x^*\|)\|x_n - x^*\| \leq \|x_n - x^*\| < r,
\end{aligned}$$

$$\begin{aligned}
z_n - x^* &= y_n - x^* - F'(y_n)^{-1}F(y_n) \\
&\quad -[([x_n, y_n; F]^{-1} - F'(x_n)^{-1}) \\
&\quad +([x_n, y_n; F]^{-1} - F'(y_n)^{-1})]F(y_n), \\
\|z_n - x^*\| &\leq \left[\frac{\int_0^1 \omega((1-\theta)\|y_n - x^*\|)d\theta}{1 - \omega_0(\|y_n - x^*\|)}\right. \\
&\quad + \left(\frac{\varphi(\|x_n - x^*\|, \|y_n - x^*\|)}{(1 - \omega_0(\|x_n - x^*\|))(1 - \varphi_0(\|x_n - x^*\|, \|y_n - x^*\|))}\right. \\
&\quad \left. + \frac{\varphi_1(\|x_n - x^*\|, \|y_n - x^*\|)}{(1 - \omega_0(\|y_n - x^*\|))(1 - \varphi_0(\|x_n - x^*\|, \|y_n - x^*\|))}\right) \\
&\quad \left. \times (1 + \int_0^1 \omega_0(\theta\|y_n - x^*\|)d\theta)\right]\|y_n - x^*\| \\
&\leq g_2(\|x_n - x^*\|)\|x_n - x^*\| \leq \|x_n - x^*\|
\end{aligned}$$

and

$$\begin{aligned}
x_{n+1} - x^* &= z_n - x^* - F'(z_n)^{-1}F(z_n) \\
&\quad -[([x_n, y_n; F]^{-1} - F'(x_n)^{-1}) + ([x_n, y_n; F]^{-1} - F'(z_n)^{-1})]F(z_n), \\
\|x_{n+1} - x^*\| &\leq \left[\frac{\int_0^1 \omega((1-\theta)\|z_n - x^*\|)d\theta}{1 - \omega_0(\|z_n - x^*\|)}\right. \\
&\quad + \left(\frac{\varphi(\|x_n - x^*\|, \|y_n - x^*\|)}{(1 - \omega_0(\|x_n - x^*\|))(1 - \varphi_0(\|x_n - x^*\|, \|y_n - x^*\|))}\right. \\
&\quad \left. + \frac{\varphi_2(\|x_n - x^*\|, \|y_n - x^*\|, \|z_n - x^*\|}{(1 - \omega_0(\|z_n - x^*\|))(1 - \varphi_0(\|x_n - x^*\|, \|y_n - x^*\|))}\right) \\
&\quad \left. \times (1 + \int_0^1 \omega_0(\theta\|z_n - x^*\|)d\theta)\right]\|z_n - x^*\| \\
&\leq g_3(\|x_n - x^*\|)\|x_n - x^*\| \leq \|x_n - x^*\|.
\end{aligned}$$

3. Semi-local Convergence

Method (25.1)

The majorizing sequence $\{\alpha_n\}$ is defined for $\alpha_0 = 0, \beta_0 \geq \|F'(x_0)^{-1}F(x_0)\|$ and each $n =$

$0,1,2,\ldots$ by

$$\begin{aligned}
a_n &= \int_0^1 \omega((1-\theta)(\beta_n-\alpha_n))d\theta(\beta_n-\alpha_n),\\
\gamma_n &= \beta_n+\frac{1}{1-v_0(\alpha_n)}(1+2\frac{\psi_0(\alpha_n,\beta_n)}{1-v_0(\alpha_n)})a_n,\\
d_n &= (1+\int_0^1 v_0(\beta_n+\theta(\gamma_n-\beta_n))d\theta)(\gamma_n-\beta_n)+a_n,\\
\alpha_{n+1} &= \gamma_n+\left(\frac{1}{1-v_0(\alpha_n)}(1+2\frac{\psi(\alpha_n,\beta_n)}{1-v_0(\alpha_n)}\right)d_n,\\
\delta_{n+1} &= (1+\int_0^1 v_0(\alpha_n+\theta(\alpha_{n+1}-\alpha_n))d\theta(\alpha_{n+1}-\alpha_n)\\
&\quad +(1+v_0(\alpha_n))(\alpha_{n+1}-\beta_n)
\end{aligned}$$

and

$$\beta_{n+1}=\alpha_{n+1}+\frac{\delta_{n+1}}{1-v_0(\alpha_{n+1})}.$$

The motivation follows in turn by the estimates:

$$\begin{aligned}
z_n-y_n &= -F'(x_n)^{-1}F(y_n)\\
&\quad -2F'(x_n)^{-1}(F'(x_n)-[x_n,y_n;F])F'(x_n)^{-1}F(y_n),\\
\|z_n-y_n\| &\leq \frac{1}{1-v_0(\alpha_n)}(1+\frac{2\psi(\alpha_n,\beta_n)}{1-v_0(\alpha_n)}a_n\\
&= \gamma_n-\beta_n,\\
F(z_n) &= F(z_n)-F(y_n)+F(y_n)\\
&= \int_0^1 F'(y_n+\theta(z_n-y_n))d\theta(z_n-y_n)\\
&\quad +\int_0^1 F'((1-\theta)(y_n-x_n))d\theta(y_n-x_n),\\
x_{n+1}-z_n &= -F'(x_n)^{-1}F(z_n)\\
&\quad -2F'(x_n)^{-1}(F'(x_n)-[x_n,y_n;F])F'(x_n)^{-1}F(y_n),\\
\|L^{-1}F(z_n)\| &\leq (1+\int_0^1 v_0(\beta_n+\theta(\gamma_n-\beta_n))d\theta)(\gamma_n-\beta_n)+a_n\\
&= d_n
\end{aligned}$$

so,

$$\begin{aligned}
\|x_{n+1}-z_n\| &\leq \alpha_{n+1}-\gamma_n,\\
F(x_{n+1}) &= F(x_{n+1})-F(x_n)-F'(x_n)(x_{n+1}-x_n)\\
&\quad +F'(x_n)(x_{n+1}-y_n),\\
\|LF(x_{n+1})\| &\leq (1+\int_0^1 v_0(\alpha_n+\theta(\alpha_{n+1}-\alpha_n))d\theta)(\alpha_{n+1}-\alpha_n)\\
&\quad +(1+v_0(\alpha_n))(\alpha_{n+1}-\beta_n)=\delta_{n+1}
\end{aligned}$$

and

$$\begin{aligned}\|y_{n+1}-x_{n+1}\| &\leq \|F'(x_{n+1})^{-1}L\|\|L^{-1}F(x_{n+1})\| \\ &\leq \frac{\delta_{n+1}}{1-v_0(\alpha_{n+1})} \\ &= \beta_{n+1}-\alpha_{n+1}.\end{aligned}$$

Method (25.2)

The majorant sequence $\{\alpha_n\}$ for $\alpha_0=0$, $\beta_0 \geq \|F'(x_0)^{-1}F(x_0)\|$ and each $n=0,1,2,\ldots$ is defined by The motivation follows in turn by the estimates:

$$\begin{aligned}z_n-y_n &= -[x_n,y_n;F]^{-1}F(y_n) \\ &= -([x_n,y_n;F]^{-1}-F'(x_n)^{-1})F'(y_n), \\ \|z_n-y_n\| &\leq \left[\frac{1}{1-\psi_0(\alpha_n,\beta_n)}\right. \\ &\quad \left.+\frac{\psi(\alpha_n,\beta_n)}{(1-v_0(\alpha_n))(1-\psi_0(\alpha_n,\beta_n))}\right]a_n \\ &\leq \gamma_n-\beta_n, \\ x_{n+1}-z_n &= -[x_n,y_n;F]^{-1}F(z_n) \\ &\quad -([x_n,y_n;F]^{-1}-F'(x_n)^{-1})F(z_n),\end{aligned}$$

$$\begin{aligned}F(z_n) &= F(z_n)-F(y_n)+F(y_n) \\ \|L^{-1}F(z_n)\| &\leq (1+\int_0^1 v_0(\beta_n+\theta(\gamma_n-\beta_n))d\theta)(\gamma_n-\beta_n)+a_n \\ &= c_n \\ \|x_{n+1}-z_n\| &\leq \left(\frac{1}{1-\psi_0(\alpha_n,\beta_n)}+\frac{\psi(\alpha_n,\beta_n)}{(1-v_0(\alpha_n))(1-\psi_0(\alpha_n,\beta_n))}\right)c_n \\ &\leq \alpha_{n+1}-\gamma_n,\end{aligned}$$

The iterate β_{n+1} is computed in method (25.1).

4. References

[1] Argyros, C. I., Regmi, S., Argyros, I. K., George, S., *Contemporary Algorithms: Theory and Applications*, Volume III, NOVA Publishers, (2023).

[2] Argyros, I. K., *Advances in the Efficiency of Computational Methods and Applications*, World Scientific Publishing, Oxon, United Kingdom, (2000).

[3] Argyros, I. K., *The theory and applications of iteration methods*, 2nd Edition, CRC Press, Engineering Series, Taylor and Francis Group, (2022).

[4] Argyros, I. K., George, S., *Ball comparison between two optimal eight-order methods under weak conditions*, SeMA Journal, 72, (2015), 1-11.

[5] Cordero, A., Jordan, C., Sanabria-Codesal, E., Torregrosa, J. R., *Highly efficient iterative algorithms for solving nonlinear systems with arbitrary order of convergence $p+3, p \geq 5$*, J. Comput. Appl. Math., 330, (2018), 748-758.

[6] Grau-Snchez, M., Grau, A., Noguera, M., *Ostrowski-type methods for solving systems of nonlinear equations*, Appl. Math. Comput., 218, (2011), 2377-2385.

[7] Hueso, J., Martnez, E., Teruel, C., *Convergence, efficiency and dynamics of new fourth and sixth order families of iterative methods for nonlinear systems*, J. Comput. Appl. Math., 275, (2015), 412-420.

[8] Ortega, J. M., Rheinbolt, W. C., *Iterative Solution of Nonlinear Equations in Several Variables*, Academic Press, New York, NY, USA, (1970).

[9] Sharma, J. R., Arora, H., *On efficient weighted Newton methods for solving systems of nonlinear equations*, Appl. Math. Comput., 222, (2013), 497-506.

[10] Soleymani, F., Lofti, T., Bakhtiari, P., *A multistep class of iterative methods for nonlinear systems*, Optim. Lett., 8, (2014), 1001-1015.

Chapter 26

Comparison of Two Fourth-Order Methods

1. Introduction

The methods are:

$$\begin{aligned} y_n &= x_n - \frac{2}{3}F'(x_n)^{-1}F(x_n), \\ A_n &= F'(x_n) + F'(y_n), \\ x_{n+1} &= x_n - 2\left[2I - \frac{7}{4}F'(x_n)^{-1}F'(y_n) + \frac{3}{4}(A_n^{-1}F'(y_n))^2\right]A_n^{-1}F(x_n) \end{aligned} \tag{26.1}$$

and

$$\begin{aligned} y_n &= x_n - \frac{2}{3}F'(x_n)^{-1}F(x_n), \\ x_{n+1} &= x_n - \left(-\frac{3}{8}I + F'(y_n)^{-1}F'(x_n) + \frac{1}{3}F'(x_n)^{-1}F'(y_n)\right. \\ &\quad \left. + \frac{1}{24}(F'(y_n)^{-1}F'(x_n))^2\right)F'(x_n)^{-1}F(x_n). \end{aligned} \tag{26.2}$$

The local convergence order four is established under the Taylor series expansion approach using assumptions reaching the fifth derivative in [5,6], respectively when $B_1 = B_2 = \mathbb{R}^k$. In this chapter their convergence is extended using only the operators on the methods in both the local as well as the semi-local convergence case and in the more general setting of a Banach space. Notice that the semi-local case is not considered in [6]. The approach can be applied to extend the applicability of other methods in a similar way [1]-[4], [7]-[10].

2. Local Convergence

Method(26.1)

The majorant functions $g_i,\ i = 1,2$ are:

$$g_1(t) = \frac{\int_0^1 w((1-\theta)t)d\theta + \frac{1}{3}(1 + \int_0^1 w_0(\theta t)d\theta)}{1 - w_0(t)}$$

and

$$\begin{aligned} g_2(t) &= \frac{\int_0^1 w((1-\theta)t)d\theta}{1-w_0(t)} \\ &+ \left(\frac{\bar{w}(t)}{(1-w_0(t))(1-w_0(g_1(t)t))} + \frac{\bar{w}(t)}{2(1-w_0(t))} + \frac{3\bar{w}(t)}{8(1-q(t))} \right. \\ &+ \left. \frac{1}{32}\left(\frac{\bar{w}(t)}{1-q(t)}\right)^2 + \frac{9}{8}\right)(1+\int_0^1 w_0(\theta t)d\theta), \end{aligned}$$

where

$$q(t) = \frac{1}{2}(w_0(t)+w_0(g_1(t)t)).$$

The motivation is:

$$\begin{aligned} y_n - x^* &= x_n - x^* - F'(x_n)^{-1}F(x_n) + \frac{1}{3}F'(x_n)^{-1}F(x_n), \\ \|y_n - x^*\| &\le \frac{[\int_0^1 w((1-\theta)\|x_n-x^*\|)d\theta + \frac{1}{3}(1+\int_0^1 w(\theta\|x_n-x^*\|)d\theta)]\|x_n-x^*\|}{1-w_0(\|x_n-x^*\|)} \\ &\le g_1(\|x_n-x^*\|)\|x_n-x^*\| \le \|x_n-x^*\| < r, \\ x_{n+1}-x^* &= x_n - x^* - F'(x_n)^{-1}F(x_n) \\ &+(F'(x_n)^{-1} - 2(F'(x_n)+F'(y_n))^{-1}F(x_n) \\ &\left\{\frac{7}{2}(I - F'(x_n)^{-1}F'(y_n)) + \frac{3}{2}\left(\frac{3}{4}I - \frac{1}{2}(2A_n^{-1}F'(y_n)-I)\right)\right. \\ &\left. -\frac{1}{4}(2A_n^{-1}F'(y_n)-I)^2\right\}F(x_n), \\ \|x_{n+1}-x^*\| &\le \left[\frac{\int_0^1 w((1-\theta)\|x_n-x^*\|)d\theta}{1-w_0(\|x_n-x^*\|)}\right. \\ &+\left(\frac{\bar{w}_n}{(1-w_0(\|x_n-x^*\|))(1-w_0(\|y_n-x^*\|))}\right. \\ &+ \left.\frac{7\bar{w}_n}{2(1-w_0)\|x_n-x^*\|))} + \frac{9}{8} + \frac{3\bar{w}_n}{8(1-q_n)} + \frac{1}{32}\left(\frac{\bar{w}_n}{1-q_n}\right)^2\right) \\ &(1+\int_0^1 w_0(\theta\|x_n-x^*\|)d\theta)]\|x_n-x^*\| \\ &\le g_2(\|x_n-x^*\|)\|x_n-x^*\| \le \|x_n-x^*\|, \end{aligned}$$

where

$$\bar{q}_n = \frac{1}{2}(w_0(\|x_n-x^*\|)+w_0(\|y_n-x^*\|)$$

$$\bar{w} = \begin{cases} w(\|x_n-x^*\|+\|y_n-x^*\|) \\ w_0(\|x_n-x^*\|)+w_0(\|y_n-x^*\|). \end{cases}$$

Method(26.2)

$$g_1(t) = \frac{\int_0^1 w((1-\theta)t)d\theta + \frac{1}{3}(1+\int_0^1 w_0(\theta t)d\theta)}{1-w_0(t)}$$

and

$$
\begin{aligned}
g_2(t) &= \frac{\int_0^1 w((1-\theta)t)d\theta}{1-w_0(t)} \\
&+\frac{1}{24}\left(\frac{8\bar{w}(t)}{1-w_0(t)}+\frac{2b\bar{w}(t)}{1-w_0(g_1(t)t)}+\left(\frac{\bar{w}(t)}{1-w_0(g_1(t)t)}\right)^2\right) \\
&\times\frac{(1+\int_0^1 w_0(\theta t)d\theta)}{1-w_0(t)}.
\end{aligned}
$$

The motivation for g_1 is given in method(26.1) and for g_2:

$$
\begin{aligned}
x_{n+1}-x^* &= x_n-x^*-F'(x_n)^{-1}F(x_n)-\left\{\left(\frac{1}{8}I+\frac{1}{3}(F'(x_n)^{-1}F'(y_n)-I)+\frac{1}{3}I\right.\right. \\
&+\frac{25}{24}I+\frac{13}{12}(F'(y_n)^{-1}F'(x_n)-I)+\frac{1}{24}(F'(y_n)^{-1}F'(x_n)-I)^2\Big\}F'(x_n)^{-1}F(x_n), \\
\|x_{n+1}-x^*\| &\leq \left[\frac{\int_0^1 w((1-\theta)\|x_n-x^*\|)d\theta}{1-w_0(\|x_n-x^*\|)}\right. \\
&+\frac{1}{24}\left(\frac{8\bar{w}_n}{1-w_0(\|x_n-x^*\|)}+\frac{26\bar{w}_n}{1-w_0(\|y_n-x^*\|)}+\left(\frac{\bar{w}_n}{1-w_0(\|y_n-x^*\|)}\right)^2\right) \\
&\left.\frac{(1+\int_0^1 w_0(\|x_n-x^*\|)d\theta)}{1-w_0(\|x_n-x^*\|)}\right]\|x_n-x^*\| \\
&\leq g_2(\|x_n-x^*\|)\|x_n-x^*\|\leq\|x_n-x^*\|.
\end{aligned}
$$

3. Semi-local Convergence

Method(26.1)

The majorant sequence $\{\alpha_m\}$ is defined for $\alpha_0=0,\beta_0\geq\frac{2}{3}\|F'(x_0)^{-1}F(x_0)\|$ and each $n=0,1,2,...$ by

$$
\begin{aligned}
p_n &= \frac{1}{2}(v_0(\alpha_n)+v_0(\beta_n)), \\
\alpha_{n+1} &= \beta_n+\frac{1}{32}\left[\frac{84\bar{v}_n}{1-v_0(\alpha_n)}+9\left(\frac{\bar{v}_n}{1-p_n}\right)^2+36\left(\frac{\bar{v}_n}{1-p_n}\right)+20\right] \\
&\times\frac{(1+v_0(\alpha_n))}{1-p_n}(\beta_n-\alpha_n),
\end{aligned}
$$

where

$$
\bar{v} = \begin{cases} v(\alpha_n-\beta_n) \\ v_0(\alpha_n)+v_0(\beta_n). \end{cases}
$$

and

$$
\begin{aligned}
\delta_{n+1} &= (1+\int_0^1 v_0(\alpha_n+\theta(\alpha_{n+1}-\alpha_n))d\theta)(\alpha_{n+1}-\alpha_n) \\
&+(1+v_0(\alpha_n))(\beta_n-\alpha_n), \\
\beta_{n+1} &= \alpha_{n+1}+\frac{2}{3}\frac{\delta_{n+1}}{1-v_0(\alpha_{n+1})}.
\end{aligned}
$$

By substituting the value of x_n from the first to the second substep we get

$$\begin{aligned} x_{n+1}-y_n &= \left[\frac{2}{3}I-4I+\frac{7}{2}F'(x_n)^{-1}F'(y_n)-\frac{3}{2}(A_n^{-1}F'(y_n))^2\right] \\ &\quad A_n^{-1}F'(x_n)F'(x_n)^{-1}F(x_n) \\ &= \frac{1}{6}[-20I+21F'(x_n)^{-1}F'(y_n)-9(A_n^{-1}F'(y_n))^2]A_n^{-1}F'(x_n) \\ &\quad \times(-\frac{3}{2}(y_n-x_n)) \\ &= -\frac{1}{4}[21(F'(x_n)^{-1}F'(y_n)-I)-\frac{5}{4} \\ &\quad -\frac{9}{4}(2A_n-I)^2-\frac{9}{2}(2A_n-I)^2]A_n^{-1}F'(x_n)(y_n-x_n), \end{aligned}$$

thus

$$\begin{aligned} \|x_{n+1}-y_n\| &\leq \frac{1}{32}\left[\frac{84\bar{v}_n}{1-v_0(\alpha_n)}+20+9\left(\frac{\bar{v}_n}{1-p_n}\right)^2+36\left(\frac{\bar{v}_n}{1-p_n}\right)\right] \\ &\quad \frac{(1+v_0(\alpha_n))}{1-p_n}(\beta_n-\alpha_n). \end{aligned}$$

Moreover,

$$\begin{aligned} F(x_{n+1}) &= F(x_{n+1})-F(x_n)-\frac{3}{2}F'(x_n)(y_n-x_n), \\ \|L^{-1}F(x_{n+1})\| &\leq (1+\int_0^1 v_0(\alpha_n+\theta(\alpha_{n+1}-\alpha_n))d\theta)(\alpha_{n+1}-\alpha_n) \\ &\quad +\frac{3}{2}(1+v_0(\alpha_n))(\beta_n-\alpha_n)=\delta_{n+1}, \\ \|y_{n+1}-x_{n+1}\| &\leq \frac{2}{3}\|F'(x_{n+1})^{-1}L\|\|L^{-1}F(x_{n+1})\| \\ &\leq \frac{2}{3}\frac{\delta_{n+1}}{1-v_0(\alpha_{n+1})}=\beta_{n+1}-\alpha_{n+1}. \end{aligned}$$

Method(26.2)

The majorant sequence $\{\alpha_n\}$ is defined for $\alpha_0=0, \beta_0 \geq \frac{2}{3}\|F'(x_0)^{-1}F(x_0)\|$ and each $n = 0,1,2,...$ by

$$\alpha_{n+1} = \beta_n+\frac{1}{16}\left(1+\frac{8\bar{v}_n}{1-v_0(\alpha_n)}+\frac{26\bar{v}_n}{1-v_0(\beta_n)}+\left(\frac{\bar{v}_n}{1-v_0(\beta_n)}\right)^2\right)(\beta_n-\alpha_n)$$

and

$$\beta_{n+1} = \alpha_{n+1}+\frac{2}{3}\frac{\delta_{n+1}}{1-v_0(\alpha_{n+1})},$$

where $\bar{v}_n, \delta_{n+1}$ are given in method(26.1). The motivation for the iterate β_{n+1} is given in method (26.1), whereas

$$\begin{aligned}
x_{n+1}-y_n &= \Big[\frac{2}{3}I+\frac{3}{8}I-F'(y_n)^{-1}F'(x_n) \\
&\quad -\frac{1}{3}F'(x_n)^{-1}F'(y_n)-\frac{1}{24}(F'(y_n)^{-1}F'(x_n))^2](-\frac{3}{2}(y_n-x_n)) \\
&= \frac{1}{16}[I-8(I-F'(x_n)^{-1}F'(y_n))+26(F'(y_n)^{-1}F'(x_n)-I) \\
&\quad +(F'(y_n)^{-1}F'(x_n)-I)^2](y_n-x_n), \\
\|x_{n+1}-y_n\| &\le \frac{1}{16}\left[1+8\frac{\bar{v}_n}{1-v_0(\alpha_n)}+26\frac{\bar{v}_n}{1-v_0(\beta_n)}+\left(\frac{\bar{v}_n}{1-v_0(\beta)}\right)^2\right](\beta_n-\alpha_n) \\
&\le \alpha_{n+1}-\beta_n.
\end{aligned}$$

4. References

[1] Abad, M. F., Cordero, A., Torregrosa, J. R., *Fourth- and fifth-order for solving nonlinear systems of equations: An application to the global positioning system*, Abstr. Appl. Anal., (2013).

[2] Argyros, I. K., *The theory and application of iteration methods*, second edition, Engineering series, Boca Raton, Florida, USA, (2022).

[3] Argyros, I. K., George, S., *Mathematical modeling for the solution of equations and systems of equations with applications*, Volume-IV, Nova Publishes, NY, (2020).

[4] Argyros, I. K., George, S., *On the complexity of extending the convergence region for Traubs method*, Journal of Complexity, 56, (2022), 101423.

[5] Hueso, J. L., Martnez, E., Teruel, C., *Convergence, efficiency and dynamics of new fourth and sixth order families of iterative methods for nonlinear systems*, J. Comput. Appl. Math., 275, (2015), 412-420.

[6] Jaiswal, J. P., *Some class of third- and fourth-order iterative methods for solving nonlinear equations*, J. Appl. Math., (2014).

[7] King, R. F., *A family of fourth order methods for nonlinear equations*, SIAM J. Numer. Anal., 10, (1973), 876-879.

[8] Ortega, J. M., Rheinboldt, W. C., *Iterative Solutions of Nonlinear Equations in Several Variables*, Academic Press, New York, (1970).

[9] Sharma, J. R., Arora, H., Petkovi, M., *An efficient derivative free family of fourth order methods for solving systems of nonlinear equations*, Appl. Math. Comput., 235, (2014), 383-393.

[10] Sharma, J. R., Guna, R. K., Sharma, R., *An efficient fourth order weighted-Newton method for systems of nonlinear equations*, Numer. Algorithm, 62, (2013), 307-323.

Chapter 27

Related Methods of Fifth, Sixth, and Eighth Order

1. Introduction

The five studied methods are in turn:

$$\begin{aligned} y_n &= x_n + F'(x_n)^{-1}F(x_n), && (27.1)\\ z_n &= y_n - F'(x_n)^{-1}F(y_n),\\ x_{n+1} &= z_n + (F'(y_n)^{-1} - 2F'(x_n)^{-1})F(z_n)\\ y_n &= x_n - F'(x_n)^{-1}F(x_n), && (27.2)\\ x_{n+1} &= y_n - [a_1 I + a_2 F'(y_n)^{-1}F'(x_n) + a_3(F'(y_n)^{-1}F'(x_n))^2]F'(y_n)^{-1}F(y_n),\\ y_n &= x_n - \frac{2}{3}F'(x_n)^{-1}F(x_n),\\ z_n &= x_n - [\frac{5}{8}I + \frac{3}{8}(F'(y_n)^{-1}F'(x_n))^2]F'(x_n)^{-1}F(x_n), && (27.3)\\ x_{n+1} &= z_n - [\frac{-9}{4}I + \frac{15}{8}F'(y_n)^{-1}F'(x_n) + \frac{11}{8}F'(x_n)^{-1}F'(y_n)]F'(y_n)^{-1}F(z_n),\\ y_n &= x_n - \frac{2}{3}F'(x_n)^{-1}F(x_n), A_n = 3F'(y_n) - F'(x_n), && (27.4)\\ z_n &= x_n - \frac{1}{2}A_n^{-1}(3F'(y_n) + F'(x_n))F'(x_n)^{-1}F(x_n),\\ x_{n+1} &= z_n - (\frac{1}{2}A_n^{-1}(3F'(y_n) + F'(x_n)))^2F'(x_n)^{-1}F(z_n),\\ y_n &= x_n - F'(x_n)^{-1}F(x_n), && (27.5)\\ z_n &= y_n - [\frac{1}{4}I + \frac{1}{2}F'(y_n)^{-1}F'(x_n) + \frac{1}{4}(F'(y_n)^{-1}F'(x_n))^2]F'(x_n)^{-1}F(y_n) \end{aligned}$$

and

$$x_{n+1} = z_n - [\frac{1}{2}I + \frac{1}{2}(F'(y_n)^{-1}F'(x_n))^2]F'(x_n)^{-1}F(z_n),$$

where $a_1 = \frac{5}{4}, a_2 = -\frac{1}{2}$ and $a_3 = \frac{1}{4}$. The methods are shown to be respectively of local order five [10], five [5], six [7], six [9], eight [5] provided the derivative of order one higher than the convergence order is bounded when $B_1 = B_2 = \mathbb{R}^k$. In this chapter the convergence is shown based only on the operators on the methods and in the more general setting of a Banach space.

Moreover, the more interesting semi-local convergence (not given in [5,7,9,10]) is studied. The process can extend the applicability of other methods in an analogous way. [1]-[4], [6,7,11].

2. Local Convergence

Method 27.1

The majorant functions $g_i, i = 1,2,3$ are defined by

$$g_1(t) = \frac{\int_0^1 w((1-\theta)t)d\theta + 2\left(1+\int_0^1 w_0(\theta t)d\theta\right)}{1-w_0(t)},$$

$$g_2(t) = \left[\frac{\int_0^1 w((1-\theta)g_1(t)t)d\theta}{1-w_0(g_1(t)t)} + \frac{\bar{w}(t)\left(1+\int_0^1 w_0(\theta g_1(t)td\theta)\right)}{(1-w_0(t))(1-w_0(g_1(t)t))}\right] g_1(t),$$

$$\bar{w}(t) = \begin{cases} w((1+g_1(t))t) \\ w_0(t)+w_0(g_1(t)t), \end{cases}$$

$$\bar{\bar{w}}(t) = \begin{cases} w((1+g_2(t))t) \\ w_0(t)+w_0(g_2(t)t), \end{cases}$$

and

$$g_3(t) = \left[\frac{\int_0^1 w((1-\theta)g_2(t)t)d\theta}{1-w_0(g_2(t)t)} + \frac{1}{1-w_0(t)}\left(\frac{\bar{\bar{w}}(t)}{1-w_0(g_2(t)t)} + \frac{\bar{w}(t)}{1-w_0(g_1(t)t)}\right)\left(1+\int_0^1 w_0(\theta g_2(t)t)d\theta\right)\right] g_2(t).$$

Assume the equations $g_2(t)-1=0$ and $g_3(t)-1=0$ have smallest positive solutions $r_2, r_3 \in (0, r_0)$, where r_0 is the smallest positive solution of the equation $w_0(t)-1=0$ (if it exists). Set $r = \min\{r_2, r_3\}$ and $R = g_1(r)r$. Moreover, assume $U[x^*, \bar{r}] \subset \Omega$, where $\bar{r} = \max\{r, R\}$. The motivation for these functions is:

$$\begin{aligned}
y_n - x^* &= x_n - x^* - F'(x_n)^{-1}F(x_n) + 2F'(x_n)^{-1}F(x_n),\\
\|y_n - x^*\| &\leq \frac{\left[\int_0^1 w((1-\theta)\|x_n - x^*\|)d\theta + 2\left(1 + \int_0^1 w_0(\theta\|x_n - x^*\|)d\theta\right)\right]\|x_n - x^*\|}{1 - w_0(\|x_n - x^*\|)}\\
&\leq g_1(\|x_n - x^*\|)\|x_n - x^*\| \leq g_1(r)r = R,\\
z_n - x^* &= y_n - x^* - F'(y_n)^{-1}F(y_n) + (F'(y_n)^{-1} - F'(x_n)^{-1})F(y_n),\\
\|z_n - x^*\| &\leq \left[\frac{\int_0^1 w((1-\theta)\|y_n - x^*\|)d\theta}{1 - w_0(\|y_n - x^*\|)}\right.\\
&\left. + \frac{\bar{w}_n\left(1 + \int_0^1 w_0(\theta\|y_n - x^*\|)d\theta\right)}{(1 - w_0(\|x_n - x^*\|))(1 - w_0(\|y_n - x^*\|))}\right]\|y_n - x^*\||\\
&\leq g_2(\|x_n - x^*\|)\|x_n - x^*\| \leq \|x_n - x^*\|,\\
x_{n+1} - x^* &= z_n - x^* - F'(z_n)^{-1}F(z_n) + (F'(z_n)^{-1} - F'(x_n)^{-1})F(z_n)\\
&\quad + (F'(y_n)^{-1} - F'(x_n)^{-1})F(z_n),\\
\|x_{n+1} - x^*\| &\leq \left[\frac{\int_0^1 w((1-\theta)\|z_n - x^*\|)d\theta}{1 - w_0(\|z_n - x^*\|)}\right.\\
&\quad + \frac{1}{1 - w_0(\|x_n - x^*\|)}\left(\frac{\bar{\bar{w}}_n}{1 - w_0(\|z_n - x^*\|)} + \frac{\bar{w}_n}{1 - w_0(\|y_n - x^*\|)}\right)\\
&\left.\quad \times \left(1 + \int_0^1 w_0(\theta\|z_n - x^*\|)d\theta\right)\right]\|z_n - x^*\|\\
&\leq g_3(\|x_n - x^*\|)\|x_n - x^*\| \leq \|x_n - x^*\|.
\end{aligned}$$

Method (27.2)

The majorant functions $g_i, i = 1, 2$ are:

$$g_1(t) = \frac{\int_0^1 w((1-\theta)t)d\theta}{1 - w_0(t)}$$

and

$$\begin{aligned}
g_2(t) &= \left[\frac{\int_0^1 w((1-\theta)g_1(t)t)d\theta}{1 - w_0(g_1(t)t)}\right.\\
&\quad + \left(|a_2 + 2a_3|\frac{\bar{w}(t)}{1 - w_0(g_1(t)t)} + |a_3|\left(\frac{\bar{w}(t)}{1 - w_0(g_1(t)t)}\right)^2\right)\\
&\left.\quad \times \frac{\left(1 + \int_0^1 w_0(\theta g_1(t)t)d\theta\right)}{1 - w_0(g_1(t)t)}\right] g_1(t).
\end{aligned}$$

The estimate leading to g_1 is given in Method (27.1). Then, for g_2 we have in turn

$$\begin{aligned}
x_{n+1} - x^* &= y_n - x^* - F'(y_n)^{-1}F(y_n)\\
&\quad -[(a_1 - 1)I + a_2F'(y_n)^{-1}F'(x_n) + a_3(F'(y_n)^{-1}F'(x_n))^2]F'(y_n)^{-1}F(y_n)\\
&= y_n - x^* - F'(y_n)^{-1}F(y_n) - [(a_1 + a_2 + a_3)I\\
&\quad + (a_2 + 2a_3)(F'(y_n)^{-1}F'(x_n) - I) + a_3(F'(y_n)^{-1}F'(x_n) - I)^2]F'(y_n)^{-1}F(y_n),
\end{aligned}$$

so since $a_1 + a_2 + a_3 = 0$,

$$\begin{aligned}
\|x_{n+1} - x_n\| &\leq \Bigg[\frac{\int_0^1 w((1-\theta)\|y_n - x^*\|)d\theta}{1 - w_0(\|y_n - x^*\|)} \\
&+ \Bigg(|a_2 + 2a_3| \frac{\bar{w}_n}{1 - w_0(\|y_n - x^*\|)} \\
&+ |a_3| \left(\frac{\bar{w}_n}{1 - w_0(\|y_n - x^*\|)} \right)^2 \Bigg) \frac{\left(1 + \int_0^1 w_0(\theta\|y_n - x^*\|)d\theta\right)}{1 - w_0(\|y_n - x^*\|)} \Bigg] \|y_n - x^*\| \\
&\leq g_2(\|x_n - x^*\|)\|x_n - x^*\| \leq \|x_n - x^*\|.
\end{aligned}$$

Method (27.3)

The majorant functions $g_i, i = 1,2$ are:

$$\begin{aligned}
g_1(t) &= \frac{\int_0^1 w((1-\theta)t)d\theta + \frac{1}{3}\left(1 + \int_0^1 w_0(\theta t)d\theta\right)}{1 - w_0(t)}, \\
g_2(t) &= \frac{\int_0^1 w((1-\theta)t)d\theta}{1 - w_0(t)} \\
&+ \frac{3}{8} \frac{\bar{w}(t)}{1 - w_0(g_1(t)t)} \left(2 + \frac{\bar{w}(t)}{1 - w_0(g_1(t)t)}\right) \frac{\left(1 + \int_0^1 w_0(\theta t)d\theta\right)}{1 - w_0(t)}
\end{aligned}$$

and

$$\begin{aligned}
g_3(t) &= \Bigg[\frac{\int_0^1 w((1-\theta)g_2(t)t)d\theta}{1 - w_0(g_2(t)t)} \\
&+ \Bigg(\frac{\bar{\bar{w}}(t)}{(1 - w_0(t))(1 - w_0(g_2(t)t))} \\
&+ \frac{3}{4}\left(1 + \frac{(1 - w_0(t))(1 - w_0(g_1(t)t))}{(1 - p(t))^2}\right)\Bigg) \frac{\left(1 + \int_0^1 w_0(\theta g_2(t)t)\right)}{1 - w_0(t)} \Bigg] g_2(t).
\end{aligned}$$

The estimated lending to the functions g_i are:

$$\begin{aligned}
y_n - x^* &= x_n - x^* - F'(x_n)^{-1}F(x_n) + \frac{1}{3}F'(x_n)^{-1}F(x_n),\\
\|y_n - x^*\| &\leq \frac{\left[\int_0^1 w((1-\theta)\|x_n - x^*\|)d\theta + \frac{1}{3}\left(1 + \int_0^1 w_0(\theta\|x_n - x^*\|)d\theta\right)\right]\|x_n - x^*\|}{1 - w_0(\|x_n - x^*\|)}\\
&\leq g_1(\|x_n - x^*\|)\|x_n - x^*\| \leq \|x_n - x^*\| < r,\\
z_n - x^* &= x_n - x^* - F'(x_n)^{-1}F(x_n)\\
&\quad - \left[-\frac{3}{8}I + \frac{3}{8}(F'(y_n)^{-1}F'(x_n))^2\right]F'(x_n)^{-1}F(x_n)\\
&= x_n - x^* - F'(x_n)^{-1}F(x_n)\\
&\quad - \frac{3}{8}(I - F'(y_n)^{-1}F'(x_n))[2I + (I - F'(y_n)^{-1}F'(x_n))]F'(x_n)^{-1}F(x_n),\\
\|z_n - x^*\| &\leq \left[\frac{\int_0^1 w((1-\theta)\|x_n - x^*\|)d\theta}{1 - w_0(\|x_n - x^*\|)}\right.\\
&\quad \left. + \frac{3}{8}\frac{\bar{w}_n}{1 - w_0(\|y_n - x^*\|)}\left(2 + \frac{\bar{w}_n}{1 - w_0(\|y_n - x^*\|)}\right)\frac{\left(1 + \int_0^1 w_0(\theta\|x_n - x^*\|)d\theta\right)}{1 - w_0(\|x_n - x^*\|)}\right]\\
&\quad \times \|x_n - x^*\|\\
&\leq g_2(\|x_n - x^*\|)\|x_n - x^*\| \leq \|x_n - x^*\|,\\
x_{n+1} - x^* &= z_n - x^* - F'(z_n)^{-1}F(z_n) + (F'(z_n)^{-1} - F'(y_n)^{-1})F(z_n)\\
&\quad -[-I - \frac{9}{4}I + \frac{15}{8}F'(y_n)^{-1}F'(x_n) + \frac{11}{8}F'(x_n)^{-1}F'(y_n)]F'(y_n)^{-1}F(z_n)\\
&= z_n - x^* - F'(z_n)^{-1}F(z_n) + (F'(z_n)^{-1} - F'(y_n)^{-1})F(z_n)\\
&\quad - \frac{1}{8}[15(F'(y_n)^{-1}F'(x_n) - I) + 11(F'(x_n)^{-1}F'(y_n) - I)]F'(y_n)^{-1}F(z_n),\\
\|x_{n+1} - x^*\| &\leq \left[\frac{\int_0^1 w((1-\theta)\|x_n - x^*\|)d\theta}{1 - w_0(\|x_n - x^*\|)}\right.\\
&\quad \left. + \frac{1}{8}\left(15\frac{\bar{\bar{w}}(t)}{1 - w_0(g_1(t)t)} + 11\frac{\bar{w}(t)}{1 - w_0(t)}\right)\frac{\left(1 + \int_0^1 w_0(\theta g_2(t)t)d\theta\right)}{1 - w_0(g_1(t)t)}\right]\|z_n - x^*\|\\
&\leq g_3(\|x_n - x^*\|)\|x_n - x^*\| \leq \|x_n - x^*\|,
\end{aligned}$$

where

$$\bar{\bar{w}}_n = \begin{cases} w(\|y_n - x^*\| + \|z_n - x^*\|)\\ w_0(\|y_n - x^*\|) + w_0(\|z_n - x^*\|), \end{cases}$$

and

$$\bar{\bar{w}}(t) = \begin{cases} w((g_1(t) + g_2(t))t)\\ w_0(g_1(t)t) + w_0(g_2(t)t). \end{cases}$$

Method (27.4)

The majorant functions $g_i, i = 1,2,3$ are:

$$\begin{aligned}
g_1(t) &= \frac{\int_0^1 w((1-\theta)td\theta) + \frac{1}{3}\left(1+\int_0^1 w_0(\theta t)d\theta\right)}{1-w_0(t)},\\
p(t) &= \frac{1}{2}(3w_0(g_1(t)t) + w_0(t)),\\
g_2(t) &= \frac{\int_0^1 w((1-\theta)t)d\theta}{1-w_0(t)} + \frac{3}{4}\frac{\bar{w}(t)\left(1+\int_0^1 w_0(\theta t)d\theta\right)}{(1-p(t))(1-w_0(t))},\\
g_3(t) &= \left[\frac{\int_0^1 w((1-\theta)g_2(t)t)d\theta}{1-w_0(g_2(t)t)}\right.\\
&+ \left(\frac{\bar{\bar{w}}(t)}{(1-w_0(t))(1-w_0(g_2(t)t))} + \frac{3}{4}\left(1+\frac{(1+w_0(t))(1+w_0(g_1(t)t))}{(1-p(t))^2}\right)\right.\\
&\times \left.\left.\frac{\left(1+\int_0^1 w_0(\theta g_2(t))d\theta\right)}{1-w_0(t)}\right)\right] g_2(t).
\end{aligned}$$

The motivation is in turn: The estimate for the function g_1 is given in method (27.3). Moreover, we have

$$\begin{aligned}
z_n - x^* &= x_n - x^* - F'(x_n)^{-1}F(x_n) + \\
&+ \left[I - \frac{1}{2}(3F'(y_n) - F'(x_n))^{-1}(3F'(y_n) + F'(x_n))\right] F'(x_n)^{-1}F(x_n)\\
&= x_n - x^* - F'(x_n)^{-1}F(x_n)\\
&+ \frac{3}{2}(3F'(y_n) - F'(x_n))^{-1}(F'(y_n) - F'(x_n))F'(x_n)^{-1}F(x_n),\\
\|z_n - x^*\| &\le \left[\frac{\int_0^1 w((1-\theta)\|x_n - x^*\|)d\theta}{1-w_0(\|x_n - x^*\|)}\right.\\
&+ \left.\frac{3\bar{w}_n\left(1+\int_0^1 w_0(\theta\|x_n - x^*\|)d\theta\right)}{4(1-p_n)(1-w_0(\|x_n - x^*\|))}\right] \|x_n - x^*\|\\
&\le g_2(\|x_n - x^*\|)\|x_n - x^*\| \le \|x_n - x^*\|,
\end{aligned}$$

$$\begin{aligned}
x_{n+1} - x^* &= z_n - x^* - F'(z_n)^{-1}F(z_n)\\
&+ (F'(z_n)^{-1} - F'(x_n)^{-1})F(z_n)\\
&+ [I - \left(\frac{1}{2}A_n^{-1}(3F'(y_n) + F'(x_n))\right)^2]F'(x_n)^{-1}F(z_n)\\
&= z_n - x^* - F'(z_n)^{-1}F(z_n) + (F'(z_n)^{-1} - F'(x_n)^{-1})F(z_n)\\
&+ [I - (\frac{1}{2}(A_n^{-1}(A_n + 2F'(x_n))))^2]F'(x_n)^{-1}F(z_n)\\
&= z_n - x^* - F'(z_n)^{-1}F(z_n) + (F'(z_n)^{-1} - F'(x_n)^{-1})F(z_n)\\
&+ [\frac{3}{4} - A_n^{-1}F'(x_n) - (A_n^{-1}F'(x_n))^2]F'(x_n)^{-1}F(z_n)\\
&= z_n - x^* - F'(z_n)^{-1}F(z_n) + (F'(z_n)^{-1} - F'(x_n)^{-1})F(z_n)\\
&+ \left[\frac{3}{4}I - A_n^{-1}F'(x_n)(I + (A_n^{-1}F'(x_n)))\right] F'(x_n)^{-1}F(z_n),
\end{aligned}$$

since

$$\begin{aligned}
& I-\left(\frac{1}{2}A_n^{-1}(A_n+2F'(x_n))\right)^2 \\
= & I-\frac{1}{4}(I^2+4(A_n^{-1}F'(x_n))^2+4A_n^{-1}F'(x_n)) \\
= & \frac{3}{4}I-A_n^{-1}F'(x_n)(I+A_n^{-1}F'(x_n)),
\end{aligned}$$

thus

$$\begin{aligned}
\|x_{n+1}-x^*\| \leq & \left[\frac{\int_0^1 w((1-\theta)\|z_n-x^*\|)d\theta}{1-w_0(\|z_n-x^*\|)}\right. \\
& +\left(\frac{\bar{\bar{w}}_n}{(1-w_0(\|x_n-x^*\|))(1-w_0(\|z_n-x^*\|))}\right. \\
& \left.\left.+\frac{3}{4}\left(1+\frac{(1+w_0(\|x_n-x^*\|))(1+w_0(\|y_n-x^*\|))}{(1-p_n)^2}\right)\frac{1+\int_0^1 w_0(\theta\|z_n-x^*\|d\theta)}{1-w_0(\|x_n-x^*\|)}\right)\right] \\
& \times\|z_n-x^*\| \\
\leq & g_3(\|x_n-x^*\|)\|x_n-x^*\|\leq\|x_n-x^*\|,
\end{aligned}$$

where

$$p_n = \frac{1}{2}(3w_0(\|y_n-x^*\|)+w_0(\|x_n-x^*\|)).$$

Method (27.5)

The majorant functions $g_i, i=1,2,3$ are:

$$\begin{aligned}
g_1(t) = & \frac{\int_0^1 w((1-\theta)t)d\theta}{1-w_0(g_1(t)t)}, \\
g_2(t) = & \left[\frac{w((1-\theta)g_1(t)t)d\theta}{1-w_0(g_1(t)t)}\right. \\
& +\left(\frac{\bar{w}(t)}{(1-w_0(t))(1-w_0(g_1(t)t))}\right. \\
& \left.\left.+\frac{1}{4}\frac{\bar{w}(t)}{1-w_0(g_1(t)t)}\left(4+\frac{\bar{w}(t)}{1-w_0(g_1(t)t)}\right)\frac{\left(1+\int_0^1 w_0(\theta g_1(t)t)d\theta\right)}{1-w_0(t)}\right)\right]g_1(t), \\
g_3(t) = & \left[\frac{\int_0^1 w((1-\theta)g_2(t)t)d\theta}{1-w_0(g_2(t)t)}\right. \\
& +\left(\frac{\bar{\bar{w}}(t)}{(1-w_0(t))(1-w_0(g_2(t)t))}\right. \\
& \left.\left.+\frac{1}{2}\frac{\bar{w}(t)}{1-w_0(g_1(t)t)}\left(2+\frac{\bar{w}(t)}{1-w_0(g_1(t)t)}\right)\frac{\left(1+\int_0^1 w_0(\theta g_2(t)t)d\theta\right)}{1-w_0(t)}\right)\right]g_2(t).
\end{aligned}$$

The motivation for g_1 is given method (27.2). Moreover, we get

$$\begin{aligned}
z_n - x^* &= y_n - x^* - F'(y_n)^{-1}F(y_n) + (F'(y_n)^{-1} - F'(x_n)^{-1})F(y_n) \\
&\quad -[-I + \frac{1}{4}I + \frac{1}{2}(F'(y_n)^{-1}F'(x_n) - I) + \frac{1}{2}I + \frac{1}{4}(F'(y_n)^{-1}F'(x_n) - I)^2]F'(x_n)^{-1}F(y_n), \\
\|z_n - x^*\| &\leq \left[\frac{\int_0^1 w((1-\theta)\|y_n - x^*\|)d\theta}{1 - w_0(\|y_n - x^*\|)}\right. \\
&\quad + \left(\frac{\bar{w}_n}{(1 - w_0(\|x_n - x^*\|))(1 - w_0(\|y_n - x^*\|))}\right) \\
&\quad \left. + \frac{1}{4}\frac{\bar{w}_n}{1 - w_0(\|y_n - x^*\|)}\left(4 + \frac{\bar{w}_n}{1 - w_0(\|y_n - x^*\|)}\right)\frac{\left(1 + \int_0^1 w_0(\theta\|y_n - x^*\|)d\theta\right)}{1 - w_0(\|x_n - x^*\|)}\right] \\
&\quad \times \|y_n - x^*\| \\
&\leq g_2(\|x_n - x^*\|)\|x_n - x^*\| \leq \|x_n - x^*\|,
\end{aligned}$$

$$\begin{aligned}
x_{n+1} - x^* &= z_n - x^* - F'(z_n)^{-1}F(z_n) + (F'(z_n)^{-1} - F'(x_n)^{-1})F(z_n) \\
&\quad -\frac{1}{2}[(F'(y_n)^{-1}F'(x_n))^2 - I]F'(x_n)^{-1}F(z_n) \\
&\quad = z_n - x^* - F'(z_n)^{-1}F(z_n) + (F'(z_n)^{-1} - F'(x_n)^{-1})F(z_n) \\
&\quad -\frac{1}{2}(F'(y_n)^{-1}F'(x_n) - I)[(F'(y_n)^{-1}F'(x_n) - I) + 2I]F'(x_n)^{-1}F(z_n), \\
\|x_{n+1} - x^*\| &\leq \left[\frac{\int_0^1 w((1-\theta)\|z_n - x^*\|)d\theta}{1 - w_0(\|z_n - x^*\|)}\right. \\
&\quad + \left(\frac{\bar{\bar{w}}_n}{(1 - w_0(\|x_n - x^*\|))(1 - w_0(\|z_n - x^*\|))}\right. \\
&\quad \left.\left. + \frac{1}{2}\frac{\bar{w}_n}{1 - w_0(\|y_n - x^*\|)}\left(2 + \frac{\bar{w}_n}{1 - w_0(\|y_n - x^*\|)}\right)\frac{\left(1 + \int_0^1 w_0(\theta\|z_n - x^*\|)d\theta\right)}{1 - w_0(\|x_n - x^*\|)}\right)\right] \\
&\quad \times \|z_n - x^*\|. \\
&\leq g_3(\|x_n - x^*\|)\|x_n - x^*\| \leq \|x_n - x^*\|.
\end{aligned}$$

3. Semi-local Convergence

The computations are similar to the ones in the local case but the role of x^*, "w" is exchanged by x_0, "r", respectively.

Method (27.1)

The motivational estimates are:

$$
\begin{aligned}
F(y_n) &= F(y_n)-F(x_n)+F'(x_n)(y_n-x_n),\\
z_n-y_n &= -F'(x_n)^{-1}F(y_n),\\
\|L^{-1}F(y_n)\| &\leq \left(1+\int_0^1 v_0(\alpha_n+\theta(\beta_n-\alpha_n))d\theta\right)(\beta_n-\alpha_n)\\
&\quad +(1+v_0(\alpha_n))(\beta_n-\alpha_n)=\lambda_n,\\
\|z_n-y_n\| &\leq \frac{\lambda_n}{1-v_0(\alpha_n)}=\gamma_n-\beta_n,\\
F(z_n) &= F(z_n)-F(y_n)+F(y_n),\\
x_{n+1}-z_n &= -F'(x_n)^{-1}F(z_n)+(F'(y_n)^{-1}-F'(x_n)^{-1})F(z_n),\\
\|L^{-1}F(z_n)\| &\leq \left(1+\int_0^1 v_0(\beta_n+\theta(\gamma_n-\beta_n))d\theta\right)(\gamma_n-\beta_n)+\lambda_n=\mu_n,\\
\|x_{n+1}-z_n\| &\leq \frac{1}{1-v_0(\alpha_n)}\left(1+\frac{\bar{v}_n}{1-v_0(\beta_n)}\right)\mu_n=\alpha_{n+1}-\gamma_n,
\end{aligned}
$$

$$
\begin{aligned}
F(x_{n+1}) &= F(x_{n+1})-F(x_n)+F'(x_n)(y_n-x_n),\\
\|L^{-1}F(x_{n+1})\| &\leq \left(1+\int_0^1 v_0(\alpha_n+\theta(\alpha_{n+1}-\alpha_n))d\theta\right)(\alpha_{n+1}-\alpha_n)\\
&\quad +(1+v_0(\alpha_n))(\beta_n-\alpha_n)=\delta_{n+1},\\
\|y_{n+1}-x_{n+1}\| &\leq \|F'(x_{n+1})^{-1}L\|\|L^{-1}F(x_{n+1})\|\\
&\leq \frac{\delta_{n+1}}{1-v_0(\alpha_{n+1})}=\beta_{n+1}-\alpha_{n+1}.
\end{aligned}
$$

where

$$
\bar{v}_n=\begin{cases} v(\beta_n-\alpha_n)\\ v_0(\alpha_n)+v_0(\beta_n).\end{cases}
$$

Thus the majorant sequence $\{\alpha_n\}$ is determined provided that $\alpha_0=0$ and $\beta_0\geq \|F'(x_0)^{-1}F(x_0)\|$.

Method (27.2)

The motivational estimates are:

$$
\begin{aligned}
x_{n+1}-y_n &= -[(a_1+a_2+a_3)I+(a_2+2a_3)(F'(y_n)^{-1}F'(x_n)-I)\\
&\quad +a_3(F'(y_n)^{-1}F'(x_n)-I)^2]F'(y_n)^{-1}F(y_n),\\
\|x_{n+1}-y_n\| &\leq \left[|a_1+a_2+a_3|+|a_2+2a_3|\frac{\bar{v}_n}{1-v_0(\beta_n)}\right.\\
&\quad \left.+|a_3|\left(\frac{\bar{v}_n}{1-v_0(\beta_n)^2}\right)\right]\frac{\int_0^1 v((1-\theta)(\beta_n-\alpha_n))d\theta(\beta_n-\alpha_n)}{1-v_0(\beta_n)}=\alpha_{n+1}-\gamma_n
\end{aligned}
$$

and the iterate β_{n+1} is as defined in the method (27.1).

Method (27.3)

The estimates are:

$$\begin{aligned}
z_n - y_n &= \left[\frac{2}{3}I - \frac{5}{8}I - \frac{3}{8}(F'(y_n)^{-1}F'(x_n))^2\right]\left(-\frac{3}{2}(y_n - x_n)\right)\\
&= -\frac{3}{2}\left[-\frac{1}{3} + \frac{3}{8}(I - (F'(y_n)^{-1}F'(x_n))^2)\right](y_n - x_n),\\
\|z_n - y_n\| &\leq \frac{3}{2}\left(\frac{1}{3} + \frac{3}{8}\frac{\bar{v}_n}{1 - v_0(\beta_n)}\left(2 + \frac{\bar{v}_n}{1 - v_0(\beta_n)}\right)\right)(\beta_n - \alpha_n) = \gamma_n - \beta_n,\\
\|x_{n+1} - z_n\| &\leq \frac{1}{8}\left(8 + 15\frac{\bar{v}_n}{1 - v_0(\beta_n)} + 11\frac{\bar{v}_n}{1 - v_0(\alpha_n)}\right)\frac{\lambda_n}{1 - v_0(\beta_n)}
\end{aligned}$$

and β_{n+1} is the same as in method (27.1) and λ_n is also as previously given.

Method (27.4)

The estimates are:

$$\begin{aligned}
z_n - y_n &= \left[\frac{2}{3}I - \frac{1}{2}A_n^{-1}(3F'(y_n) + F'(x_n))\right]\left(-\frac{3}{2}(y_n - x_n)\right)\\
&= -\frac{1}{4}A_n^{-1}\left[3(F'(y_n) - F'(x_n)) - 4F'(x_n)\right](y_n - x_n),\\
\|z_n - y_n\| &\leq \frac{(3\bar{v}_n + 4 + 4v_0(\alpha_n))(\beta_n - \alpha_n)}{8(1 - q_n)} = \gamma_n - \beta_n,\\
\|x_{n+1} - z_n\| &\leq \frac{1}{4}\left(1 + \frac{3(1 + v_0(\alpha_n))(1 + v_0(\beta_n))}{(1 - q_n)^2 2}\right)\frac{\bar{\lambda}_n}{1 - v_0(\alpha_n)} = \alpha_{n+1} - \gamma_n,\\
\bar{\delta}_{n+1} &= \left(1 + \int_0^1 v_0(\alpha_n + \theta(\alpha_{n+1} - \alpha_n))d\theta\right)(\alpha_{n+1} - \alpha_n)\\
&\quad + \frac{3}{2}(1 + v_0(\alpha_n))(\beta_n - \alpha_n)\\
\|y_{n+1} - x_{n+1}\| &\leq \frac{2}{3}\|F'(x_{n+1})^{-1}L\|\|L^{-1}F(x_{n+1})\|\\
&\leq \frac{2}{3}\frac{\bar{\delta}_{n+1}}{1 - v_0(\alpha_{n+1})} = \beta_{n+1} - \alpha_{n+1},
\end{aligned}$$

for $\alpha_0 = 0$ and $\beta_0 \geq \frac{2}{3}\|F'(x_0)^{-1}F(x_0)\|$, where we also used

$$\begin{aligned}
F(z_n) &= F(z_n) - F(x_n) - \frac{3}{2}F'(x_n)(y_n - x_n),\\
\|L^{-1}F(z_n)\| &\leq \bar{\lambda}_n,\\
F(x_{n+1}) &= F(x_{n+1}) - F(x_n) - \frac{3}{2}F'(x_n)(y_n - x_n)
\end{aligned}$$

and

$$\|L^{-1}F(x_{n+1})\| \leq \bar{\delta}_{n+1}.$$

4. References

[1] Argyros, I. K., *Convergence and Applications of Newton-Type Iterations*, Springer-Verlag, New York, (2008).

[2] Argyros, I. K., *The theory and application of iteration methods*, second edition, Engineering series, Boca Raton, Florida, USA, (2022).

[3] Argyros, I. K., George, S., *Mathematical modeling for the solution of equations and systems of equations with applications*, Volume-IV, Nova Publishes, NY, (2020).

[4] Argyros, I. K., George, S., *On the complexity of extending the convergence region for Traubs method*, Journal of Complexity, 56, (2022), 101423.

[5] Cordero, A., Gomez, E., Torregrosa, J. R., *Efficient high-order iterative methods for solving nonlinear systems and their application on heat conduction problems*, Hindawi Complexity, (2017), Article ID 6457532, https://doi.org/10.1155/2017/6457532.

[6] Cordero, A., Hueso, J. L., Martinez, E., Torregrosa, J. R., *A modified Newton-Jarratt's composition*, Numerical Algorithms, 55, 1, (2010), 87-99.

[7] Hueso, J. L., Martinez, E., Teruel, C., *Convergence, efficiency and dynamics of new fourth and sixth order families of iterative methods for nonlinear systems*, Journal of Computational and Applied Mathematics, 275, (2015), 412-420.

[8] Ostrowski, A. M., *Solution of equations and Systems of equations*, Prentice Hall, Englewood Cliffs, NY, USA, (1964).

[9] Sharma, J. R., Arora, H., *Efficient Jarratt-like methods for solving systems of nonlinear equations*, Calcolo, 51, 1, (2014), 193-210.

[10] Soleymani, F., Lofti, T., Bhaktiari, P., *A multi-step class of iterative methods for nonlinear systems*, Optimization Letters, 8, 3, (2014), 1001-1015.

[11] Xiao, X. Y., Yin, H. W., *Increasing the order of convergence for iterative methods to solve nonlinear systems*, Calcolo, 53, 3, (2016), 285-300.

Chapter 28

Quadrature Formula-Based Methods

1. Introduction

The method studied in [10] is defined for $x_0 \in \Omega$ and each $n = 0, 1, 2, \ldots$ by

$$\begin{aligned} y_n &= x_n - F'(x_n)^{-1}F(x_n), \\ A_n &= \sum_{i=1}^{k} c_i F'(x_n + \theta i(y_n - x_n)), \\ z_n &= y_n - A_n^{-1}F(y_n), \\ B_n &= \sum_{i=1}^{k} c_i F'(x_n + \theta i(z_n - x_n)) \end{aligned} \tag{28.1}$$

and

$$x_{n+1} = z_n - B_n^{-1}F(z_n),$$

where $c_i, \theta i \in [0, 1]$ and $\sum_{i=1}^{k} c_i = 1$ where k is a natural number. By specializing the parameters methods already in the literature can be obtained. The local convergence order four is shown by assuming the existence of the ninth derivative not on the method. In this chapter, the local as well as the semi-local convergence case (not studied in [10]) are studied using only the operators on the method. The same methodology applies to extend the applicability of other methods [1]-[9], [11,12].

2. Local Convergence

The majorant functions $g_i, i = 1,2,3$ are:

$$\begin{aligned}
g_1(t) &= \frac{\int_0^1 w((1-\theta)t)d\theta}{1-w_0(t)},\\
\lambda(t) &= \sum_{i=1}^{k} c_i w((1-\theta i)(1+g_1(t))t),\\
p(t) &= \sum_{i=1}^{k} c_i w_0((1-\theta i+\theta i g_1(t))t),\\
g_2(t) &= \left[\frac{\int_0^1 w((1-\theta)g_1(t)t)d\theta}{1-w_0(g_1(t)t)} + \frac{\lambda(t)\left(1+\int_0^1 w_0(\theta g_1(t)t)d\theta\right)}{(1-w_0(g_1(t)t))(1-p(t))}\right] g_1(t),\\
\mu(t) &= \sum_{i=1}^{k} c_i w((1-\theta i)(1+g_2(t))t),\\
q(t) &= \sum_{i=1}^{k} c_i w_0((1-\theta i+\theta i g_2(t))t)
\end{aligned}$$

and

$$g_3(t) = \left[\frac{\int_0^1 w((1-\theta)g_2(t)t)d\theta}{1-w_0(g_2(t)t)} + \frac{\mu(t)\left(1+\int_0^1 w_0(\theta g_2(t)t)d\theta\right)}{(1-w_0(g_2(t)t))(1-q(t))}\right] g_2(t).$$

The motivation is in turn:

$$\begin{aligned}
y_n - x^* &= x_n - x^* - F'(x_n)^{-1}F(x_n),\\
\|y_n - x^*\| &\le \frac{\int_0^1 w((1-\theta)\|x_n-x^*\|)d\theta\|x_n-x^*\|}{1-w_0(\|x_n-x^*\|)}\\
&\le g_1(\|x_n-x^*\|)\|x_n-x^*\| \le \|x_n-x^*\| < r,\\
z_n - x^* &= y_n - x^* - F'(y_n)^{-1}F(y_n) + F'(y_n)^{-1}(A_n - F'(y_n))A_n^{-1}F(y_n).
\end{aligned}$$

But

$$\begin{aligned}
\|L^{-1}(A_n - \sum_{i=1}^{k} c_i L)\| &\le \sum_{i=1}^{k} c_i \|L^{-1}(F'(x_n+\theta i(y_n-x_n)) - L)\|\\
&\le \sum_{i=1}^{k} c_i w_0((1-\theta i)\|x_n-x^*\| + \theta i\|y_n-x^*\|) = p_n,\\
\|z_n - x^*\| &\le \left[\frac{\int_0^1 w((1-\theta)\|y_n-x^*\|)d\theta}{1-w_0(\|y_n-x^*\|)}\right.\\
&\quad \left. + \frac{\lambda_n\left(1+\int_0^1 w_0(\theta\|y_n-x^*\|)d\theta\right)}{(1-w_0(\|y_n-x^*\|))(1-p_n)}\right]\|y_n-x^*\|,\\
&\le g_2(\|x_n-x^*\|)(\|x_n-x^*\|) < \|x_n-x^*\|,
\end{aligned}$$

where we also used

$$\|L^{-1}(A_n - F'(y_n))\| \leq \sum_{i=1}^{k} w((1-\theta i)\|y_n - x_n\|) = \lambda_n.$$

Moreover, similarly

$$\begin{aligned}\|L^{-1}(B_n - \sum_{i=1}^{k} L)\| &\leq \sum_{i=1}^{k} c_i w_0((1-\theta i) + \theta i\|z_n - x^*\|) = q_n,\\ \|L^{-1}(B_n - F'(z_n))\| &\leq \sum_{i=1}^{k} w((1-\theta i)\|z_n - x_n\|) = \mu_n,\end{aligned}$$

thus

$$x_{n+1} - x^* = z_n - x^* - F'(z_n)^{-1}F(z_n) + F'(z_n)^{-1}(B_n - F'(z_n))B_n^{-1}F(z_n)$$

and consequently

$$\begin{aligned}\|x_{n+1} - x^*\| &\leq \left[\frac{\int_0^1 w((1-\theta)\|z_n - x^*\|)d\theta}{1 - w_0(\|z_n - x^*\|)}\right.\\ &\quad \left.+ \frac{\mu_n\left(1 + \int_0^1 w_0(\theta\|z_n - x^*\|)d\theta\right)}{(1 - w_0(\|z_n - x^*\|)(1 - q_n))}\right]\|z_n - x^*\|\\ &\leq g_3(\|x_n - x^*\|)\|x_n - x^*\| \leq \|x_n - x^*\|.\end{aligned}$$

3. Semi-local Convergence

The majorant sequence $\{\alpha_n\}$ is defined for $\alpha_0 = 0, \beta_0 \geq \|F'(x_0)^{-1}F(x_0)\|$ and each $n = 0, 1, 2, \ldots$ by

$$\begin{aligned}\xi_n &= \int_0^1 v((1-\theta)(\beta_n - \alpha_n))d\theta(\beta_n - \alpha_n),\\ \bar{p}_n &= \sum_{i=1}^{k} c_i v_0(\alpha_n + \theta i(\beta_n - \alpha_n)),\\ \gamma_n &= \beta_n + \frac{\xi_n}{1 - \bar{p}_n},\\ d_n &= \left(1 + \int_0^1 v_0(\alpha_n + \theta(\gamma_n - \alpha_n))d\theta\right)(\gamma_n - \alpha_n) + \xi_n,\\ \bar{q}_n &= \sum_{i=1}^{k} c_i v_0(\alpha_n + \theta i(\gamma_n - \alpha_n)),\\ \alpha_{n+1} &= \gamma_n + \frac{d_n}{1 - \bar{q}_n},\\ \delta_{n+1} &= \int_0^1 v((1-\theta)(\alpha_{n+1} - \alpha_n))d\theta(\alpha_{n+1} - \alpha_n) + (1 + v_0(\alpha_n))(\alpha_{n+1} - \beta_n)\end{aligned}$$

and

$$\beta_{n+1} = \alpha_{n+1} + \frac{\delta_{n+1}}{1 - v_0(\alpha_{n+1})}.$$

The motivation follows using similar calculations to the local case:

$$\begin{aligned}
F(y_n) &= F(y_n)-F(x_n)-F'(x_n)(y_n-x_n)\\
&= \int_0^1 (F'(x_n+\theta(y_n-x_n))d\theta-F'(x_n))(y_n-x_n),\\
\|L^{-1}F(y_n)\| &\leq \int_0^1 v((1-\theta)(\beta_n-\alpha_n))d\theta(\beta_n-\alpha_n)=\xi_n,\\
\|z_n-y_n\| &\leq \frac{\xi_n}{1-p_n}=\gamma_n-\beta_n,\\
F(z_n) &= F(z_n)-F(y_n)+F(y_n),\\
\|L^{-1}F(z_n)\| &\leq d_n,\\
\|x_{n+1}-z_n\| &\leq \frac{d_n}{1-\bar{q}_n}=\alpha_{n+1}-\gamma_n,\\
F(x_{n+1}) &= F(x_{n+1})-F(x_n)-F'(x_n)^{-1}(x_{n+1}-x_n)+F'(x_n)(x_{n+1}-y_n),\\
\|L^{-1}F(x_{n+1})\| &\leq \delta_{n+1},\\
\|y_{n+1}-x_{n+1}\| &\leq \|F'(x_{n+1})^{-1}L\|\|L^{-1}F(x_{n+1})\|\\
&\leq \frac{\delta_{n+1}}{1-v_0(\alpha_{n+1})}=\beta_{n+1}-\alpha_{n+1}.
\end{aligned}$$

4. References

[1] Abbasbandy, S., *Improving Newton-Raphson method for nonlinear equations modified Adomian decomposition method*, Appl. Math. Comput., 145, (2003), 887-893.

[2] Adomian, G., *Solving Frontier Problems of Physics: The Decomposition Method*, Kluwer Academic Publishers, Dordrecht, (1994).

[3] Argyros, I. K., *Convergence and Applications of Newton-Type Iterations*, Springer-Verlag, New York, (2008).

[4] Argyros, I. K., George, S., *Mathematical modeling for the solution of equations and systems of equations with applications*, Volume-IV, Nova Publishes, NY, (2020).

[5] Argyros, I. K., George, S., *On the complexity of extending the convergence region for Traubs method*, Journal of Complexity, 56, (2022), 101423.

[6] Bhalekhar, S., Daftardar-Gejji, V., *Convergence of the new iterative method*, Int. J. Differ. Eqs. 2011, (2011), 1-10. Article ID-989065.

[7] Chun, C., *Iterative methods improving Newton's method by the decomposition method*, Comput. Math. Appl., 50, (2005), 1559-1568.

[8] Daftardar-Gejji, V., Jafari, H., *An iterative method for solving nonlinear functional equations*, J. Math. Anal. Appl., 316, (2006), 753-763.

[9] He, J. H., *A new iteration method for solving algebraic equations*, Appl. Math. Comput., 135, (2003), 81-84.

[10] Noor, M. A., Noor, K. I., Waseem, M., *Higher order iterative algorithms for solving nonlinear equations*, World Appl. Sci. J., 16, (2012), 1657-1663.

[11] Noor, M. A., Waseem, M., Noor, K. I., Ali, M. A., *New iterative technique for solving nonlinear equations*, Appl. Math. Comput., 265, (2015), 1115-1125.

[12] Traub, J., *Iterative methods for solution of equations*, Prentice-Hall, Englewood Cliffs, New Jersey, USA, (1964).

Chapter 29

An Extension of Chebyshev's Method

1. Introduction

The method is

$$\begin{aligned} y_n &= x_n - F'(x_n)^{-1}F(x_n) \\ z_n &= x_n + a(y_n - x_n) \\ x_{n+1} &= x_n - \frac{1}{a^2}F'(x_n)^{-1}((a^2+a-1)F(x_n) + F(z_n)), \end{aligned} \tag{29.1}$$

where $a \in (0,1]$ is a given parameter. The method is studied in [6] as a simplification of Chebyshev's method which can be given in a similar form as

$$\begin{aligned} a_n &= -F'(x_n)^{-1}F(x_n) \\ b_n &= -\frac{1}{2}F'(x_n)^{-1}F''(x_n)a_n a_n \\ x_{n+1} &= x_n + \delta_n + \gamma_n. \end{aligned} \tag{29.2}$$

The local convergence is shown to be three for method (29.1) in [6] which is the same as with the method (29.2). However it is required that $F^{(4)}$ not on the method exists and is bounded when $B_1 = B_2 = \mathbb{R}^k$. The new local convergence analysis relies on the operators on the method (29.1) (i.e., F'). The same is true for the semi-local convergence analysis. Moreover, the convergence in both cases is presented in the more general Banach space setting. Similar works can be done for the methods in [1]-[5], [7]-[11].

2. Local Convergence

The majorant functions $g_i = 1,2,3$ are:

$$\begin{aligned} g_1(t) &= \frac{\int_0^1 w((1-\theta)t)d\theta}{1-w_0(t)}, \\ g_2(t) &= |1-a| + a g_1(t) \end{aligned}$$

and

$$g_3(t) = \frac{\int_0^1 w((1-\theta)t)d\theta}{1-w_0(t)} + \frac{1-a}{a^2}\frac{1+\int_0^1 w_0(\theta t)d\theta}{1-w_0(t)} + \frac{\left(1+\int_0^1 w_0(\theta g_2(t)t)d\theta\right) g_2(t)}{a^2(1-w_0(t))}.$$

The motivation follows from the estimates:

$$\begin{aligned}
y_n - x^* &= x_n - x^* - F'(x_n)^{-1}F(x_n),\\
\|y_n - x^*\| &\le \frac{\int_0^1 w((1-\theta)\|x_n-x^*\|)d\theta\|x_n-x^*\|}{1-w_0(\|x_n-x^*\|)}\\
&\le g_1(\|x_n-x^*\|)\|x_n-x^*\| \le \|x_n-x^*\| < r,\\
z_n - x^* &= x_n - x^* + a(y_n - x^* + x^* - x_n)\\
&= (1-a)(x_n-x^*) + a(y_n-x^*),\\
\|z_n - x^*\| &\le [|1-a| + ag_1(\|x_n-x^*\|)]\|x_n-x^*\|\\
&\le g_2(\|x_n-x^*\|)\|x_n-x^*\| \le \|x_n-x^*\|,\\
x_{n+1} &= x_n - x^* - F'(x_n)^{-1}F(x_n) + \frac{1-a}{a^2}F'(x_n)^{-1}F(x_n) - \frac{1}{a^2}F'(x_n)^{-1}F(z_n)
\end{aligned}$$

and

$$\begin{aligned}
\|x_{n+1} - x^*\| \le & \left[\frac{\int_0^1 w((1-\theta)\|x_n-x^*\|)d\theta}{1-w_0(\|x_n-x^*\|)}\right.\\
& + \frac{(1-a)\left(1+\int_0^1 w_0(\theta\|x_n-x^*\|)d\theta\right)}{a^2(1-w_0(\|x_n-x^*\|))}\\
& \left.\frac{\left(1+\int_0^1 w_0(\theta\|z_n-x^*\|)d\theta\right) g_2(\|x_n-x^*\|)}{a^2(1-w_0(\|x_n-x^*\|))}\right] \|x_n-x^*\|\\
\le & \; g_3(\|x_n-x^*\|)\|x_n-x^*\| \le \|x_n-x^*\|.
\end{aligned}$$

3. Semi-Local Analysis

The majorant sequence $\{\alpha_n\}$ for $\alpha_0 = 0, \beta_0 \ge \|F'(x_0)^{-1}F(x_0)\|$ and each $n = 0, 1, 2, \ldots$ by

$$\begin{aligned}
\gamma_n &= \beta_n + (1-a)(\beta_n - \alpha_n),\\
\lambda_n &= \left(1+\int_0^1 v_0(\beta_n + \theta(\gamma_n - \beta_n))d\theta\right)(\gamma_n - \beta_n) + \int_0^1 v((1-\theta)(\beta_n-\gamma_n))d\theta(\beta_n - \alpha_n),\\
\alpha_{n+1} &= \gamma_n + \frac{1-a}{a^2}(\beta_n - \alpha_n) + \frac{\lambda_n}{a^2(1-v_0(\alpha_n))},\\
\delta_{n+1} &= \int_0^1 v((1-\theta)(\alpha_{n+1}-\alpha_n))d\theta(\alpha_{n+1}-\alpha_n) + (1+v_0(\alpha_n))(\alpha_{n+1}-\beta_n)
\end{aligned}$$

and

$$\beta_{n+1} = \alpha_{n+1} + \frac{\delta_{n+1}}{1-v_0(\alpha_{n+1})}.$$

The motivation follows from

$$
\begin{aligned}
z_n - y_n &= x_n - y_n + a(y_n - x_n) = (a-1)(y_n - x_n),\\
\|z_n - y_n\| &\leq (1-a)\|y_n - x_n\| \leq (1-a)(\beta_n - \alpha_n) = \gamma_n - \beta_n,\\
F(y_n) &= F(y_n) - F(x_n) - F'(x_n)(y_n - x_n)\\
&= \int_0^1 (F'(x_n + \theta(y_n - x_n))d\theta - F'(x_n))(y_n - x_n),\\
\|L^{-1}F(y_n)\| &\leq \int_0^1 v((1-\theta)(\beta_n - \alpha_n))d\theta(\beta_n - \alpha_n) = \mu_n,
\end{aligned}
$$

$$
\begin{aligned}
F(z_n) &= F(z_n) - F(y_n) + F(y_n),\\
\|L^{-1}F(z_n)\| &\leq \left(1 + \int_0^1 v_0(\beta_n + \theta(\gamma_n - \beta_n))d\theta\right)(\gamma_n - \beta_n) + \mu_n = \lambda_n,\\
x_{n+1} - z_n &= -\frac{1}{a^2}(a^2 + a - 1)F'(x_n)^{-1}F(x_n)\\
&\quad -\frac{1}{a^2}F'(x_n)^{-1}F(z_n) - a(y_n - x_n)\\
&= \frac{a-1}{a^2}(y_n - x_n) - \frac{1}{a^2}F'(x_n)^{-1}F(z_n),\\
\|x_{n+1} - z_n\| &\leq \frac{1-a}{a^2}(\beta_n - \alpha_n) + \frac{\lambda_n}{a^2(1 - v_0(\alpha_n))} = \alpha_{n+1} - \gamma_n.\\
F(x_{n+1}) &= F(x_{n+1}) - F(x_n) - F'(x_n)(y_n - x_n)\\
&= F(x_{n+1}) - F(x_n) - F'(x_n)(x_{n+1} - x_n) + F'(x_n)(x_{n+1} - y_n),\\
\|L^{-1}F(x_{n+1})\| &\leq \int_0^1 v((1-\theta)\|x_{n+1} - x_n\|)d\theta\|x_{n+1} - x_n\|\\
&\quad +(1 + v_0(\|x_n - x_0\|))\|x_{n+1} - y_n\| \leq \delta_{n+1},\\
\|y_{n+1} - x_{n+1}\| &\leq \|F'(x_{n+1})^{-1}L\|\|L^{-1}F(x_{n+1})\|\\
&\leq \frac{\delta_{n+1}}{1 - v_0(\alpha_{n+1})} = \beta_{n+1} - \alpha_{n+1}.
\end{aligned}
$$

The rest of the analysis follows as in previous chapters.

4. References

[1] Argyros, I. K., *Convergence and Applications of Newton-Type Iterations*, Springer-Verlag, New York, (2008).

[2] Argyros, I. K., *The theory and application of iteration methods*, second edition, Engineering series, Boca Raton, Florida, USA, (2022).

[3] Argyros, I. K., Chen, D., *Results on the Chebyshev method in Banach spaces*, Proyecciones, 12(2), (1993), 119-128.

[4] Argyros, I. K., George, S., *Mathematical modeling for the solution of equations and systems of equations with applications*, Volume-IV, Nova Publishes, NY, (2020).

[5] Argyros, I. K., George, S., *On the complexity of extending the convergence region for Traubs method*, Journal of Complexity, 56, (2022), 101423.

[6] Ezquerro, J. A., Hernndez, M. A., *An optimization of Chebyshev's method*, Journal of Complexity, 25, (2009), 343-361.

[7] Hernndez, M. A., *Chebyshev's approximation algorithms and applications*, Comput. Math. Appl., 41, (2001), 433-445.

[8] Kantorovich, L. V., Akilov, G. P., *Functional Analysis*, Pergamon Press, Oxford, (1982).

[9] Traub, J. F., *Iterative Methods for the Solution of Equations*, Prentice Hall, Englewood Cliffs, NJ, (1964).

[10] Weerakoon, S., Fernando, T. G. I., *A variant of Newton's method with accelerated third-order convergence*, Appl. Math. Lett., 13, (2000), 87-93.

Chapter 30

Newton-Simpson-Type Method

1. Introduction

The method is defined for $x_0 \in \Omega$ and each $n = 0, 1, 2...$ by

$$\begin{aligned} y_n &= x_n - F'(x_n)^{-1}F(x_n) \\ A_n &= F'(y_n) + 3F'\left(\frac{x_n + 2y_n}{3}\right) + 3F'\left(\frac{y_n + 2x_n}{3}\right) + F'(x_n) \qquad (30.1) \\ x_{n+1} &= x_n - 8A_n^{-1}F(x_n). \end{aligned}$$

The local order three is shown in [22] under the Taylor series expansion approach when $B_1 = B_2 = \mathbb{R}^k$ and by assuming the existence and boundedness of $F^{(4)}$. In this chapter only the operators on the method are used to show the local convergence. Moreover, the semilocal convergence (not studied in [22]) is also presented. Our technique can be used to apply the applicability of other methods[1]-[21], [23,24][.

2. Local Convergence

The majorant functions $g_i, i = 1, 2$ are:

$$\begin{aligned} g_1(t) &= \frac{\int_0^1 w((1-\theta)t)d\theta}{1 - w_0(t)}, \\ \bar{w}(t) &= \begin{cases} w((1+g_1(t))t) \\ w_0(t) + w_0(g_1(t)t), \end{cases} \\ \bar{\bar{w}}(t) &= \begin{cases} w\left(\frac{(2+g_1(t))t}{3}\right) \\ w_0(t) + w_0\left(\frac{(1+2g_1(t))t}{3}\right), \end{cases} \\ \bar{\bar{\bar{w}}}(t) &= \begin{cases} w\left(\frac{(1+g_1(t))t}{3}\right) \\ w_0(t) + w_0\left(\frac{(2+g_1(t))t}{3}\right), \end{cases} \\ q(t) &= \frac{1}{8}(w_0(t) + w_0(g_1(t)t) + 3w_0\left(\frac{(1+2g_1(t))t}{3}\right) + 3w_0\left(\frac{(2+g_1(t))t}{3}\right) \end{aligned}$$

and

$$g_2(t) = \frac{\int_0^1 w((1-\theta)t)d\theta}{1-w_0(t)} + \frac{(\bar{w}(t)+3\bar{\bar{w}}(t)+3\bar{\bar{\bar{w}}}(t))(1+\int_0^1 w_0(\theta t)d\theta)}{8(1-q(t))(1-w_0(t))}.$$

The calculations leading to these functions are:

$$\begin{aligned} y_n - x^* &= x_n - x^* - F'(x_n)^{-1}F(x_n), \\ \|y_n - x^*\| &\leq \frac{\int_0^1 w((1-\theta)\|x_n-x^*\|)d\theta\|x_n-x^*\|}{1-w_0(\|x_n-x^*\|)} \\ &\leq g_1(\|x_n-x^*\|)\|x_n-x^*\|)\|x_n-x^*\| \leq \|x_n-x^*\| < r, \end{aligned}$$

$$\begin{aligned} \|(8L)^{-1}(A_n-8L)\| &\leq \frac{1}{8}[\|L^{-1}(F'(x_n)-L)\| \\ &\quad +\|L^{-1}(F'(y_n)-L)\|+3\|L^{-1}\left(\left(\frac{x_n+2y_n}{3}\right)-L\right)\|] \\ &\leq \frac{1}{8}[w_0(\|x_n-x^*\|)+w_0(\|y_n-x^*\|)] \end{aligned}$$

$$+3w_0\left(\frac{\|x_n-x^*\|+2\|y_n-x^*\|}{3}\right)+3w_0\left(\frac{2\|x_n-x^*\|+\|y_n-x^*\|}{3}\right)=q_n,$$

$$\begin{aligned} \|A_n^{-1}L\| &\leq \frac{1}{8(1-q_n)}, \\ \|L^{-1}(A_n-8F'(x_n))\| &\leq \|L^{-1}(F'(y_n)-F'(x_n))\| \\ &\quad +3\|L^{-1}\left(F'\left(\frac{x_n+2y_n}{3}\right)-F'(x_n)\right)\| \\ &\quad +3\|L^{-1}\left(F'\left(\frac{2x_n+y_n}{3}\right)-F'(x_n)\right)\| \\ &\leq \bar{w}_n+3\bar{\bar{w}}_n+3\bar{\bar{\bar{w}}}_n \end{aligned}$$

and

$$\begin{aligned} \|x_{n+1}-x^*\| &\leq \left[\frac{\int_0^1 w((1-\theta)\|x_n-x^*\|)d\theta}{1-w_0(\|x_n-x^*\|)}\right. \\ &\quad \left.+\frac{(\bar{w}_n+3\bar{\bar{w}}_n+3\bar{\bar{\bar{w}}}_n)(1+\int_0^1 w_0(\theta\|x_n-x^*\|)d\theta)}{8(1-q_n)(1-w_0(\|x_n-x^*\|))}\right]\|x_n-x^*\| \\ &\leq g_2(\|x_n-x^*\|)\|x_n-x^*\| \leq \|x_n-x^*\|. \end{aligned}$$

3. Semi-local Convergence

The majorant sequence $\{\alpha_n\}$ is defined for $\alpha_0 = 0, \beta_0 \geq \|F'(x_0)^{-1}F(x_0)\|$ and each $n = 0,1,2,...$ by

$$\begin{aligned}
\bar{v}(t) &= \begin{cases} v(\beta_n - \alpha_n) \\ v_0(\alpha_n) + v_0(\beta_n), \end{cases} \\
\bar{\bar{v}}(t) &= \begin{cases} v\left(\dfrac{2(\beta_n - \alpha_n)}{3}\right) \\ v_0(\alpha_n) + v_0\left(\dfrac{\alpha_n + 2\beta_n}{3}\right), \end{cases} \\
\bar{\bar{\bar{v}}}(t) &= \begin{cases} v\left(\dfrac{\beta_n - \alpha_n}{3}\right) \\ v_0(\alpha_n) + v_0\left(\dfrac{2\alpha_n + \beta_n}{3}\right), \end{cases} \\
p_n &= \frac{1}{8}(v_0(\alpha_n) + v_0(\beta_n) + 3v_0\left(\frac{\alpha_n + 2\beta_n}{3}\right) + 3v_0\left(\frac{2\alpha_n + \beta_n}{3}\right)], \\
\alpha_{n+1} &= \beta_n + \frac{(\bar{v}(t) + \bar{\bar{v}}(t) + \bar{\bar{\bar{v}}}(t))(\beta_n - \alpha_n)}{8(1 - p_n)} \\
\delta_{n+1} &= \int_0^1 v((1-\theta)(\beta_n - \alpha_n))d\theta(\beta_n - \alpha_n) + (1 + v_0(\alpha_n))(\alpha_{n+1} - \beta_n),
\end{aligned}$$

and

$$\beta_{n+1} = \alpha_{n+1} + \frac{\delta_{n+1}}{1 - v_0(\alpha_{n+1})},$$

The estimates are:

$$\begin{aligned}
x_{n+1} - y_n &= (F'(x_n)^{-1} - 8A_n^{-1})F(x_n) \\
&= -A_n^{-1}(8F'(x_n) - A_n)F'(x_n)^{-1}F(x_n) \\
&= A_n^{-1}(8F'(x_n) - A_n)(y_n - x_n), \\
\|x_{n+1} - y_n\| &\leq \|A_n^{-1}L\|\|L^{-1}(8F'(x_n) - A_n)\|\|y_n - x_n\| \\
&\leq \frac{(\bar{v}_n + 3\bar{\bar{v}}_n + 3\bar{\bar{\bar{v}}}_n)(\beta_n - \alpha_n)}{8(1 - p_n)} = \alpha_{n+1} - \beta_n, \\
\|L^{-1}F(x_{n+1})\| &= \|L^{-1}(F(x_{n+1}) - F(x_n) - F'(x_n)(x_{n+1} - x_n) \\
&\quad + F'(x_n)(x_{n+1} - x_n))\| \leq \bar{\delta}_{n+1} \leq \delta_{n+1}, \\
\|y_{n+1} - x_{n+1}\| &\leq \|F'(x_{n+1})^{-1}L\|\|L^{-1}F(x_{n+1})\| \\
&\leq \frac{\delta_{n+1}}{1 - v_0(\alpha_{n+1})} = \beta_{n+1} - \alpha_{n+1}.
\end{aligned}$$

4. References

[1] Argyros, I.K., *Convergence and Application of Newton-type Iterations*, Springer, Berlin (2008).

[2] Argyros, I.K., Cho, Y.J., George, S., *Local convergence for some third-order iterative methods under weak conditions*, J. Korean Math. Soc., 53(4), (2016), 781793.

[3] Argyros, I.K., Cho, Y.J., Hilout, S., *Numerical Methods for Equations and its Applications*, Taylor & Francis, CRC Press, New York (2012).

[4] Argyros, I.K., George, S., *Extending the applicability of a seventh-order method without inverses of derivatives under weak conditions*, Int. J. Appl. Comput. Math., 6(4), (2019). https://doi.org/10.1007/s40819-019-0760-6.

[5] Argyros, I.K., George, S., *Local convergence for an almost sixth-order method for solving equations under weak conditions*, SeMA J., 75(2), (2017), 163171.

[6] Argyros, I.K., George, S., *Local convergence of deformed Halley method in Banach space under Hlder continuity conditions*, J. Nonlinear Sci. Appl., 8, (2015), 246254.

[7] Argyros, I.K., George, S., *Local convergence of modified Halley-like methods with less computation of inversion*, Novi Sad J. Math., 45(2), (2015), 4758.

[8] Argyros, I.K., George, S., *On the complexity of extending the convergence region for Traubs method*, J. Complex., 56, 101423 (2020). https://doi.org/10.1016/j.jco.2019.101423.

[9] Argyros, I.K., George, S., Magren, .A., *Local convergence for multi-point-parametric Chebyshev-Halley-type methods of higher convergence order*, J. Comput. Appl. Math., 282, (2015), 215224.

[10] Argyros, I.K., González, D., *Local convergence for an improved Jarratt-type method in Banach space*, Int. J. Interact. Multimed. Artif. Intell., 3, (2015), 2025.

[11] Argyros, I.K., Hilout, S., *Computational Methods in Nonlinear Analysis*, World Scientific Publishing, New Jersey (2013).

[12] Argyros, I.K., Hilout, S., *On the local convergence of fast two-step Newton-like methods for solving nonlinear equations*, J. Comput. Appl. Math., 245, (2013), 19.

[13] Argyros, I.K., Magren, .A., *A study on the local convergence and the dynamics of Chebyshev-Halley-type methods free from second derivative*, Numer. Algor., 71(1), (2015), 123.

[14] Argyros, I.K., Sharma, D., Parhi, S.K., *On the local convergence of Weerakoon-Fernando method with continuous derivative in Banach spaces*, SeMA J., (2020). https://doi.org/10.1007/s40324-020-00217-y.

[15] Cordero, A., Ezquerro, J.A., Hernandez-Veron, M.A., *On the local convergence of a fifth-order iterative method in Banach spaces*, Appl. Math. Comput., 251, (2014), 396403.

[16] Kou, J., Li, Y., Wang, X., *A composite fourth-order iterative method for solving nonlinear equations*, Appl. Math. Comput., 184, (2007), 471475.

[17] Magren, .A., Argyros, I.K., *On the local convergence and the dynamics of Chebyshev-Halley methods with six and eight orders of convergence*, J. Comput. Appl. Math., 298, (2016), 236251.

[18] Martínez, E., Singh, S., Hueso, J.L., Gupta, D.K., *Enlarging the convergence domain in local convergence studies for iterative methods in Banach spaces*, Appl. Math. Comput., 281, (2016), 252265.

[19] Potra, F.A., Ptak, V., *Nondiscrete induction and iterative processes*, Research Notes in Mathematics, vol. 103, Pitman Publ., Boston (1984).

[20] Sharma, J.R., Argyros, I.K., *Local convergence of a Newton-Traub composition in Banach spaces*, SeMA J., 75(1), (2017), 5768.

[21] Singh, M.K., Singh, A.K., *Variant of Newtons method using Simpsons 3/8th rule*, Int. J. Appl. Comput. Math., 6, 20 (2020). https://doi.org/10.1007/s40819-020-0770-4.

[22] Traub, J.F., *Iterative Methods for Solution of Equations*, Prentice-Hall, Englewood Cliffs (1964).

[23] Weerakoon, S., Fernando, T.G.I., *A variant of Newtons method with accelerated third-order convergence*, Appl. Math. Lett., 13(8), (2000), 8793.

Chapter 31

Chebyshev-Type Three-Step Method

1. Introduction

The method is defined for $x_0, u_0 \in \Omega$ and each $n = 0, 1, 2, ...$ by

$$\begin{aligned} y_n &= x_n - A_n^{-1}F(x_n), \\ z_n &= x_n + a(y_n - x_n) \end{aligned} \tag{31.1}$$

and

$$x_{n+1} = x_n - A_n^{-1}(bF(x_n) + cF(z_n)),$$

where $a, b, c \in \mathbb{R}$ and $A : \Omega \times \Omega \longrightarrow f(B_1)$. The method is studied in [2]. In the present chapter the applicability of the method is extended. The methodology can be applied on other methods [1,3].

2. Local Convergence

The conditions are:

(C_1) There exists a continuous and nondecreasing function $w_0 : T = [0, \infty) \longrightarrow T$ such that the equation $w_0(t) - 1 = 0$ has a smallest positive solution ρ_0. Let $T_0 = [0, \rho_0)$.

(C_2) There exist continuous and nondecreasing functions $w : T_0 \longrightarrow T$, $f : T_0 \longrightarrow T$ and $w_1 : T_0 \times T_0 \longrightarrow T$. Moreover, the equation $q(t) - 1 = 0$ has a smallest solution $\rho \in (0, \rho_0)$, where $q(t) = w_1(t, f(t))$.

(C_3) There exists an invertible operator L such that for all $x \in \Omega$

$$\|L^{-1}(F'(x) - L)\| \leq w_0(\|x - x^*\|).$$

Let $\Omega_0 = \Omega \cap U(x^*, \rho)$.

(C_4) The equations $g_i(t) - 1 = 0, i = 1, 2, 3$ have smallest solutions $r_i \in (0, \rho)$, respectively, where

$$\begin{aligned} p(t) &= w_1(t, f(t)) + w_0(t), \\ g_1(t) &= \frac{\int_0^1 w((1-\theta)t)d\theta}{1 - w_0(t)} + \frac{p(t)(1 + \int_0^1 w_0(\theta t)d\theta)}{(1 - w_0(t))(1 - q(t))}, \\ g_2(t) &= (|1 - a| + |a|g_1(t))t \end{aligned}$$

and

$$g_1(t) = \frac{\int_0^1 w((1-\theta)t)d\theta}{1-w_0(t)} + \frac{p(t)(1+\int_0^1 w_0(\theta t)d\theta)}{(1-w_0(t))(1-q(t))}, \\ \frac{|b-1|(1+\int_0^1 w_0(\theta t)d\theta)}{1-q(t)} + \frac{|c|(1+\int_0^1 w_0(\theta g_2(t)t)d\theta)}{1-q(t)}.$$

The motivational calculations for the functions g_i in turn are:

$$\begin{aligned}
\|L^{-1}(A(x_n,u_n)-L)\| &\leq w_1(\|x_n-x^*\|, \|u_n-x^*\|) \\
&\leq w_1(\|x_n-x^*\|, f(\|x_n-x^*\|)) = q_n, \\
\|A(x_n,u_n)^{-1}L\| &\leq \frac{1}{1-q_n}, \\
y_n-x^* &= x_n-x^*-F'(x_n)^{-1}F(x_n)+(F'(x_n)^{-1}-A_n^{-1})F(x_n), \\
\|L^{-1}(A_n-F'(x_n))\| &\leq \|L^{-1}(A_n-F'(x_n))\|+\|L^{-1}(F'(x_n)-L)\| \\
&\leq w_1(\|x_n-x^*\|, \|u_n-x^*\|)+w_0(\|x_n-x^*\|) \\
&\leq w_1(\|x_n-x^*\|, f(\|x_n-x^*\|))+w_0(\|x_n-x^*\|) = p_n,
\end{aligned}$$

$$\begin{aligned}
\|y_n-x^*\| &\leq \left[\frac{\int_0^1 w((1-\theta)\|x_n-x^*\|)d\theta}{1-w_0(\|x_n-x^*\|)} + \frac{p_n(1+\int_0^1 w_0(\theta\|x_n-x^*\|)d\theta)}{(1-w_0(\|x_n-x^*\|))(1-q_n)}\right] \\
&\quad \times \|x_n-x^*\| \\
z_n-x^* &= (1-a)(x_n-x^*)+a(y_n-x^*), \\
\|z_n-x^*\| &\leq (|1-a|+|a|g_1(\|x_n-x^*\|))\|x_n-x^*\| \\
&\leq g_2(\|x_n-x^*\|)\|x_n-x^*\| \leq \|x_n-x^*\|, \\
x_{n+1}-x^* &= x_n-x^*-F'(x_n)^{-1}F(x_n)+(F'(x_n)^{-1}-A_n^{-1})F(x_n) \\
&\quad -(b-1)A_n^{-1}F(x_n)-cA_n^{-1}F(z_n), \qquad (31.2)
\end{aligned}$$

$$\begin{aligned}
\|x_{n+1}-x^*\| &\leq \left[\frac{\int_0^1 w((1-\theta)\|x_n-x^*\|)d\theta}{1-w_0(\|x_n-x^*\|)}\right. \\
&\quad + \frac{p_n(1+\int_0^1 w_0(\theta\|x_n-x^*\|)d\theta)}{(1-w_0(\|x_n-x^*\|))(1-q_n)} \\
&\quad + \frac{|b-1|(1+\int_0^1 w_0(\theta\|x_n-x^*\|)d\theta)}{1-q_n} \\
&\quad \left. + \frac{|c|(1+\int_0^1 w_0(\theta g_2(\|x_n-x^*\|)\|x_n-x^*\|)d\theta)}{1-q_n}\right] \|x_n-x^*\| \\
&\leq g_3(\|x_n-x^*\|)]\|x_n-x^*\| \leq \|x_n-x^*\|.
\end{aligned}$$

3. Semi-local Convergence

The conditions are:

(H_1) There exist a CNF $v_0 : T \longrightarrow T$ such that the equation $v_0(t) - 1 = 0$ has a smallest solution $s_0 \in T - \{0\}$.
Let $T_2 = [0, s_0)$.

(H_2) There exist CNF functions $g : T_2 \longrightarrow T$, $v_1 : T_2 \times T_2 \longrightarrow T$. Moreover, the equation $v_0(t, g(t)) - 1 = 0$ has a smallest solution $s_1 \in T_2 - \{0\}$.

(H_3) There exists an invertible operator L such that for each $x, u \in \Omega$

$$\|L^{-1}(A(x,u) - L)\| \leq v_0(\|x - x_0\|, \|u - x_0\|). \tag{31.3}$$

Let $\Omega_2 = \Omega \cap U(x_0, s_1)$

(H_4)

$$\begin{aligned} \|u - x_0\| &\leq g(\|x - x_0\|) \\ \|L^{-1}(A(x,u) - L)\| &\leq v_1(\|x - x_0\|, \|u - x_0\|) \end{aligned}$$

for each $x^u \in \Omega_2$
Notice that if $x_0 \in \Omega$, $v_0(\|x_0 - x_0\|, g(\|x_0 - x_0\|)) < 1$, so $A(x_0, u)^{-1}$ exists. Let $\|A_0^{-1}F(x_0)\| \leq \beta_0$.

Define the majorant sequence $\{\alpha_n\}$ for $\alpha_0 = 0, \beta_0 \geq 0$ and each $n = 0, 1, 2, \ldots$ by

$$\begin{aligned} \gamma_n &= \beta_n + |1 - a|(\beta_n - a_n), \\ \lambda_n &= (1 + v_0(\alpha_n, \gamma_n)d\theta)(\gamma_n - \alpha_n) \\ &\quad + (1 + v_1(\alpha_n, g(\alpha_n)))(\beta_n - \alpha_n), \\ \bar{q}_n &= v_1(\alpha_n, g(\alpha_n)), \\ \alpha_{n+1} &= \gamma_n + |b - a|(\beta_n - \alpha_n) + \frac{|c|\lambda_n}{1 - \bar{q}_n}, \\ \delta_{n+1} &= (1 + v_0(\alpha_n, \alpha_{n+1})))(\alpha_{n+1} - \alpha_n) \\ &\quad + (1 + v_0(\alpha_n, g(\alpha_n)))(\beta_n - \alpha_n) \end{aligned}$$

and

$$\beta_{n+1} = \alpha_{n+1} + \frac{\delta_{n+1}}{1 - \bar{q}_{n+1}}$$

The calculations leading to the majorizing sequence $\{\alpha_n\}$ are in turn:

$$\begin{aligned} z_n - y_n &= A_n^{-1}F(x_n) + (y_n - x_n) \\ &= -(y_n - x_n) + a(y_n - x_n) = (a - 1)(y_n - x_n), \\ \|z_n - y_n\| &\leq |1 - a|\|y_n - x_n\| \leq |1 - a|(\beta_n - \alpha_n) = \gamma_n - \beta_n, \\ x_{n+1} - z_n &= -a(y_n - x_n) - bA_n^{-1}F(x_n) - cA_n^{-1}F(z_n), \\ &= -a(y_n - x_n) + b(y_n - x_n) - cA_n^{-1}F(z_n) \\ &= (b - a)(y_n - x_n) - cA_n^{-1}F(z_n) \end{aligned}$$

But

$$\begin{aligned} F(z_n) &= F(z_n) - F(x_n) + F(x_n) = F(z_n) - F(x_n) - A_n(y_n - x_n), \\ \|L^{-1}F(z_n)\| &\leq (1 + v_0(\alpha_n, \gamma_n))(\gamma_n - \alpha_n) \\ &\quad + (1 + v_1(\alpha_n, g(\alpha_n)))(\beta_n - \alpha_n) = \lambda_n, \end{aligned}$$

so

$$
\begin{aligned}
\|x_{n+1}-z_n\| &\leq |b-a|(\beta_n-\alpha_n)+\frac{|c|\lambda_n}{1-\bar{q}_n}=\alpha_{n+1}-\gamma_n,\\
F(x_{n+1}) &= F(x_{n+1})-F(x_n)-A_n(y_n-x_n),\\
\|L^{-1}F(x_{n+1})\| &\leq (1+v_0(\alpha_n,\alpha_{n+1}))(\alpha_{n+1}-\alpha_n)\\
&\quad +(1+v_0(\alpha_n,g(\alpha_n)))(\beta_n-\alpha_n)=\delta_{n+1},
\end{aligned}
$$

thus

$$
\begin{aligned}
\|y_{n+1}-x_{n+1}\| &\leq \|A_{n+1}^{-1}L\|\|L^{-1}F(x_{n+1})\|\\
&\leq \frac{\delta_{n+1}}{1-\bar{q}_{n+1}}=\beta_{n+1}-\alpha_{n+1}.
\end{aligned}
$$

Remark. If $L=F'(x^*)$(local case) or $L=[x_{-1},x_0;F]$ (semi-local case) the results can be reduced to the ones in [2]. Otherwise, the results in this chapter generalize and improve the earlier ones. The new technique does not require x^* to be a simple solution, i.e $F'(x^*)^{-1}$ to exist. Hence, it can also be applied to solve equations with solutions of multiplicity greater than one.

4. References

[1] Argyros, C.I., Argyros, M.I., Argyros, I.K., Magren, .A., Sarria, I., *Local and Semi-local convergence for Chebyshev two-point-like methods with applications in different fields*, CMMSE, (2020).

[2] Argyros, I.K., *The theory and applications of iteration methods*, 2nd Edition, Engineering Series, CRC Press, Taylor and Francis Group, (2022).

[3] Traub, J.F., *Iterative Methods for the Solution of Equations*, Chelsea Publishing Company, New York, (1982).

Chapter 32

Two Comparable Sixth-Order Methods

1. Introduction

The methods are defined for $x_0 \in \Omega$ and each $n = 0,1,2,...$ by

$$\begin{aligned}
y_n &= x_n - aF'(x_n)^{-1}F(x_n), \\
z_n &= x_n - \left[\frac{23}{8}I - 3F'(x_n)^{-1}F'(y_n) + \frac{9}{8}(F'(x_n)^{-1}F'(y_n))^2\right] \\
&\quad \times F'(x_n)^{-1}F(x_n), \qquad (32.1) \\
x_{n+1} &= z_n - \left(\frac{5}{2}I - \frac{3}{2}F'(x_n)^{-1}F'(y_n)\right) F'(x_n)^{-1}F(z_n)
\end{aligned}$$

and

$$\begin{aligned}
y_n &= x_n - aF'(x_n)^{-1}F(x_n), \\
z_n &= x_n - \left(I - \frac{21}{8}F'(x_n)^{-1}F'(y_n) - \frac{9}{2}(F'(x_n)^{-1}F'(y_n))^2\right. \\
&\quad \left. + \frac{15}{8}(F'(x_n)^{-1}F'(y_n))^3\right)F'(x_n)^{-1}F(x_n), \qquad (32.2) \\
x_{n+1} &= z_n - \left[3I - \frac{5}{2}F'(x_n)^{-1}F'(y_n) + (\frac{1}{2}F'(x_n)^{-1}F'(y_n))^2\right] \\
&\quad \times F'(x_n)^{-1}F(z_n).
\end{aligned}$$

The local convergence order six is shown in [11,14], respectively under the Taylor series expansion approach, using the seventh derivative when $B_1 = B_2 = \mathbb{R}^k$. In this chapter, only conditions on the operators involved in the methods are used to establish both local and semi-local convergence (not studied in [11,14]), within the more general setting of a Banach space. This approach can also be used to extend the applicability of other methods in an analogous manner. [1]-[10], [12,13].

2. Local Convergence

Method (32.1)
The majorant functions $g_i, i = 1,2,3$ are:

$$g_1(t) = \frac{\int_0^1 w((1-\theta)t)d\theta + |1-a|(1+\int_0^1 w_0(\theta t)d\theta)}{1-w_0(t)},$$

$$g_2(t) = \frac{\int_0^1 w((1-\theta)t)d\theta}{1-w_0(t)} + \frac{3}{8}\frac{\bar{w}(t)}{(1-w_0(t))^2}\left(2+\frac{\bar{w}(t)}{1-w_0(t)}\right)(1+\int_0^1 w_0(\theta t)d\theta)$$

and

$$g_3(t) = \left[\frac{\int_0^1 w((1-\theta)g_2(t)t)d\theta}{1-w_0(g_2(t)t)} + \left(\frac{\bar{\bar{w}}(t)}{(1-w_0(t))(1-w_0(g_2(t)t))} + \frac{3}{2}\frac{\bar{w}(t)}{1-w_0(t)}\right)\frac{(1+\int_0^1 w_0(\theta t)d\theta)}{1-w_0(t)}\right] g_2(t).$$

The motivation is:

$$y_n - x^* = x_n - x^* - F'(x_n)^{-1}F(x_n) + (1-a)F'(x_n)^{-1}F(x_n),$$

$$\begin{aligned}\|y_n - x^*\| &\le \frac{[\int_0^1 w((1-\theta)\|x_n-x^*\|)d\theta + |1-a|(1+\int_0^1 w_0(\theta\|x_n-x^*\|)d\theta]}{1-w_0(\|x_n-x^*\|)} \\ &\quad \times\|x_n-x^*\| \\ &\le g_1(\|x_n-x^*\|)\|x_n-x^*\| \le \|x_n-x^*\| < r,\end{aligned}$$

$$\begin{aligned}z_n - x^* &= x_n - x^* - F'(x_n)^{-1}F(x_n) \\ &\quad -\left(\frac{23}{8}I - I - 3F'(x_n)^{-1}F'(y_n) + \frac{9}{8}(F'(x_n)^{-1}F'(y_n))^2\right) \\ &\quad F'(x_n)^{-1}F(x_n) \\ &= x_n - x^* - F'(x_n)^{-1}F(x_n) \\ &\quad -\frac{3}{8}(F'(x_n)^{-1}F'(y_n) - I)[3(F'(x_n)^{-1}F'(y_n) - I) - 2I] \\ &\quad F'(x_n)^{-1}F(x_n),\end{aligned}$$

$$\begin{aligned}\|z_n - x^*\| &\le \left[\frac{\int_0^1 w((1-\theta)\|x_n-x^*\|)d\theta}{1-w_0(\|x_n-x^*\|)}\right. \\ &\quad +\frac{3}{8}\frac{\bar{w}_n}{(1-w_0(\|x_n-x^*\|))^2}\left(\frac{3\bar{w}_n}{1-w_0(\|x_n-x^*\|)}+2\right) \\ &\quad \left.\times(1+\int_0^1 w_0(\theta\|x_n-x^*\|)d\theta)\right]\|x_n-x^*\| \\ &\le g_2(\|x_n-x^*\|)\|x_n-x^*\| \le \|x_n-x^*\|,\end{aligned}$$

$$\begin{aligned} x_{n+1}-x^* &= z_n-x^*-F'(z_n)^{-1}F(z_n)+(F'(z_n)^{-1}-F'(x_n)^{-1})F(z_n) \\ &\quad -\frac{3}{2}(I-F'(x_n)^{-1}F'(y_n))F'(x_n)^{-1}F(z_n), \\ \|x_{n+1}-x^*\| &\leq \left[\frac{\int_0^1 w((1-\theta)\|z_n-x^*\|)d\theta}{1-w_0(\|z_n-x^*\|)}\right. \\ &\quad +\left(\frac{\bar{\bar{w}}_n}{(1-w_0(\|x_n-x^*\|))(1-w_0(\|z_n-x^*\|)}\right. \\ &\quad \left.+\frac{3}{2}\frac{\bar{w}_n}{1-w_0(\|x_n-x^*\|)}\right) \\ &\quad \left.\times\frac{(1+\int_0^1 w_0(\theta\|z_n-x^*\|)d\theta}{1-w_0(\|x_n-x^*\|)}\right]\|z_n-x^*\| \\ &\leq g_3(\|x_n-x^*\|)\|x_n-x^*\| \leq \|x_n-x^*\|. \end{aligned}$$

Method (32.2)
This time the functions g_i are:

$$g_1(t) = \frac{\int_0^1 w((1-\theta)t)d\theta+|1-a|(1+\int_0^1 w_0(\theta t)d\theta)}{1-w_0(t)},$$

$$g_2(t) = \frac{\int_0^1 w((1-\theta)t)d\theta}{1-w_0(t)}+\frac{3}{8}\left(\frac{\bar{w}(t)}{1-w_0(t)}\right)^2\left(2+5\frac{\bar{w}(t)}{1-w_0(t)}\right)\frac{(1+\int_0^1 w_0(\theta t)d\theta)}{1-w_0(t)}$$

and

$$\begin{aligned} g_3(t) &= \left[\frac{\int_0^1 w((1-\theta)g_2(t)t)d\theta}{1-w_0(g_2(t)t)}\right. \\ &\quad +\frac{\bar{\bar{w}}(t)(1+\int_0^1 w_0(\theta g_2(t)t)d\theta)}{(1-w_0(t))(1-w_0(g_2(t)t))} \\ &\quad \left.+\frac{1}{4}\frac{\bar{w}(t)}{1-w_0(t)}\left(3+\frac{\bar{w}(t)}{1-w_0(t)}\right)\frac{(1+\int_0^1 w_0(\theta g_2(t)t)d\theta)}{1-w_0(t)}\right]g_2(t). \end{aligned}$$

The motivation for the function g_1 is already given in method (32.2). Moreover, we can write in turn:

$$\begin{aligned} z_n-x^* &= x_n-x^*-F'(x_n)^{-1}F(x_n) \\ &\quad -\left(\frac{21}{8}F'(x_n)^{-1}F'(y_n)-\frac{9}{2}(F'(x_n)^{-1}F'(y_n))^2\right. \\ &\quad \left.+\frac{15}{8}(F'(x_n)^{-1}F'(y_n))^3\right) \\ &\quad \times F'(x_n)^{-1}F(x_n) \\ &= x_n-x^*-F'(x_n)^{-1}F(x_n)-\frac{3}{8}(F'(x_n)^{-1}F'(y_n)-I)^2 \\ &\quad (5(F'(x_n)^{-1}F'(y_n)-I)-2I)F'(x_n)^{-1}F(x_n), \end{aligned}$$

$$\begin{aligned}
\|z_n - x^*\| &\leq \Bigg[\frac{\int_0^1 w((1-\theta)\|x_n - x^*\|)d\theta}{1 - w_0(\|x_n - x^*\|)} \\
&\quad + \frac{3}{8}\left(\frac{\bar{w}_n}{1 - w_0(\|x_n - x^*\|)}\right)^2 \left(2 + 5\frac{\bar{w}_n}{1 - w_0(\|x_n - x^*\|)}\right) \\
&\quad \times \frac{(1 + \int_0^1 w_0(\theta\|x_n - x^*\|)d\theta)}{1 - w_0(\|x_n - x^*\|)}\Bigg] \|x_n - x^*\| \\
&\leq g_2(\|x_n - x^*\|)\|x_n - x^*\| \leq \|x_n - x^*\|, \\
x_{n+1} - x^* &= z_n - x^* - F'(z_n)^{-1}F(z_n) + (F'(z_n)^{-1} - F'(x_n)^{-1})F(z_n) \\
&\quad -(2I - \frac{5}{2}F'(x_n)^{-1}F'(y_n) + (\frac{1}{2}F'(x_n)^{-1}F'(y_n))^2)F'(x_n)^{-1}F(z_n) \\
&= z_n - x^* - F'(z_n)^{-1}F(z_n) + (F'(z_n)^{-1} - F'(x_n)^{-1})F(z_n) \\
&\quad -\frac{1}{4}(F'(x_n)^{-1}F'(y_n) - I)[(F'(x_n)^{-1}F'(y_n) - I) - 3I]F'(x_n)^{-1}F(z_n),
\end{aligned}$$

$$\begin{aligned}
\|x_{n+1} - x^*\| &\leq \Bigg[\frac{\int_0^1 w((1-\theta)\|z_n - x^*\|)d\theta}{1 - w_0(\|z_n - x^*\|)} \\
&\quad + \frac{\bar{\bar{w}}_n(1 + \int_0^1 w_0(\theta\|z_n - x^*\|)d\theta)}{(1 - w_0(\|x_n - x^*\|))(1 - w_0(\|z_n - x^*\|))} \\
&\quad + \frac{1}{4}\frac{\bar{w}_n}{1 - w_0(\|x_n - x^*\|)}\left(3 + \frac{\bar{w}_n}{1 - w_0(\|x_n - x^*\|)}\right) \\
&\quad \times \frac{(1 + \int_0^1 w_0(\theta\|z_n - x^*\|)d\theta)}{1 - w_0(\|x_n - x^*\|)}\Bigg] \\
&\quad \times \|z_n - x^*\| \\
&\leq g_3(\|x_n - x^*\|)\|x_n - x^*\| \leq \|x_n - x^*\|.
\end{aligned}$$

3. Semi-local Convergence

Method (32.1)

The majorant sequence $\{\alpha_n\}$ is defined for $\alpha_0 = 0, \beta_0 \geq |a|\|F'(x_0)^{-1}F(x_0)\|$ and each $n = 0, 1, 2, \ldots$ by

$$\begin{aligned}
\gamma_n &= \beta_n + \frac{1}{8|a|}\left(9\left(\frac{\bar{v}_n}{1 - v_0(\alpha_n)}\right)^2 + 6\left(\frac{\bar{v}_n}{1 - v_0(\alpha_n)}\right) + 8|1 - a|\right)(\beta_n - \alpha_n), \\
\lambda_n &= (1 + \int_0^1 v_0(\alpha_n + \theta(\gamma_n - \alpha_n))d\theta)(\gamma_n - \alpha_n) \\
&\quad + \frac{1}{|a|}(1 + v_0(\alpha_n))(\beta_n - \alpha_n), \\
\alpha_{n+1} &= \gamma_n + \frac{1}{1 - v_0(\alpha_n)}(1 + \frac{3}{2}\bar{v}_n)\lambda_n,
\end{aligned}$$

$$\begin{aligned}
\delta_{n+1} &= (1 + \int_0^1 v_0(\alpha_n + \theta(\alpha_{n+1} - \alpha_n))d\theta)(\alpha_{n+1} - \alpha_n) \\
&\quad + \frac{1}{|a|}(1 + v_0(\alpha_n))(\beta_n - \alpha_n)
\end{aligned}$$

and

$$\beta_{n+1} = \alpha_{n+1} + \frac{|a|\delta_{n+1}}{1 - v_0(\alpha_{n+1}}.$$

The role of x^*, "w" is exchanged by x_0, "v", respectively in the calculations. In particular, we have in turn:

$$z_n - y_n = [(a - \frac{23}{8})I + 3F'(x_n)^{-1}F'(x_n) - \frac{9}{8}(F'(x_n)^{-1}F'(y_n))^2](-\frac{1}{a}(y_n - x_n)),$$

Method (32.2)

The majorant sequence $\{\alpha_n\}$ for $\alpha_0 = 0, \beta \geq |a|\|F'(x_0)^{-1}F(x_0)\|$ and each $n = 0, 1, 2, \ldots$ is given by

$$\begin{aligned}
\gamma_n &= \beta_n + \frac{1}{8|a|}\left(|1-a| + 6\frac{\bar{v}_n}{1 - v_0(\alpha_n)} + 9\left(\frac{\bar{v}_n}{1 - v_0(\alpha_n)}\right)^2 + 15\left(\frac{\bar{v}_n}{1 - v_0(\alpha_n)}\right)^3\right) \\
&\quad \times (\beta_n - \alpha_n), \\
\alpha_{n+1} &= \gamma_n + \frac{1}{4}\left(13 + 8\frac{\bar{v}_n}{1 - v_0(\alpha_n)} + \left(\frac{\bar{v}_n}{1 - v_0(\alpha_n)}\right)^2\right)\frac{\lambda_n}{1 - v_0(\alpha_n)},
\end{aligned}$$

and

$$\beta_{n+1} = \alpha_{n+1} + |a|\frac{\delta_{n+1}}{1 - v_0(\alpha_{n+1})},$$

where the sequence $\{\bar{v}_n\}, \{\lambda_n\}$ and $\{\delta_{n+1}\}$ are is given in the method (32.1). The motivation is:

$$\begin{aligned}
z_n - y_n &= \frac{1}{8a}[(1-a)I - 6(F'(x_n)^{-1}F'(y_n) - I) \\
&\quad + 9(F'(x_n)^{-1}F'(y_n) - I)^2 + 15(F'(x_n)^{-1}F'(y_n) - I)^3](y_n - x_n), \\
\|z_n - y_n\| &\leq \frac{1}{8|a|}\left(|1-a| + 6\left(\frac{\bar{v}_n}{1 - v_0(\alpha_n)}\right) + 9\left(\frac{\bar{v}_n}{1 - v_0(\alpha_n)}\right)^2 + 15\left(\frac{\bar{v}_n}{1 - v_0(\alpha_n)}\right)^3\right) \\
&\quad \times (\beta_n - \alpha_n) \\
&= \gamma_n - \beta_n, \\
&= \frac{1}{8a}[9(F'(x_n)^{-1}F'(y_n) - I)^2 - 6(F'(x_n)^{-1}F'(y_n) - I) - 8(a-1)](y_n - x_n), \\
\|z_n - y_n\| &\leq \frac{1}{8|a|}\left[9\left(\frac{\bar{v}_n}{1 - v_0(\alpha_n)}\right)^2 + 6\left(\frac{\bar{v}_n}{1 - v_0(\alpha_n)}\right) + 8|1-a|\right](\beta_n - \alpha_n) \\
&= \gamma_n - \beta_n,
\end{aligned}$$

$$\begin{aligned}
F(z_n) &= F(z_n) - F(x_n) + F(x_n) = F(z_n) - F(x_n) - \frac{1}{a}F'(x_n)(y_n - x_n), \\
\|L^{-1}F(z_n)\| &\leq (1 + \int_0^1 v_0(\alpha_n + \theta(\gamma_n - \alpha_n))d\theta)(\gamma_n - \alpha_n) \\
&\quad + \frac{1}{|a|}(1 + v_0(\alpha_n))(\beta_n - \alpha_n) = \lambda_n, \\
x_{n+1} - z_n &= -F'(x_n)^{-1}F(z_n) - \frac{3}{2}(I - F'(x_n)^{-1}F'(y_n))F'(x_n)^{-1}F(z_n), \\
\|x_{n+1} - z_n\| &\leq \left[\frac{1}{1 - v_0(\alpha_n)} + \frac{3}{2}\frac{\bar{v}_n}{1 - v_0(\alpha_n)}\right]\lambda_n = \alpha_{n+1} - \gamma_n,
\end{aligned}$$

$$\begin{aligned}
F(x_{n+1}) &= F(x_{n+1})-F(x_n)-\frac{1}{a}F'(x_n)(y_n-x_n),\\
\|L^{-1}F(x_{n+1})\| &\leq (1+\int_0^1 v_0(\alpha_n+\theta(\alpha_{n+1}-\alpha_n))d\theta)(\alpha_{n+1}-\alpha_n)\\
&\quad +\frac{1}{|a|}(1+v_0(\alpha_n))(\beta_n-\alpha_n)=\delta_{n+1},\\
\|y_{n+1}-x_{n+1}\| &\leq |a|\|F(x_{n+1})^{-1}L\|\|L^{-1}F(x_{n+1})\|\\
&\leq |a|\frac{\delta_{n+1}}{1-v_0(\alpha_{n+1}}=\beta_{n+1}-\alpha_{n+1}.\\
x_{n+1}-z_n &= -[3I-\frac{5}{2}F'(x_n)^{-1}F'(y_n)+(\frac{1}{2}F'(x_n)^{-1}F'(y_n))^2]\\
&\quad \times F'(x_n)^{-1}F(z_n)\\
&= -\frac{1}{4}[13I-8(F'(x_n)^{-1}F'(y_n)-I)\\
&\quad +(F'(x_n)^{-1}F'(y_n)-I)^2]F'(x_n)^{-1}F(z_n),\\
\|x_{n+1}-z_n\| &\leq \frac{1}{4}\left(13+8\frac{\bar{v}_n}{1-v_0(\alpha_n)}+\left(\frac{\bar{v}_n}{1-v_0(\alpha_n)}\right)^2\right)\frac{\lambda_n}{1-v_0(\alpha_n)}\\
&= \alpha_{n+1}-\gamma_n.
\end{aligned}$$

The calculations for the iterate β_{n+1} is given in the method (32.1). The rest of the chapter follows as before.

4. References

[1] Grau-Snchez, M., Noguera, M., Amat, S., *On the approximation of derivatives using divided difference operators preserving the local convergence order of iterative methods*, J. Comput. Appl. Math., 237(1), (2013), 363-372.

[2] Homeier, H.H.H., *A modified Newton method with cubic convergence: the multivariate case*, J. Comput. Appl. Math., 169, (2004), 161-169.

[3] Hueso, J.L., Martnez, E., Teruel, C., *Convergence, efficiency, and dynamics of new fourth and sixth order families of iterative methods for nonlinear systems*, J. Comput. Appl. Math., 275, (2015), 412-420.

[4] Jarratt, P., *Some fourth-order multipoint iterative methods for solving equations*, Math. Comp., 20(95), (1966), 434-437.

[5] Kou, J., Li, Y., *An improvement of the Jarratt method*, Appl. Math. Comput., 189(2), (2007), 1816-1821.

[6] Magren, .A., *Different anomalies in a Jarratt method*, Appl. Math. Comput., 189(2), (2014), 1816-1821.

[7] Noor, M.A., Waseem, M., *Some iterative methods for solving a system of nonlinear equations*, Comput. Math. Appl., 57, (2009), 101-106.

[8] Petkovi, M.S., Neta, B., Petkovi, L., Duni, D., *Multipoint methods for solving nonlinear equations*, Elsevier, Amsterdam, Netherlands, (2013).

[9] Rheinboldt, W.C., *An adaptive continuation process for solving systems of nonlinear equations*, In Mathematical Models and Numerical Methods (A.N. Tikhonov et al. eds.), pub. 3, pp. 129-142, Banach Center, Warsaw, Poland, (1978).

[10] Sharma, D., Parhi, S.K., *On the local convergence of higher-order methods*, Fixed Point Theory, 22(2), (2021), 855-870.

[11] Sharma, J.R., Arora, H., *Efficient Jarratt-like methods for solving systems of nonlinear equations*, Calcolo, 51(1), (2014), 193-210.

[12] Sharma, J.R., Guha, R.K., Sharma, R., *An efficient fourth-order weighted-Newton method for systems of nonlinear equations*, Numer. Algor., 62(2), (2013), 307-323.

[13] Traub, J.F., *Iterative Methods for Solution of Equations*, Prentice Hall, Upper Saddle River, New Jersey, USA, (1964).

[14] Xiao, X., Yin, H., *A simple and efficient method with high-order convergence for solving systems of nonlinear equations*, Comput. Math. Appl., 69(10), (2015), 1220-1231.

Chapter 33

Local Convergence of a Seventh-Order Method with One Parameter

1. Introduction

The method is defined for $x_0 \in \Omega$ and each $n = 0, 1, 2, \ldots$ by

$$\begin{aligned}
y_n &= x_n - F(x_n)^{-1}F(x_n), \\
z_n &= y_n - \frac{1}{a}F'(x_n)^{-1}F(y_n), \\
v_n &= z_n - F'(x_n)^{-1}[(2 - \frac{1}{a} - a)F(y_n) + aF(z_n)], \\
u_n &= I - \frac{1}{a}F'(x_n)^{-1}[y_n, x_n; F], \\
A(t) &= (1 - \frac{1}{a})I + a(t - (1 - \frac{1}{a}I) \\
&\quad + \frac{1}{2}(6a^2 - a)(t - (1 - \frac{1}{a})I)^2
\end{aligned} \tag{33.1}$$

and

$$x_{n+1} = z_n - A(u_n)F'(z_n)^{-1}F(v_n),$$

where $a \in \mathbb{R} - \{0\}$ is a given parameter and $[.,.;F] : \Omega \times \Omega \longrightarrow L(B_1, B_2)$.

The seventh order of convergence shown in [1] provided that $B_1 = B_2 = \mathbb{R}^k$, where $F^{(8)}$ exists and is bounded. The present Chapter provides a technique for expanding the convergence via means of the operators on the method (see also [2]-[16]). Notice also that the semi-local convergence of the method (33.1) is given in [7].

2. Convergence

The majorant functions g_i, $i = 1, 2, 3, 4$ are defined as:

$$g_1(t) = \frac{\int_0^1 \omega((1-\theta)t)d\theta}{1 - \omega_0(t)},$$

$$\bar{\omega}(t) = \begin{cases} \omega((1+g_1(t))t), \\ \omega_0(t)+\omega_0(g_1(t)t), \end{cases}$$

$$g_2(t) = \left[\frac{\int_0^1 \omega((1-\theta)g_1(t)t)d\theta}{1-\omega_0(g_1(t)t)} + \frac{\bar{\omega}(t)}{(1-\omega_0(t))(1-\omega_0(g_1(t)t))} + |1-\frac{1}{a}|\frac{1}{1-\omega_0(t)}\right) \times (1+\int_0^1 \omega_0(\theta g_1(t)t)d\theta\right] g_1(t),$$

$$\bar{\bar{\omega}}(t) = \begin{cases} \omega((1+g_2(t))t), \\ \omega_0(t)+\omega_0(g_2(t)t), \end{cases}$$

$$g_3(t) = \frac{\int_0^1 \omega((1-\theta)g_2(t)t)d\theta}{1-\omega_0(g_2(t)t)} + \left(\frac{\bar{\bar{\omega}}(t)}{(1-\omega_0(t))(1-\omega_0(g_2(t)t))} + \frac{|1-a|}{1-\omega_0(t)}\right) \times (1+\int_0^1 \omega_0(\theta g_2(t)t)d\theta) + |2-\frac{1}{a}-a|\frac{1+\int_0^1 \omega_0(\theta g_1(t)t)d\theta}{1-\omega_0(t)} g_1(t),$$

$$\mu(t) = \frac{\omega_1(t, g_1(t)t)}{|a|(1-\omega_0(t))} + |1-\frac{1}{a}|,$$

$$c = \frac{(6a-1)(1-a)^2+2-a}{a},$$

$$p(t) = |c| + 2|\frac{(6a-1)(1-a)}{a}|\mu(t) + \frac{1}{2}|\frac{6a-1}{a}|\mu(t)^2,$$

$$\bar{\bar{\bar{\omega}}}(t) = \begin{cases} \omega((g_2(t)+g_3(t))t), \\ \omega_0(g_2(t)t)+\omega_0(g_3(t)t), \end{cases}$$

and

$$g_4(t) = \left[\frac{\int_0^1 \omega((1-\theta)g_3(t)t)d\theta}{1-\omega_0(g_3(t)t)} + \left(\frac{\bar{\bar{\bar{\omega}}}(t)}{(1-\omega_0(g_2(t)t))(1-\omega_0(g_3(t)t))} + \frac{p(t)}{1-\omega_0(g_2(t)t)}\right) \times (1+\int_0^1 \omega_0(\theta g_3(t)t)d\theta)\right] g_3(t).$$

The motivation for the development of the functions g_i follows in turn by the calculations

$$\begin{aligned}
\|y_n - x^*\| &\leq \frac{\int_0^1 \omega((1-\theta)\|x_n - x^*\|)d\theta}{1-\omega_0(\|x_n - x^*\|)} \\
&\leq g_1(\|x_n - x^*\|)\|x_n - x^*\| \\
&\leq \|x_n - x^*\| < r, \\
z_n - x^* &= y_n - x^* - F'(y_n)^{-1}F(y_n) + F'(y_n)^{-1}F(y_n) - \frac{1}{a}F'(x_n)^{-1}F(y_n) \\
&= y_n - x^* - F'(y_n)^{-1} + (F'(y_n)^{-1} - F'(x_n)^{-1})F(y_n) \\
&\quad +(1-\frac{1}{a})F'(x_n)^{-1}F(y_n),
\end{aligned}$$

$$\begin{aligned}
\|z_n - x^*\| &\leq \Bigg[\frac{\int_0^1 \omega((1-\theta)\|y_n - x^*\|)d\theta}{1-\omega_0(\|y_n - x^*\|)} \\
&\quad + \Bigg(\frac{\bar{\omega}_n}{(1-\omega_0(\|y_n - x^*\|))(1-\omega_0(\|x_n - x^*\|))} \\
&\quad + |1-\frac{1}{a}|\frac{1}{1-\omega_0(\|x_n - x^*\|)}\Bigg) \\
&\quad \times (1 + \int_0^1 \omega_0(\theta\|y_n - x^*\|)d\theta)\Bigg]\|y_n - x^*\| \\
&\leq g_2(\|x_n - x^*\|)\|x_n - x^*\| \leq \|x_n - x^*\|,
\end{aligned}$$

$$\begin{aligned}
v_n - x^* &= z_n - x^* - F'(z_n)^{-1}F(z_n) + F'(z_n)^{-1}F(z_n) \\
&\quad -aF'(x_n)^{-1}F(z_n) - (2 - \frac{1}{a} - a)F'(x_n)^{-1}F(y_n) \\
&= z_n - x^* - F'(z_n)^{-1}F(z_n) + (F'(z_n)^{-1} - F'(x_n)^{-1})F(z_n) \\
&\quad (1-a)F'(x_n)^{-1}F(z_n) - (2 - \frac{1}{a} - a)F'(x_n)^{-1}F(y_n),
\end{aligned}$$

$$\begin{aligned}
\|v_n - x^*\| &\leq \frac{\int_0^1 \omega((1-\theta)\|z_n - x^*\|)d\theta\|z_n - x^*\|}{1-\omega_0(\|z_n - x^*\|)} \\
&\quad + \left(\frac{\bar{\bar{\omega}}_n}{(1-\omega_0(\|x_n - x^*\|))(1-\omega_0(\|z_n - x^*\|))} + \frac{|1-a|}{1-\omega_0(\|x_n - x^*\|)}\right) \\
&\quad \times (1 + \int_0^1 \omega_0(\theta\|z_n - x^*\|)d\theta)\|z_n - x^*\| \\
&\quad + |2 - \frac{1}{a} - a|\frac{1 + \int_0^1 \omega_0(\theta\|y_n - x^*\|)d\theta}{1-\omega_0(\|x_n - x^*\|)}\|y_n - x^*\| \\
&\leq g_3(\|x_n - x^*\|)\|x_n - x^*\| \leq \|x_n - x^*\|,
\end{aligned}$$

$$\begin{aligned}
u_n &= I - \frac{1}{a}F'(x_n)^{-1}[y_n, x_n; F] \\
&= (1 - \frac{1}{a}I + \frac{1}{a}(I - F'(x_n)^{-1}[y_n, x_n; F]), \\
\|u_n\| &\leq |1 - \frac{1}{a}| + \frac{1}{|a|}\|F'(x_n)^{-1}(F'(x_n) - [y_n, x_n; F])\| \\
&\leq |1 - \frac{1}{a}| + \frac{1}{|a|}\frac{\omega_1(\|x_n - x^*\|, \|y_n - x^*\|)}{1 - \omega_0(\|x_n - x^*\|)} \\
&= \mu_n, \\
I - A(u_n) &= I - (1 - \frac{1}{a})I - a(u_n - (1 - \frac{1}{a})I) \\
&\quad - \frac{1}{2}(-a + 6a^2)(u_n - (1 - \frac{1}{a})I)^2 \\
&= c + \frac{2(6a-1)(a-1)}{a}\mu_n + \frac{1}{2a}(6a-1)\mu_n^2, \\
\|I - A_n\| &\leq |c| + |\frac{2(6a-1)(a-1)}{a}|\frac{\omega_1(\|x_n - x^*\|, \|y_n - x^*\|)}{1 - \omega_0(\|x_n - x^*\|)} \\
&\quad + |\frac{1}{2a}(6a-1)|\left(\frac{\omega_1(\|x_n - x^*\|, \|y_n - x^*\|)}{1 - \omega_0(\|x_n - x^*\|)}\right)^2 \\
&= p_n,
\end{aligned}$$

$$\begin{aligned}
x_{n+1} - x^* &= v_n - x^* - F'(v_n)^{-1}F(v_n) \\
&\quad + (F'(v_n)^{-1} - F'(z_n)^{-1})F(v_n) \\
&\quad + (I - A(u_n))F'(z_n)^{-1}F(v_n),
\end{aligned}$$

$$\begin{aligned}
\|x_{n+1} - x^*\| &\leq \Bigg[\frac{\int_0^1 \omega((1-\theta)\|v_n - x^*\|)d\theta}{1 - \omega_0(\|v_n - x^*\|)} \\
&\quad + \left(\frac{\bar{\bar{\bar{\omega}}}_n}{(1 - \omega_0(\|z_n - x^*\|))(1 - \omega_0(\|v_n - x^*\|))}\right. \\
&\quad \left. + \frac{p_n}{1 - \omega_0(\|z_n - x^*\|)}\right) \\
&\quad \times (1 + \int_0^1 \omega_0(\theta\|v_n - x^*\|)d\theta)\Bigg]\|v_n - x^*\| \\
&\leq g_4(\|x_n - x^*\|)\|x_n - x^*\| \leq \|x_n - x^*\|,
\end{aligned}$$

where we also used the condition

$$\|L^{-1}(F'(x) - [x, y; F])\| \leq \omega_1(\|x - x^*\|, \|y - x^*\|)$$

for each $x, y \in \Omega_0$.

3. References

[1] Amiri, A., Cordero, A., Darvishi, M.T., Torregrosa, J.R., *Stability analysis of a parametric family of seventh-order iterative methods for solving nonlinear systems*, Appl. Math. Comput., 323, (2018), 43–57.

[2] Argyros, C. I., Argyros, I.K., Regmi, S., John, J.A., Jayaraman, J., *Semi-local convergence of a seventh-order method with one parameter for solving non-linear equations*, Foundations, (2023).

[3] Argyros, C. I., Regmi, S., Argyros, I.K., George, S., *Contemporary Algorithms: Theory and Applications*, Volume III, NOVA Publishers, (2023).

[4] Argyros, I.K., *Advances in the Efficiency of Computational Methods and Applications*, World Scientific Publishing, Oxon, United Kingdom, (2000).

[5] Argyros, I.K., *The theory and applications of iteration methods*, 2nd Edition, Engineering Series, CRC Press, Taylor and Francis Group, (2022).

[6] Argyros, I.K., *Unified Convergence Criteria for Iterative Banach Space Valued Methods with Applications*, Mathematics, 2021, 9(16), 1942; https://doi.org/10.3390/math9161942.

[7] Argyros, I.K., George, S., *Ball comparison between two optimal eight-order methods under weak conditions*, SeMA Journal, 72, (2015), 1–11.

[8] Cordero, A., Villalba, E.G., Torregrosa, J.R., Triguero-Navarro, P., *Convergence and Stability of a Parametric Class of Iterative Schemes for Solving Nonlinear Systems*, Mathematics, 9(1), Article Number: 86, (2021), 1–19.

[9] Grau-Snchez, M., Gutirrez, J.M., *Zero-finder methods derived from Obreshkovs techniques*, Appl. Math. Comput., 215, (2009), 2992–3001.

[10] Kumar, D., Sharma, J.R., Jntschi, L., *Convergence Analysis and Complex Geometry of an Efficient Derivative-Free Iterative Method*, Mathematics, 7(10), Article Number: 919, (2019), 1–11.

[11] Magren, .A., *Different anomalies in a Jarratt family of iterative root-finding methods*, Appl. Math. Comput., 233, (2014), 29–38.

[12] Ortega, J.M., Rheinholdt, W.C., *Iterative Solution of Nonlinear Equations in Several Variables*, Academic Press, New York, USA, (1970).

[13] Rheinboldt, W.C., *An adaptive continuation process for solving systems of nonlinear equations*, In: Mathematical Models and Numerical Methods (A.N. Tikhonov et al. eds.), pub. 3, (1978), 129–142, Banach Center, Warsaw, Poland.

[14] Sharma, D., Parhi, S.K., *Extending the Applicability of a Newton-Simpson-Like Method*, Int. J. Appl. Comput. Math., 6(3), Article Number: 79, (2020). DOI: https://doi.org/10.1007/s40819-020-00832-3.

[15] Sharma, J.R., Arora, H., *Improved Newton-like methods for solving systems of nonlinear equations*, SeMA J., 74(2), (2016), 1–7.

[16] Traub, J.F., *Iterative Methods for Solution of Equations*, Prentice-Hall, Upper Saddle River, New Jersey, USA, (1964).

Chapter 34

Generalized Eighth-Order Method with a Parameter

1. Introduction

The three step method is defined for $x_0 \in \Omega$ and each $n = 0, 1, 2, \ldots$ by

$$\begin{aligned} y_n &= x_n - F(x_n)^{-1}F(x_n), \\ z_n &= A_4(x_n, y_n) \end{aligned} \tag{34.1}$$

and

$$x_{n+1} = B_8(x_n, y_n, z_n),$$

$a \in \mathbb{R}$ is a given parameter, where $A_4 : \Omega \times \Omega \longrightarrow L(B_1, B_2)$, $B_8 : \Omega \times \Omega \times \Omega \longrightarrow L(B_1, B_2)$ are iterations of order four and eight, respectively. The Taylor series expansion approach is used to show local convergence order eight in [4], provided that $a = 1, B_1 = B_2 = \mathbb{R}^k$ and $F^{(9)}$ exists and is bounded. These restrictions limit the applicability of the method to solving equations where the operator F is at least nine times differentiable. But the method may converge if these conditions are not necessarily satisfied. That is why in this Chapter only the operators on the method are utilized to show the local convergence of the method. Moreover, the more interesting semi-local convergence (not studied in [4]) is also presented (see also [1,2,3,5]).

2. Local Convergence

The following conditions are needed provided that $x^* \in \Omega$ with $F(x^*) = 0$.

Suppose:

(C1) There exists a continuous and nondecreasing function $\omega_0 : [0, \infty) \longrightarrow [0, \infty)$ such that the equation $\omega_0(t) - 1 = 0$ has a smallest positive solution ρ.

(C2) There exists continuous and nondecreasing functions $\omega : [0, \rho) \longrightarrow [0, \infty)$, $g_2 : S_1 \longrightarrow [0, \infty)$ and $g_2 : S \longrightarrow [0, \infty)$ such that the equations $g_i(t) - 1 = 0$, where

(C3) There exists an invertible operator L such that

$$\|L^{-1}(F'(x) - L)\| \le \omega_0(\|x - x^*\|)$$

for each $x \in \Omega$.

(C4) $\|A_4(x,y)-x^*\| \leq g_2(\|x-x^*\|)\|x-x^*\|$ for each $x,y \in \Omega_1$ and

$$\|B_8(x,y,z)-x^*\| \leq g_3(\|x-x^*\|, \|y-x^*\|, \|z-x^*\|)$$

for each $x,y,z \in \Omega_2$ and

(C5) $U[x^*,r] \subset \Omega$, where $r = \min\{r_i\}$.

The estimates leading to the introduction of the functions g_i are in turn:

$$\begin{aligned}
y_n - x^* &= x_n - x^* - F'(x_n)^{-1}F(x_n) + (1-a)F'(x_n)^{-1}F(x_n),\\
\|y_n - x^*\| &\leq \frac{1}{1-\omega_0(\|x_n-x^*\|)}[\int_0^1 \omega((1-\theta)\|x_n-x^*\|)d\theta\\
&+|1-a|(1+\int_0^1 \omega_0(\theta\|x_n-x^*\|)d\theta)]\|x_n-x^*\|\\
&\leq g_1(\|x_n-x^*\|)\|x_n-x^*\| \leq \|x_n-x^*\|,\\
\|z_n-x^*\| &= \|A_4(x_n,y_n)-x^*\|\\
&\leq g_2(\|x_n-x^*\|)\|x_n-x^*\| \leq \|x_n-x^*\|
\end{aligned}$$

and

$$\begin{aligned}
\|x_{n+1}-x^*\| &= \|B_8(x_n,y_n,z_n)-x^*\|\\
&\leq g_3(\|x_n-x^*\|)\|x_n-x^*\| \leq \|x_n-x^*\|
\end{aligned}$$

provided that $x_0 \in U(x^*,r)$. Thus $\lim_{n\longrightarrow\infty} x_n = x^*$ and $\{x_n\} \subset U(x^*,r)$.

3. Semi-local Convergence

The role of x^*, "ω" is exchanged by x_0, "v" in the semi-local case. The conditions in this case are:

(H1) There exists a continuous and nondecreasing function $v_0 : [0,\infty) \longrightarrow [0,\infty)$ such that the equation $v_0(t)-1=0$ has a smallest positive solution ρ_3.

(H2) There exists an invertible operator L and some $x_0 \in \Omega$ such that

$$\|L^{-1}(F'(x)-L)\| \leq v_0(\|x-x^*\|)$$

for each $x \in \Omega$. Notice that

$$\|L^{-1}(F'(x_0)-L)\| \leq v_0(0) < 1,$$

thus $F'(x_0)^{-1} \in L(B_2,B_1)$. Set $|a|\|F'(x_0)^{-1}F(x_0)\| \leq \beta_0$.

(H3) The iterates x_n, y_n, z_n are well defined and there exist nonnegative sequences $\{\alpha_n\}$, $\{\beta_n\}$, $\{\gamma_n\}$ such that for each $n = 0,1,2,\ldots$

$$\|z_n-y_n\| = \|x_n - aF'(x_n)^{-1}F9x_n) - A_4(x_n,y_n)\| \leq \gamma_n - \beta_n$$

and

$$\|x_{n+1}-z_n\| + \|B_4(x_n,y_n,z_n)-z_n\| \leq \alpha_{n+1} - \gamma_n.$$

(H4) There exists $\rho_4 \in [0, \rho_3)$ such that $v_0(\alpha_n) < 1$ and $\alpha_n \leq \rho_4$.
It follows that the sequence $\{\alpha_n\}$ defined for $\alpha_0 = 0$, $\beta_0 \geq 0$ and each $n = 0, 1, 2, \ldots$ defined by

$$\begin{aligned}
\gamma_n &= \beta_n + \|z_n - y_n\|, \\
\alpha_{n+1} &= \gamma_n + \|x_{n+1} - z_n\|, \\
\delta_{n+1} &= (1 + \int_0^1 v_0(\alpha_n + \theta(\alpha_{n+1} - \alpha_n)d\theta)(\alpha_{n+1} - \alpha_n) \\
&\quad + \frac{1}{|a|}(1 + v_0(\alpha_n))(\beta_n - \alpha_n)
\end{aligned}$$

and

$$\beta_{n+1} = \alpha_{n+1} + |a| \frac{\delta_{n+1}}{1 - v_0(\alpha_{n+1})},$$

satisfies $0 \leq \alpha_n \leq \beta_n \leq \gamma_n \leq \alpha_{n+1} \leq \rho_3$ and there exists ρ^* such that $\lim_{n \longrightarrow \infty} \alpha_n = \rho^* \in [0, \rho_3]$.
and

(H5) $U[x_0, \rho^*] \subset \Omega$.

The motivation is in turn as follows:

$$\begin{aligned}
\|z_n - x_n\| &= \|x_n - aF'(x_n)^{-1}F(x_n) - A_4(x_n, y_n)\| \\
&\leq \alpha_{n+1} - \beta_n, \\
\|x_{n+1} - z_n\| &= \|B_8(x_n, y_n, z_n) - z_n\| \\
&\leq \alpha_{n+1} - \alpha_n, \\
F(x_{n+1}) &= F(x_{n+1}) - F(x_n) - \frac{1}{a}F'(x_n)(y_n - x_n), \\
\|L^{-1}F(x_{n+1})\| &\leq (1 + \int_0^1 v_0(\alpha_n + \theta(\alpha_{n+1} - \alpha_n))d\theta)(\alpha_{n+1} - \alpha_n) \\
&\quad + \frac{1}{|a|}(1 + v_0(\alpha_n))(\beta_n - \alpha_n) = \delta_{n+1},
\end{aligned} \tag{34.2}$$

and

$$\begin{aligned}
\|y_{n+1} - x_{n+1}\| &\leq |a| \|F'(x_{n+1})^{-1}L\| \|L^{-1}F(x_{n+1})\| \\
&\leq |a| \frac{\delta_{n+1}}{1 - v_0(\alpha_{n+1})} = \beta_{n+1} - \alpha_{n+1}.
\end{aligned}$$

If the operators A_4, B_8 are further specialized then the same is true for the sequences $\{\alpha_n\}, \{\beta_n\}$ and $\{\gamma_n\}$. Thus, the sequence $\{\alpha_n\}$ is Cauchy in the Banach space B_1, thus there exists $x^* \in U[x_0, \rho^*]$ such that $\lim_{n \longrightarrow \infty} x_n = x^*$. Finally, by letting $n \longrightarrow \infty$ in (34.2), we deduce $F(x^*) = 0$. The rest follows as in the previous Chapters.

4. References

[1] Argyros, C. I., Argyros, I.K., Regmi, S., John, J.A., Jayaraman, J., *Semi-local convergence of a seventh-order method with one parameter for solving non-linear equations*, Foundations, (2023).

[2] Argyros, C. I., Regmi, S., Argyros, I.K., George, S., *Contemporary Algorithms: Theory and Applications*, Volume III, NOVA Publishers, (2023).

[3] Argyros, I.K., *Advances in the Efficiency of Computational Methods and Applications*, World Scientific Publishing, Oxon, United Kingdom, (2000).

[4] Argyros, I.K., *The theory and applications of iteration methods*, 2nd Edition, CRC Press, Engineering Series, Taylor and Francis Group, (2022).

[5] Argyros, I.K., George, S., *Ball comparison between two optimal eight-order methods under weak conditions*, SeMA Journal, 72, (2015), 1–11.

Chapter 35

Three-Step Kurchatov-Type Methods

1. Introduction

The method is defined for $x_{-1}, x_0 \in \Omega$ and each $n = 0, 1, 2, \dots$ by

$$\begin{aligned} u_n &= x_n - [2x_n - x_{n-1}, x_{n-1}; F]^{-1} F(x_n), \\ y_n &= x_n - [u_n, x_n; F]^{-1} F(x_n), \\ x_{n+1} &= y_n - [u_n, y_n; F]^{-1} F(y_n), \end{aligned} \tag{35.1}$$

where $[., .; F] : \Omega \times \Omega \longrightarrow L(B)$ is a divided difference of order one. The local convergence of order five is established for $B = \mathbb{R}^k$ in [6] by means of the Taylor series expansion approach and the boundedness of the sixth derivative, which is not required by the method itself. In this chapter, local convergence is established based solely on the operators involved in the method. Moreover, the more interesting semi-local convergence analysis (not studied in [6]) is also developed in this chapter (see also [1]-[5], [7, 8]).

2. Local Convergence

The majorant functions $g_i, i = 1, 2, 3$ are:

$$g_1(t) = \frac{\omega(t, t, 3t)}{1 - \omega_0(t, 3t)},$$

$$g_2(t) = \frac{\omega_1(t, g_1(t)t)}{1 - \omega_0(t, g_1(t)t)}$$

and

$$g_3(t) = \frac{\omega_1(g_1(t)t, g_2(t)t) g_2(t)}{1 - \omega_0(g_1(t)t, g_2(t)t)}.$$

The motivation arises from the calculations

$$\begin{aligned} u_n - x^* &= x_n - x^* - [2x_n - x_{n-1}, x_{n-1}; F]^{-1} F(x_n) \\ &= [2x_n - x_{n-1}, x_{n-1}; F]^{-1} ([2x_n - x_{n-1}, x_{n-1}; F] - [x_n, x^*; F])(x_n - x^*), \\ \|u_n - x^*\| &\leq \frac{w(\|2x_n - x_{n-1} - x^*\|, \|x_{n-1} - x^*\|, \|x_n - x^*\|) \|x_n - x^*\|}{1 - \omega_0(\|x_n - x^*\|, \|2x_n - x_{n-1} - x^*\|)} \\ &\leq g_1(\|x_n - x^*\|) \|x_n - x^*\| \leq \|x_n - x^*\| < r, \end{aligned}$$

$$\begin{aligned}
y_n - x^* &= [u_n, x_n; F]^{-1}([u_n, x_n; F] - [x_n, x_{n-1}; F])(x_n - x^*), \\
\|y_n - x^*\| &\leq \frac{\omega_1(\|x_n - x^*\|, \|u_n - x^*\|)\|x_n - x^*\|}{1 - \omega_0(\|x_n - x^*\|, \|u_n - x^*\|)} \\
&\leq g_2(\|x_n - x^*\|) \leq \|x_n - x^*\|
\end{aligned}$$

and

$$\begin{aligned}
\|x_{n+1} - x^*\| &= \|[u_n, y_n; F]^{-1}([u_n, y_n; F] - [y_n, x^*; F])(y_n - x^*)\| \\
&\leq \frac{\omega_1(\|y_n - x^*\|, \|u_n - x^*\|)\|y_n - x^*\|}{1 - \omega_0(\|u_n - x^*\|, \|y_n - x^*\|)} \\
&\leq g_3(\|x_n - x^*\|)\|x_n - x^*\| \leq \|x_n - x^*\|.
\end{aligned}$$

Notice that the additional condition is assumed $U(x^*, R) \subset \Omega$, where $R = \max\{r, 3r\}$.

3. Semi-local Convergence

The majorant sequence $\{\alpha_n\}$ is defined for $\alpha_{-1} = 0$, $\alpha_0 \geq 0$, $\beta_0 \geq \bar{\beta}_0 \geq \alpha_0$ and each $n = 0, 1, 2, \ldots$

$$\begin{aligned}
\beta_n &= \bar{\beta}_n + \frac{v(2d_n + \alpha_{n-1}, \alpha_{n-1}, \alpha_n, \bar{\beta}_n)(\beta_n - \alpha_n)}{1 - v_0(\alpha_n, \bar{\beta}_n)}, \\
\lambda_n &= (1 + v_0(\alpha_n, \beta_n))(\beta_n - \alpha_n) + (1 + v_0(\alpha_n, \bar{\beta}_n))(\beta_n - \alpha_n), \\
\alpha_{n+1} &= \beta_n + \frac{\lambda_n}{1 - v_0(\bar{\beta}_n, \beta_n)}, \\
\delta_{n+1} &= (1 + v_0(\alpha_n, \alpha_{n+1}))(\alpha_{n+1} - \alpha_n) \\
&\quad + (1 + v_0(\alpha_{n-1}, 2\alpha_n + \alpha_{n-1}))(\bar{\beta}_n - \alpha_n)
\end{aligned}$$

and

$$\bar{\beta}_{n+1} = \alpha_{n+1} + \frac{\delta_{n+1}}{1 - v_0(\alpha_n, 2\alpha_{n+1} + \alpha_n)}.$$

The calculations are:

$$\begin{aligned}
y_n - u_n &= ([2x_n - x_{n-1}; F]^{-1} - [u_n, x_n; F]^{-1})F(x_n) \\
&= -([u_n, x_n; F]^{-1} - [2x_n - x_{n-1}, x_{n-1}; F]^{-1})F(x_n) \\
&= [u_n, x_n; F]^{-1}([2x_n - x_{n-1}, x_{n-1}; F] - [u_n, x_n; F])(y_n - x_n),
\end{aligned}$$

$$\begin{aligned}
\|y_n - u_n\| &\leq \frac{v(2\alpha_n + \alpha_{n-1}, \alpha_{n-1}, \alpha_n, \bar{\beta}_n)(\beta_n - \alpha_n)}{1 - v_0(\alpha_n, \bar{\beta}_n)} \\
&= \beta_n - \bar{\beta}_n \qquad (35.2) \\
F(y_n) &= F(y_n) - F(x_n) - [u+_n, x_n; F](y_n - x_n), \\
\|L^{-1}F(y_n)\| &\leq (1 + v_0(\alpha_n + \beta_n))(\beta_n - \alpha_n) \\
&\quad + (1 + v_0(\alpha_n, \bar{\beta}_n))(\beta_n - \alpha_n) = \lambda_n,
\end{aligned}$$

and

$$\|x_{n+1} - y_n\| \leq \frac{\lambda_n}{1 - v_0(\bar{\beta}_n, \beta_n)} = \alpha_{n+1} - \beta_n.$$

$$\begin{aligned}
F(x_{n+1}) &= F(x_{n+1}) - F(x_n) - [2x_n - x_{n-1}, x_{n-1}; F](u_n - x_n), \\
\|L^{-1}F(x_{n+1})\| &\leq (1 + v_0(\alpha_n, \alpha_{n+1}))(\alpha_{n+1} - \alpha_n) \\
&\quad + (1 + v_0(\alpha_{n-1}, 2\alpha_n + \alpha_{n-1}))(\bar{\beta}_n - \alpha_n) = \delta_{n+1}, \\
\|u_{n+1} - x_{n+1}\| &\leq \|[2x_{n+1} - x_n, x_n; F]L\| \|L^{-1}F(x_{n+1})\| \\
&\leq \frac{\delta_{n+1}}{1 - v_0(\alpha_n, 2\alpha_{n+1} + \alpha_n)} \\
&= \bar{\beta}_{n+1} - \alpha_{n+1}.
\end{aligned}$$

4. References

[1] Argyros, C. I., Regmi, S., Argyros, I. K., George, S., *Contemporary Algorithms: Theory and Applications*, Volume III, NOVA Publishers, (2023).

[2] Argyros, I. K., *Advances in the Efficiency of Computational Methods and Applications*, World Scientific Publishing, Oxon, United Kingdom, (2000).

[3] Argyros, I. K., *The theory and applications of iteration methods*, 2nd Edition, CRC Press, Engineering Series, Taylor and Francis Group, (2022).

[4] Argyros, I. K., George, S., *Ball comparison between two optimal eight-order methods under weak conditions*, SeMA Journal, 72, (2015), 1–11.

[5] Chicharro, F. I., Cordero, A., Garrido, N., Torregrosa, J. R., *On the improvement of the order of convergence of iterative methods for solving nonlinear systems by means of memory*, Applied Mathematics Letters, 104, (2020), 106277.

[6] Cordero, A., Torregrosa, J. R., *Variants of Newtons method using fifth order quadrature formulas*, Appl. Math. Comput., 190, (2007), 686–698.

[7] Narang, M., Bathia, S., Alshomrani, A. S., Kanwar, V., *General efficient class of Steffensen type methods with memory for solving systems on nonlinear equations*, J. Comput. Appl. Math., 352, (2019), 23–39.

[8] Samanskii, V., *On a modification of the Newton method*, Ukrain. Math., 19, (1967), 133–138.

Chapter 36

Two Competing Fifth-Order Methods

1. Introduction

The method is defined for $x_0 \in \Omega$ and each $n = 0, 1, 2, \ldots$ by

$$\begin{aligned} y_n &= x_n - F'(x_n)^{-1}F(x_n), \\ z_n &= x_n - \frac{1}{2}(3I - F'(x_n)^{-1}F'(y_n))F'(x_n)^{-1}F(x_n), \\ x_{n+1} &= z_n - (2I - F'(x_n)^{-1}F'(y_n))F'(x_n)^{-1}F(z_n), \end{aligned} \tag{36.1}$$

and

$$\begin{aligned} y_n &= x_n - F'(x_n)^{-1}F(x_n), \\ z_n &= x_n - \frac{1}{2}(F'(x_n)^{-1} + F'(y_n)^{-1})F(x_n), \\ x_{n+1} &= z_n - F'(y_n)^{-1}F(z_n). \end{aligned} \tag{36.2}$$

The local convergence order five is derived when $B_1 = B_2 = \mathbb{R}^k$ in [10,11], respectively under the Taylor series expansion approach and by assuming the existence of the sixth derivative not on the methods. Other limitations with this approach have been reported in earlier Chapters. The new local convergence relies only on the operators on the methods and the concept of generalized continuity. Moreover, the more interesting semi-local convergence analysis (not studied in [10,11]) is provided via means of majorizing sequences. The approach can be used to extend the applicability of other methods [1]-[9], [12,13,14]).

2. Local Convergence

Method (36.1)
The majorant functions $g_i, i = 1, 2, 3$ are:

$$g_1(t) = \frac{\int_0^1 \omega((1-\theta)t)d\theta}{1 - \omega_0(t)},$$

$$g_2(t) = \frac{\int_0^1 \omega((1-\theta)t)d\theta}{1 - \omega_0(t)} + \frac{\bar{\omega}(t)(1 + \int_0^1 \omega_0(\theta t)d\theta)}{2(1 - \omega_0(t))^2},$$

$$g_3(t) = \left[\frac{\int_0^1 \omega((1-\theta)g_2(t)t)d\theta}{1-\omega_0(g_2(t)t)} + \frac{\bar{\bar{\omega}}(1+\int_0^1 \omega_0(\theta g_2(t)t)d\theta)}{(1-\omega_0(t))(1-\omega_0(g_2(t)t))} + \frac{\bar{\omega}(1+\int_0^1 \omega_0(\theta g_2(t)t)d\theta)}{(1-\omega_0(t))^2}\right] g_2(t),$$

where $\bar{\omega}(t) = \begin{cases} \omega((1+g_1(t))t) \\ or \\ \omega_0(t)+\omega_0(g_1(t)t) \end{cases}$ and $\bar{\bar{\omega}}(t) = \begin{cases} \omega((1+g_2(t))t) \\ or \\ \omega_0(t)+\omega_0(g_2(t)t) \end{cases}$.

The motivation arises from the calculations:

$$\begin{aligned} y_n - x^* &= x_n - x^* - F'(x_n)^{-1}F(x_n) \\ \|y_n - x^*\| &\leq \frac{\int_0^1 w((1-\theta)\|x_n - x^*\|)d\theta \|x_n - x^*\|}{1-\omega_0(\|x_n - x^*\|)} \\ &\leq g_1(\|x_n - x^*\|)\|x_n - x^*\| \leq \|x_n - x^*\| < r, \end{aligned}$$

$$\begin{aligned} z_n - x^* &= x_n - x^* - F'(x_n)^{-1}F(x_n) \\ &\quad +[I - \frac{3}{2}I + \frac{1}{2}F'(x_n)^{-1}F'(y_n)]F'(x_n)^{-1}F(x_n) \\ &= x_n - x^* - F'(x_n)^{-1}F(x_n) \\ &\quad +\frac{1}{2}F'(x_n)^{-1}(F'(y_n) - F'(x_n))F'(x_n)^{-1}F(x_n), \\ \|z_n - x^*\| &\leq \left[\frac{\int_0^1 \omega((1-\theta)\|x_n - x^*\|)d\theta}{1-\omega(\|x_n - x^*\|)} + \frac{\bar{\omega}_n(1+\int_0^1 \omega_0(\theta\|x_n - x^*\|)d\theta)}{2(1-\omega_0(\|x_n - x^*\|))^2}\right]\|x_0 - x^*\| \\ &\leq g_2(\|x_n - x^*\|) \leq \|x_n - x^*\|, \end{aligned}$$

$$\begin{aligned} x_{n+1} - x^* &= z_n - x^* - F'(z_n)^{-1}F(z_n) + (F'(z_n)^{-1} - F'(x_n)^{-1})F(z_n) \\ &\quad -F'(x_n)^{-1}(F'(x_n) - F'(y_n))F'(x_n)^{-1}F(z_n), \end{aligned}$$

$$\begin{aligned} \|x_{n+1} - x^*\| &\leq \left[\frac{\int_0^1 \omega((1-\theta)\|z_n - x^*\|)d\theta}{1-\omega_0(\|z_n - x^*\|)} \right. \\ &\quad \frac{\bar{\bar{\omega}}_n(1+\int_0^1 \omega_0(\theta\|z_n - x^*\|)d\theta}{(1-\omega_0(\|x_n - x^*\|))(1-\omega_0(\|z_n - x^*\|))} \\ &\quad \left.\frac{\bar{\omega}_n(1+\int_0^1 \omega_0(\theta(\|z_n - x^*\|)d\theta)}{(1-\omega_0(\|x_n - x^*\|))^2}\right]\|z_n - x^*\| \\ &\leq g_3(\|x_n - x^*\|)\|x_n - x^*\| \leq \|x_n - x^*\|. \end{aligned}$$

Method (36.2) The corresponding majorant functions g_i are:

$$g_1(t) = \frac{\int_0^1 \omega((1-\theta)t)d\theta}{1-\omega_0(t)},$$

$$g_2(t) = \frac{\int_0^1 \omega((1-\theta)t)d\theta}{1-\omega_0(t)} + \frac{\bar{\omega}(t)(1+\int_0^1 \omega_0(\theta t)d\theta)}{2(1-\omega_0(t))(1-\omega_0(g_1(t)t))},$$

$$\begin{aligned} g_3(t) &= \left[\frac{\int_0^1 \omega((1-\theta)g_2(t)t)d\theta}{1-\omega_0(g_2(t)t)}\right. \\ &\left. + \frac{\bar{\bar{\bar{\omega}}}(1+\int_0^1 \omega_0(\theta g_2(t)t)d\theta)}{(1-\omega_0(g_1(t)t))(1-\omega_0(g_2(t)t))}\right] g_2(t), \end{aligned}$$

where $\bar{\bar{\bar{\omega}}}(t) = \begin{cases} \omega((g_1(t)+g_2(t))t) \\ or \\ \omega_0(g_1(t)t)+\omega_0(g_2(t)t) \end{cases}.$

The calculation leading to the functions g_1 is given in method (36.1). Moreover,

$$\begin{aligned} z_n - x^* &= x_n - x^* - F'(x_n)^{-1}F(x_n) \\ &+\frac{1}{2}(F'(x_n)^{-1} - F'(x_n)^{-1})F(x_n), \\ \|z_n - x^*\| &\le \left[\frac{\int_0^1 \omega((1-\theta)\|x_n - x^*\|)d\theta}{1-\omega(\|x_n - x^*\|)}\right. \\ &\left. + \frac{\bar{\omega}_n(1+\int_0^1 \omega_0(\theta\|x_n - x^*\|)d\theta)}{2(1-\omega_0(\|x_n - x^*\|))(1-\omega_0(\|y_n - x^*\|))}\right] \|x_0 - x^*\| \\ &\le g_2(\|x_n - x^*\|) \le \|x_n - x^*\|, \end{aligned}$$

$$\begin{aligned} x_{n+1} - x^* &= z_n - x^* - F'(z_n)^{-1}F(z_n) + (F'(z_n)^{-1} - F'(y_n)^{-1})F(z_n) \\ &-F'(x_n)^{-1}(F'(x_n) - F'(y_n))F'(x_n)^{-1}F(z_n), \end{aligned}$$

and

$$\begin{aligned} \|x_{n+1} - x^*\| &\le \left[\frac{\int_0^1 \omega((1-\theta)\|z_n - x^*\|)d\theta}{1-\omega_0(\|z_n - x^*\|)}\right. \\ &\left. \frac{\bar{\bar{\bar{\omega}}}_n(1+\int_0^1 \omega_0(\theta\|z_n - x^*\|)d\theta)}{(1-\omega_0(\|y_n - x^*\|))(1-\omega_0(\|z_n - x^*\|))}\right] \|z_n - x^*\| \\ &\le g_3(\|x_n - x^*\|)\|x_n - x^*\| \le \|x_n - x^*\|. \end{aligned}$$

3. Semi-local Convergence

Method (36.1)

The majorant sequence $\{\alpha_n\}$ is defined for $\alpha_0 \ge 0$, $\beta_0 \ge \|F'(x_0)^{-1}F(x_0)\|$ and each $n = 0, 1, 2, \ldots$ by

$$\bar{v}_n = \begin{cases} v(\beta_n - \alpha_n) \\ or \\ v_0(\alpha_n) + v_0(\beta_n) \end{cases}$$

$$\begin{aligned}
\gamma_n &= \beta_n + \frac{\bar{v}_n(\beta_n - \alpha_n)}{1 - v_0(\alpha_n)}, \\
\lambda_n &= (1 + \int_0^1 v_0(\alpha_n + \theta(\gamma_n - \alpha_n) d\theta)(\gamma_n - \alpha_n) \\
&\quad + (1 + v_0(\alpha_n))(\beta_n - \alpha_n) \\
\alpha_{n+1} &= \gamma_n + \frac{1}{1 - v_0(\alpha_n)}(1 + \frac{\bar{v}_n}{1 - v_0(\alpha_n)}), \\
\delta_{n+1} &= \int_0^1 v((1-\theta)(\alpha_{n+1} - \alpha_n)) d\theta(\alpha_{n+1} - \alpha_n) \\
&\quad + (1 + v_0(\alpha_n))(\alpha_{n-1} - \alpha_n)
\end{aligned}$$

and

$$\beta_{n+1} = \alpha_{n+1} + \frac{\delta_{n+1}}{1 - v_0(\alpha_{n+1})}.$$

The motivation follows from the calculations:

$$\begin{aligned}
z_n - y_n &= [I - \frac{1}{2}(3I - F'(x_n)^{-1})F(y_n)](-(y_n - x_n)) \\
&= \frac{1}{2}F'(x_n)^{-1}(F'(y_n) - F'(x_n))(y_n - x_n),
\end{aligned}$$

$$\|z_n - y_n\| \leq \frac{\bar{v}_n(\beta_n - \alpha_n)}{2(1 - v_0(\alpha_n))} = \gamma_n - \beta_n,$$

$$F(z_n) = F(z_n) - F(x_n) - F'(x_n)(y_n - x_n),$$

$$\begin{aligned}
\|L^{-1}F(z_n)\| &\leq (1 + \int_0^1 v_0(\alpha_n + \theta(\gamma_n - \alpha_n)) d\theta)(\gamma_n - \alpha_n) \\
&\quad (1 + v_0(\alpha_n))(\beta_n - \alpha_n) = \lambda_n, \\
x_{n+1} - z_n &= -F'(x_n)^{-1}F(z_n) - F'(x_n)^{-1}(F'(x_n) - F'(y_n))F'(x_n)^{-1}F(z_n), \\
\|x_{n+1} - z_n\| &\leq \left[\frac{1}{1 - v_0(\alpha_n)} + \frac{\bar{v}_n}{(1 - v_0(\alpha_n))^2}\right]\lambda_n \\
&= \alpha_{n+1} - \gamma_n.
\end{aligned}$$

$$\begin{aligned}
F(x_{n+1}) &= F(x_{n+1}) - F(x_n) - F'(x_n)(x_{n+1} - x_n) \\
&\quad + F'(x_n)(x_{n+1} - y_n) \\
\|L^{-1}F(x_{n+1})\| &\leq \int_0^1 v((1-\theta)(\alpha_{n+1} - \alpha_n)) d\theta(\alpha_{n+1} - \alpha_n) \\
&\quad + (1 + v_0(\alpha_n))(\alpha_{n+1} - \beta_n) = \delta_{n+1}
\end{aligned}$$

and

$$\begin{aligned}
\|y_{n+1} - x_{n+1}\| &\leq \frac{\delta_{n+1}}{1 - v_0(\alpha_{n+1})} \\
&= \beta_{n+1} - \alpha_{n+1}.
\end{aligned}$$

Method (36.2)

The corresponding majorant sequence $\{x_n\}$ is defined for $\alpha_0 = 0, \beta_0 \geq \|F'(x_0)^{-1}F(x_0)\|$ and each $n = 0,1,2,\ldots$ by

$$\gamma_n = \beta_n + \frac{\bar{v}_n(\beta_n - \alpha_n)}{2(1 - v_0(\beta_n))},$$

$$\alpha_{n+1} = \gamma_n + \frac{\lambda_n}{1 - v_0(\beta_n)}$$

and

$$\beta_{n+1} = \alpha_{n+1} + \frac{\delta_{n+1}}{1 - v_0(\alpha_{n+1})}.$$

The motivation follows from the calculations

$$\begin{aligned} z_n - y_n &= (F'(x_n)^{-1} - \frac{1}{2}F'(x_n)^{-1} - \frac{1}{2}F'(y_n)^{-1})F(x_n) \\ &= \frac{1}{2}F'(y_n)^{-1}(F'(x_n) - F'(y_n))(y_n - x_n), \\ \|z_n - y_n\| &\leq \frac{\bar{v}_n(\beta_n - \alpha_n)}{2(1 - v_0(\beta_n))} = \gamma_n - \beta_n, \\ x_{n+1} - z_n &= -F'(y_n)^{-1}F(z_n), \end{aligned}$$

thus

$$\|x_{n+1} - z_n\| \leq \frac{\lambda_n}{1 - v_0(\beta_n)} = \alpha_{n+1} - \gamma_n.$$

The calculation for the sequences λ_n and β_{n+1} is given in method (36.1). The rest of the convergence analysis for the local as well as the semi-local case is provided in earlier Chapters.

4. References

[1] Argyros, C. I., Regmi, S., Argyros, I. K., George, S., *Contemporary Algorithms: Theory and Applications*, Volume III, NOVA Publishers, (2023).

[2] Argyros, I. K., *Advances in the Efficiency of Computational Methods and Applications*, World Scientific Publishing, Oxon, United Kingdom, (2000).

[3] Argyros, I. K., *The theory and applications of iteration methods*, 2nd Edition, CRC Press, Engineering Series, Taylor and Francis Group, (2022).

[4] Argyros, I. K., George, S., *Ball comparison between two optimal eight-order methods under weak conditions*, SeMA Journal, 72, (2015), 1–11.

[5] Cordero, A., Hueso, J. L., Martnez, E., Torregrosa, J. R., *A modified Newton-Jarratt's composition*, Numerical Algorithms, 55, (2010), 87–99.

[6] Darvishi, M. T., Barati, A., *A fourth-order method from quadrature formulae to solve systems of nonlinear equations*, Applied Mathematics and Computation, 188, 1, (2007), 257–261.

[7] Frontini, M., Sormani, E., *Third-order methods from quadrature formulae for solving systems of nonlinear equations*, Appl. Math. Comput., 149, (2004), 771–782.

[8] Grau-Snchez, M., Grau, A., Noguera, M., *Ostrowski type methods for solving systems of nonlinear equations*, Applied Mathematics and Computation, 218, 6, (2011), 2377–2385.

[9] Gutirrez, J. M., Hernndez, M. A., *A family of Chebyshev-Halley type methods in Banach spaces*, Bulletin of the Australian Mathematical Society, 55, 1, (1997), 113–130.

[10] Homeier, H. H. H., *A modified Newton method with cubic convergence: the multivariable case*, J. Comput. Appl. Math., 169, (2004), 161–169.

[11] Ortega, J. M., Rheinboldt, W. C., *Iterative Solution of Nonlinear Equations in Several Variables*, Academic Press, New York, (1970).

[12] Ostrowski, A. M., *Solutions of Equations and System of Equations*, Academic Press, New York, (1960).

[13] Potra, F. A., Patak, V., *Nondiscrete induction and iterative processes*, Pitman, Boston, Mass, USA, (1984).

[14] Sharma, J. R., Gupta, P., *An efficient fourth-order weighted-Newton method for systems of nonlinear equations*, Numerical Algorithm, 62, 2, (2013), 307–323.

Chapter 37

Five Comparable Third-Order Methods

1. Introduction

The methods are defined for $x_0 \in \Omega$ and each $n = 0, 1, 2, \ldots$ by

$$\begin{aligned} y_n &= x_n - F'(x_n)^{-1}F(x_n), \\ x_{n+1} &= y_n - F'(x_n)^{-1}F(y_n), \end{aligned} \tag{37.1}$$

$$\begin{aligned} y_n &= x_n - \frac{1}{2}F'(x_n)^{-1}F(x_n), \\ x_{n+1} &= x_n - F'(y_n)^{-1}F(x_n), \end{aligned} \tag{37.2}$$

$$\begin{aligned} y_n &= x_n - F'(x_n)^{-1}F(x_n), \\ A_n &= F'(x_n) + 3F'\left(\frac{x_n + 2y_n}{3}\right), \\ x_{n+1} &= x_n - 4A_n^{-1}F(x_n), \end{aligned} \tag{37.3}$$

$$\begin{aligned} y_n &= x_n - F'(x_n)^{-1}F(x_n), \\ B_n &= 2F'\left(\frac{3x_n + y_n}{4}\right) + 2F'\left(\frac{x_n + 3y_n}{4}\right) - F'\left(\frac{x_n + y_n}{2}\right), \\ x_{n+1} &= x_n - 3B_n^{-1}F(x_n) \end{aligned} \tag{37.4}$$

and

$$\begin{aligned} y_n &= x_n - F'(x_n)^{-1}F(x_n), \\ x_{n+1} &= x_n - \frac{1}{2}(3I - F'(x_n)^{-1}F'(y_n))F'(x_n)^{-1}F(x_n). \end{aligned} \tag{37.5}$$

The local convergence order three is determined in [14, 9, 12, 8] for $B_1 = B_2 = \mathbb{R}^k$ under the Taylor series expansion technique and assuming the existence and boundedness of the fourth derivative not on the methods. Other limitations are reported in earlier Chapters. The local convergence relies only on the operators on these methods. Moreover, the more interesting semi-local convergence analysis (not given before) is provided based on majorizing sequences and generalized conditions of F'. The same technique can be employed to extend the applicability of other methods [1]-[7],[10,11,12]).

2. Local Convergence

Method (37.1)
The majorant functions $g_i, i = 1,2$ are:

$$g_1(t) = \frac{\int_0^1 \omega((1-\theta)t)d\theta}{1-\omega_0(t)},$$

and

$$\begin{aligned} g_2(t) &= \left[\frac{\int_0^1 \omega((1-\theta)g_1(t)t)d\theta}{1-\omega_0(g_1(t)t)}\right. \\ &\left. + \frac{\bar{\bar{\omega}}(1+\int_0^1 \omega_0(\theta t)d\theta)}{(1-\omega_0(t))(1-\omega_0(g_1(t)t))}\right] g_1(t). \end{aligned}$$

where

$$\bar{\omega} = \begin{cases} \omega((1+g_1(t))t) \\ or \\ \omega_0(t)+\omega_0(g_1(t)t). \end{cases}$$

The motivation arises from the calculations:

$$\begin{aligned} y_n - x^* &= x_n - x^* - F'(x_n)^{-1}F(x_n) \\ \|y_n - x^*\| &\leq \frac{\int_0^1 w((1-\theta)\|x_n-x^*\|)d\theta\|x_n-x^*\|}{1-\omega_0(\|x_n-x^*\|)} \\ &\leq g_1(\|x_n-x^*\|)\|x_n-x^*\| \leq \|x_n-x^*\| < r, \end{aligned}$$

$$x_{n+1} - x^* = y_n - x^* - F'(y_n)^{-1}F(y_n) + (F'(y_n)^{-1} - F'(x_n)^{-1})F(y_n)$$

and

$$\begin{aligned} \|x_{n+1}-x^*\| &\leq \left[\frac{\int_0^1 \omega((1-\theta)\|y_n-x^*\|)d\theta}{1-\omega_0(\|y_n-x^*\|)}\right. \\ &\left. \frac{\bar{\omega}_n(1+\int_0^1 \omega_0(\theta\|y_n-x^*\|)d\theta}{(1-\omega_0(\|x_n-x^*\|))(1-\omega_0(\|y_n-x^*\|))}\right]\|y_n-x^*\| \\ &\leq g_2(\|x_n-x^*\|)\|x_n-x^*\| \leq \|x_n-x^*\|. \end{aligned}$$

Method (37.2)

The corresponding majorant functions g_i are:

$$g_1(t) = \frac{\int_0^1 \omega((1-\theta)t)d\theta + \frac{1}{2}(1+\int_0^1 \omega_0(\theta t)d\theta)}{1-\omega_0(t)}$$

and

$$\begin{aligned} g_2(t) &= \frac{\int_0^1 \omega((1-\theta)t)d\theta}{1-\omega_0(t)} \\ &+ \frac{\bar{\bar{\omega}}(1+\int_0^1 \omega_0(\theta t)d\theta)}{(1-\omega_0(t))(1-\omega_0(g_1(t)t))}. \end{aligned}$$

where

$$\bar{\bar{\omega}}(t) = \begin{cases} \omega((g_1(t)+g_2(t))t) \\ or \\ \omega_0(g_1(t)t)+\omega_0(g_2(t)t) \end{cases}.$$

The calculation are:

$$\begin{aligned}
y_n - x^* &= x_n - x^* - F'(x_n)^{-1}F(x_n) \\
&\quad +\frac{1}{2}F'(x_n)^{-1}F(x_n), \\
\|y_n - x^*\| &\leq \frac{1}{1-\omega(\|x_n - x^*\|)}\Big[\int_0^1 \omega((1-\theta)\|x_n - x^*\|)d\theta \\
&\quad +\frac{1}{2}(1+\int_0^1 \omega_0(\theta\|x_n - x^*\|)d\theta)\Big]\|x_0 - x^*\| \\
&\leq g_1(\|x_n - x^*\|) \leq \|x_n - x^*\| < r,
\end{aligned}$$

$$x_{n+1} - x^* = x_n - x^* - F'(x_n)^{-1}F(x_n) + (F'(x_n)^{-1} - F'(y_n)^{-1})F(x_n),$$

and

$$\begin{aligned}
\|x_{n+1} - x^*\| &\leq \Bigg[\frac{\int_0^1 \omega((1-\theta)\|x_n - x^*\|)d\theta}{1-\omega_0(\|x_n - x^*\|)} \\
&\quad \frac{\bar{\bar{\omega}}_n(1+\int_0^1 \omega_0(\theta\|x_n - x^*\|)d\theta)}{(1-\omega_0(\|y_n - x^*\|))(1-\omega_0(\|x_n - x^*\|))}\Bigg]\|x_n - x^*\| \\
&\leq g_2(\|x_n - x^*\|)\|x_n - x^*\| \leq \|x_n - x^*\|.
\end{aligned}$$

Method (37.3)

The function g_2 is defined (g_1 is in method (37.1)) by

$$q(t) = \frac{1}{4}\left(\omega_0(t) + 3\omega_0\left(\frac{(1+2g_1(t))t}{3}\right)\right)$$

and

$$a(t) = \begin{cases} \omega_0\left(\frac{2(1+g_1(t))t}{3}\right) \\ or \\ \omega_0(t) + \omega_0\left(\frac{2(1+g_1(t))t}{3}\right) \end{cases}$$

and

$$g_2(t) = \frac{\int_0^1 \omega((1-\theta)t)d\theta}{1-\omega_0(t)} + \frac{3a(t)(1+\int_0^1 \omega_0(\theta t)d\theta}{4(1-q(t))(1-w_0(t))}.$$

The motivation for the function g_2 is:

$$\begin{aligned}
\|(4L)^{-1}(A_n - 4L)\| &\leq \frac{1}{4}[\omega_0(\|x_n - x^*\| + 3\omega_0\left(\|\frac{x_n + 2y_n}{3} - x^*\|\right)] \\
&= \bar{q}_n \leq q_n < 1, \\
x_{n+1} - x^* &= x_n - x^* - F'(x_n)^{-1}F(x_n) + (F(x_n)^{-1} - A_n^{-1})F(x_n) \\
&= x_n - x^* - F'(x_n)^{-1}F(x_n) \\
&\quad -A_n^{-1}(4F'(x_n) - A_n)F'(x_n)^{-1}F(x_n)
\end{aligned}$$

and

$$\begin{aligned}\|x_{n+1}-x^*\| &\leq \left[\frac{\int_0^1 \omega((1-\theta)\|x_n-x^*\|)d\theta}{1-\omega_0(\|x_n-x^*\|)}\right.\\ &\quad \left.+\frac{3a_n(1+\int_0^1 \omega_0(\theta\|x_n-x^*\|)d\theta)}{4(1-q_n)(1-\omega_0(\|x_n-x^*\|))}\right]\|x_n-x^*\|\\ &\leq g_2(\|x_n-x^*\|)\|x_n-x^*\| \leq \|x_n-x^*\|.\end{aligned}$$

Method (37.4)

The function g_1 is given in method (37.2), where as

$$g_2(t)=\frac{\int_0^1 \omega((1-\theta)t)d\theta}{1-\omega_0(t)}+\frac{\lambda(t)(1+\int_0^1 \omega_0(\theta t)d\theta}{3(1-\mu(t))(1-w_0(t))},$$

where

$$\mu(t)=\frac{1}{3}\left[2\omega_0\left(\frac{(3+g_1(t))t}{4}\right)+2\omega_0\left(\frac{(1+3g_1(t))t}{4}\right)+\omega_0\left(\frac{(1+g_1(t))t}{2}\right)\right],$$

$$\lambda(t)=3\omega_0\left(\frac{(1+g_1(t))t}{4}\right)+\omega_0\left(\frac{3(1+g_1(t))t}{4}\right)$$

or

$$\lambda(t)=3\omega_0(t)+\omega_0\left(\frac{(1+g_1(t))t}{2}\right)+2\omega_0\left(\frac{(3+g_1(t))t}{4}\right)+2\omega_0\left(\frac{(1+3g_1(t))t}{4}\right).$$

The calculations are:

$$\begin{aligned}\|(3L)^{-1}(B_n-3L)\| &\leq \frac{1}{3}\left[2\omega_0\left(\|\frac{3x_n+y_n}{4}-x^*\|\right)+2\omega_0\left(\|\frac{x_n+3y_n}{4}-x^*\|\right)\right.\\ &\quad \left.+2\omega_0\left(\|\frac{x_n+y_n}{2}-x^*\|\right)\right.\\ &= \bar{\lambda}_n \leq \lambda_n < 1,\\ \|B_n^{-1}L\| &\leq \frac{1}{3(1-\lambda_n)}.\end{aligned}$$

$$x_{n+1}-x^*=x_n-x^*-F'(x_n)^{-1}F(x_n)+(F'(x_n)^{-1}-A_n^{-1})F(x_n)$$

and

$$\begin{aligned}\|x_{n+1}-x^*\| &\leq \left[\frac{\int_0^1 \omega((1-\theta)\|x_n-x^*\|)d\theta}{1-\omega_0(\|x_n-x^*\|)}\right.\\ &\quad \left.+\frac{\lambda_n(1+\int_0^1 \omega_0(\theta\|x_n-x^*\|)d\theta)}{3(1-\omega_0(\|x_n-x^*\|))(1-\mu_n)}\right]\|x_n-x^*\|\\ &\leq g_2(\|x_n-x^*\|)\|x_n-x^*\| \leq \|x_n-x^*\|,\end{aligned}$$

where we also used

$$\begin{aligned}B_n-3F'(x_n) &= 2\left(F'\left(\frac{3x_n+y_n}{4}\right)-F'(x_n)\right)\\ &\quad +\left(F'\left(\frac{x_n+3y_n}{4}\right)-F'\left(\frac{x_n+y_n}{2}\right)\right)\\ &\quad +\left(F'\left(\frac{x_n+3y_n}{4}\right)-F'(x_n)\right),\end{aligned}$$

$$\|L^{-1}(B_n-3F'(x_n))\| \leq \bar{\lambda}_n \leq \lambda_n.$$

Method (37.5)

The function g_1 is given in method (37.1), where as the function g_2 is defined as

$$g_2(t) = \frac{\int_0^1 \omega((1-\theta)t)d\theta}{1-\omega_0(t)} + \frac{\bar{\omega}(t)(1+\int_0^1 \omega_0(\theta t)d\theta}{2(1-w_0(t))^2}.$$

The calculation is

$$x_{n+1}-x^* = x_n - x^* - F'(x_n)^{-1}F(x_n) - \frac{1}{2}(F'(x_n)^{-1}(F'(x_n)-F'(y_n))F'(x_n)^{-1})F(x_n),$$

and

$$\begin{aligned}\|x_{n+1}-x^*\| &\leq \left[\frac{\int_0^1 \omega((1-\theta)\|x_n-x^*\|)d\theta}{1-\omega_0(\|x_n-x^*\|)}\right.\\ &\left.+\frac{\bar{\omega}_n(1+\int_0^1 \omega_0(\theta\|x_n-x^*\|)d\theta)}{2(1-\omega_0(\|x_n-x^*\|))^2}\right]\|x_n-x^*\|\\ &\leq g_2(\|x_n-x^*\|)\|x_n-x^*\| \leq \|x_n-x^*\|.\end{aligned}$$

3. Semi-local Convergence

The role of x^*, ω is exchanged by x_0, v but the calculations are the same. We have $\alpha_0 \geq 0$, $\beta_0 \geq \frac{1}{2}\|F'(x_0)^{-1}F(x_0)\|$. Then, the majorant sequence for each method is defined for each $n = 0, 1, 2, \ldots$ as:

Method (37.1)

$$\begin{aligned}\alpha_{n+1} &= \beta_n + \frac{\int_0^1 v((1-\theta)(\beta_n-\alpha_n))d\theta(\beta_n-\alpha_n)}{1-v_0(\alpha_n)},\\ \delta_{n+1} &= \int_0^1 v((1-\theta)(\alpha_{n+1}-\alpha_n))d\theta(\alpha_{n+1}-\alpha_n)\\ &+(1+v_0(\alpha_n))(\alpha_{n-1}-\beta_n)\end{aligned}$$

and

$$\beta_{n+1} = \alpha_{n+1} + \frac{\delta_{n+1}}{1-v_0(\alpha_{n+1})}.$$

The motivational calculations are:

$$\begin{aligned}F(y_n) &= F(y_n)-F(x_n)-F'(x_n)(y_n-x_n)\\ &= \int_0^1 (F'(x_n+\theta(y_n-x_n))d\theta - F'(x_n))(y_n-x_n),\\ \|L^{-1}F(y_n)\| &\leq (1+\int_0^1 v_0((1-\theta)(\beta_n-\alpha_n))d\theta)(\beta_n-\alpha_n)\end{aligned}$$

$$\begin{aligned}F(x_{n+1}) &= F(x_{n+1})-F(x_n)-F'(x_n)(x_{n+1}-x_n)\\ &+F'(x_n)(x_{n+1}-y_n)\\ \|L^{-1}F(x_{n+1})\| &\leq \bar{\delta}_{n+1} \leq \delta_{n+1}\end{aligned}$$

and

$$\begin{aligned}\|y_{n+1}-x_{n+1}\| &\leq \|F'(x_{n+1})^{-1}L\|\|L^{-1}F(x_{n+1})\|\\ &\leq \frac{\delta_{n+1}}{1-v_0(\alpha_{n+1})}\\ &= \beta_{n+1}-\alpha_{n+1}.\end{aligned}$$

Method (37.2)

$$\begin{aligned}\xi_n &= 2(1+v_0(\alpha_n))(\beta_n-\alpha_n)\\ \alpha_{n+1} &= \beta_n+\frac{1}{2(1-v_0(\beta_n))}[\bar{v}_n(\beta_n-\alpha_n)+\xi_n]\\ d_{n+1} &= (1+\int_0^1 v_0(\alpha_n+\theta(\alpha_{n+1}-\alpha_n))d\theta)(\alpha_{n+1}-\alpha_n)\\ &\quad +2(1+v_0(\alpha_n))(\beta_n-\alpha_n)\end{aligned}$$

and

$$\beta_{n+1}=\alpha_{n+1}+\frac{d_{n+1}}{2(1-v_0(\alpha_{n+1}))}.$$

The calculations are in turn:

$$\begin{aligned}x_{n+1}-y_n &= \frac{1}{2}F'(x_n)^{-1}F(x_n)-F'(y_n)^{-1}F(x_n)\\ &= \frac{1}{2}F'(y_n)^{-1}(F'(x_n)-F'(y_n))(y_n-x_n)-\frac{1}{2}F'(y_n)^{-1}F(x_n),\\ \|x_{n+1}-y_n\| &\leq \frac{\bar{v}_n(\beta_n-\alpha_n)}{2(1-v_0(\beta_n))}+\frac{\xi_n}{2(1-v_0(\beta_n))}\\ &= \alpha_{n+1}-\beta_n,\end{aligned}$$

$$\begin{aligned}F(x_{n+1}) &= F(x_{n+1})-F(x_n)-2F'(x_n)(y_n-x_n),\\ \|L^{-1}F(x_{n+1})\| &\leq \bar{d}_{n+1}\leq d_{n+1}\end{aligned}$$

and

$$\begin{aligned}\|y_{n+1}-x_{n+1}\| &\leq \frac{1}{2}\|F'(x_{n+1})^{-1}L\|\|L^{-1}F(x_{n+1})\|\\ &\leq \frac{d_{n+1}}{2(1-v_0(\alpha_{n+1}))}\\ &= \beta_{n+1}-\alpha_{n+1}.\end{aligned}$$

Method (37.3)

$$\begin{aligned}p_n &= \frac{1}{4}(v_0(\alpha_n)+v_0(\frac{\alpha_n+2\beta_n}{3})),\\ b_n &= \begin{cases} v(\frac{2(\beta_n-\alpha_n)}{3}) \\ v_0(\alpha_n)+v_0(\frac{\alpha_n+2\beta_n}{3}) \end{cases}\\ \alpha_{n+1} &= \beta_n+\frac{3b_n(\beta_n-\alpha_n)}{4(1-p_n)},\end{aligned}$$

and the iterate β_{n+1} is given in method (37.1).

The calculations are:

$$\begin{aligned} \|x_{n+1}-y_n\| &\leq \frac{3b_n(\beta_n-\alpha_n)}{4(1-p_n)} \\ &= \alpha_{n+1}-\beta_n, \end{aligned}$$

Method (37.4)

$$\begin{aligned} s_n &= \frac{1}{3}\left(2v_0\left(\frac{3\alpha_n+\beta_n}{4}\right)\right. \\ &\quad \left.+2v_0\left(\frac{\alpha_n+3\beta_n}{4}\right)+v_0\left(\frac{\alpha_n+\beta_n}{2}\right)\right) \\ c_n &= 4v_0\left(\frac{\beta_n-\alpha_n}{4}\right)+v_0\left(\frac{3(\beta_n-\alpha_n)}{4}\right) \end{aligned}$$

or

$$\begin{aligned} c_n &= 3v_0(\alpha_n)+v_0(\frac{\alpha_n+\beta_n}{2} \\ &\quad +2v_0(\frac{\alpha_n+3\beta_n}{4}+v_0(\frac{3\alpha_n+\beta_n}{4}, \\ \alpha_{n+1} &= \beta_n+\frac{c_n(\beta_n-\alpha_n)}{3(1-s_n)}, \end{aligned}$$

whereas the iterate β_{n+1} is given in method (37.1).

The calculations for α_{n+1} is::

$$\begin{aligned} x_{n+1}-y_n &= (F'(x_n)^{-1}-3B_n^{-1})F(x_n) \\ &= B_n^{-1}(3F'(x_n)-B_n)(y_n-x_n) \end{aligned}$$

and

$$\begin{aligned} \|x_{n+1}-y_n\| &\leq \frac{c_n(\beta_n-\alpha_n)}{3(1-s_n)} \\ &= \alpha_{n+1}-\beta_n. \end{aligned}$$

The calculations for β_{n+1} is provided in method (37.1).

Method (37.5)

$$\alpha_{n+1} = \beta_n+\frac{\bar{v}_n(\beta_n-\alpha_n)}{2(1-v_0(\alpha_n))},$$

and the iterate β_{n+1} is given in method (37.1).

The calculations for α_{n+1} is::

$$\begin{aligned} x_{n+1}-y_n &= \frac{1}{2}(I-F'(x_n)^{-1}F'(y_n))(y_n-x_n) \\ &= \frac{1}{2}F'(x_n)^{-1}(F'(x_n)-F'(y_n))(y_n-x_n) \end{aligned}$$

and

$$\|x_{n+1}-y_n\| \leq \frac{\bar{v}_n(\beta_n-\alpha_n)}{2(1-v_0(\alpha_n))},$$

where $\bar{v}_n = \begin{cases} v(\|y_n-x_n\|) \\ or \\ v_0(\|x_n-x_0\|)+v_0(\|y-n-x_0\|) \end{cases}$ and $\bar{v}_n \leq v_n = \begin{cases} v(\beta_n-\alpha_n) \\ or \\ v_0(\alpha_n)+v_0(\beta_n). \end{cases}$

The rest of the convergence analysis for both the local as well as the semi-local convergence has appeared in earlier Chapters.

4. References

[1] Argyros, C. I., Regmi, S., Argyros, I. K., George, S., *Contemporary Algorithms: Theory and Applications*, Volume III, NOVA Publishers, (2023).

[2] Argyros, I. K., *Advances in the Efficiency of Computational Methods and Applications*, World Scientific Publishing, Oxon, United Kingdom, (2000).

[3] Argyros, I. K., *The theory and applications of iteration methods*, 2nd Edition, CRC Press, Engineering Series, Taylor and Francis Group, (2022).

[4] Argyros, I. K., George, S., *Ball comparison between two optimal eight-order methods under weak conditions*, SeMA Journal, 72, (2015), 1–11.

[5] Cordero, A., Hueso, J. L., Martnez, E., Torregrosa, J. R., *A modified Newton-Jarratt's composition*, Numerical Algorithms, 55, (2010), 87–99.

[6] Darvishi, M. T., Barati, A., *A fourth-order method from quadrature formulae to solve systems of nonlinear equations*, Applied Mathematics and Computation, 188, 1, (2007), 257–261.

[7] Frontini, M., Sormani, E., *Third-order methods from quadrature formulae for solving systems of nonlinear equations*, Appl. Math. Comput., 149, (2004), 771–782.

[8] Grau-Snchez, M., Grau, A., Noguera, M., *Ostrowski type methods for solving systems of nonlinear equations*, Applied Mathematics and Computation, 218, 6, (2011), 2377–2385.

[9] Gutirrez, J. M., Hernndez, M. A., *A family of Chebyshev-Halley type methods in Banach spaces*, Bulletin of the Australian Mathematical Society, 55, 1, (1997), 113–130.

[10] Homeier, H. H. H., *A modified Newton method with cubic convergence: the multivariable case*, J. Comput. Appl. Math., 169, (2004), 161–169.

[11] Noor, M. A., Waseem, M., *Some iterative methods for solving a system of nonlinear equations*, Computers and Mathematics with Applications, 57, 1, (2009), 101–106.

[12] Ortega, J. M., Rheinboldt, W. C., *Iterative Solution of Nonlinear Equations in Several Variables*, Academic Press, New York, (1970).

[13] Potra, F. A., Patak, V., *Nondiscrete induction and iterative processes*, Pitman, Boston, Mass, USA, (1984).

[14] Sharma, J. R., Gupta, P., *An efficient fourth-order weighted-Newton method for systems of nonlinear equations*, Numerical Algorithm, 62, 2, (2013), 307–323.

Chapter 38

Extended Convergence of Newton's Method under Kantorovich Conditions for k-Differentiable Operators

1. Introduction

Let $\mathcal{B}_1, \mathcal{B}_2$ be Banach spaces, $\Omega \subset \mathcal{B}_1$ be a nonempty open set and $F : \Omega \longrightarrow \mathcal{B}_2$ be a Fréchet-differentiable operator. Applications from different disciplines of sciences can be formulated like nonlinear equation

$$F(x) = 0. \tag{38.1}$$

To find a solution $x_* \in \Omega$ in closed form for (38.1) is achieved only in special cases. So, iterative procedures are mostly utilized to obtain iterates that approximate x_*. The Newton's method for (38.1) is defined as:

$$x_{n+1} = x_n - F'(x_n)^{-1}F(x_n),\ x_0 \in \Omega. \tag{38.2}$$

The Kantorovich theorem for Newton's method [2]-[6] is extended in [7] provided that $F^{(k)}$ satisfies some conditions for some fixed natural number $k = 2,3,\ldots$. In this chapter, more precise majorizing sequences are developed for Newton's method, resulting in a tighter semi-local convergence analysis under the same or weaker conditions. This new idea can be applied to other methods [1]-[10].

2. Background

Majorizing sequences play a central role in the study of iterative processes [2]-[6].

Definition 12. [9] If $\{y_n\}$ is a real sequence in a Banach space E_1 and $\{z_n\}$ is a real sequence, then $\{z_n\}$ is majorizing sequence for $\{y_n\}$, if $\|y_n - y_{n-1}\| \leq z_n - z_{n-1}$. for each $n = 1,2,\ldots$. It follows by this definition that for each $n = 0,1,2,\ldots$, $z_{n-1} \leq z_n$.

That is the sequence $\{z_n\}$ is non-decreasing. Additional condition is needed to establish the convergence of the sequence $\{y_n\}$.

Lemma 6. [9] Let $\{z_n\}$ be majorizing for a sequence $\{y_n\}$ in a Banach space E_1. If $\{z_n\}$ converges to some $z_* < +\infty$, then there exists $y_* \in B_1$ so that $y_* = \lim_{n \longrightarrow +\infty} y_n$ and $\|y_* - y_n\| \leq z_* - z_n$ for each $n = 0,1,2,\ldots$.

The following balls are used:

$$S(w,\lambda) = \{x \in B_1 : \|x - w\| < \lambda,\ \lambda > 0\}$$

and

$$S[w,\lambda] = \{x \in B_1 : \|x - w\| \leq \lambda,\ \lambda > 0\}.$$

Next, we present the results in [7]. These results are given in non-affine invariant form. However, we reproduce them in non-affine invariant form. The advantages of affine versus non affine invariant form are well explained in [2]-[6]. Later, in Section 3 we show how to improve further the semi-local results in [7]. Let $f \in C^{(k)}([t_0, \bar{t}_0])$ with $t_0 \leq \bar{t}_0$ and $k = 3, 4, \ldots$.
The conditions are:

(C1) There exists $x_0 \in D$ so that $F'(x_0)^{-1} \in L(B_2, B_1)$ with $\|F'(x_0)^{-1}F(x_0)\| \leq -\dfrac{f(t_0)}{f'(t_0)}$ and

$$\|F'(x_0)^{-1}F(x_0)\| \leq f^{(i)}(t_0)$$

for $i = 2, 3, \ldots k$.

(C2) $\|F'(x_0)^{-1}F^{(k)}(x)\| \leq f^{(k)}(t)$ if $x \in S[x_0, t - t_0]$ and $t \in [t_0, \bar{t}_0]$.

Define the scalar sequence

$$t_n = N(t_{n-1}),\ n = 1, 2, \ldots, \tag{38.3}$$

where $N(t) = t - f'(t)^{-1}f(t)$.The following auxiliary result can be found in [7, Theorem 5 and Theorem 6].

PROPOSITION 4. *Assume $f \in C^{(k)}([t_0, \bar{t}_0]), k \geq 3$ be so that $f(t_0) > 0$, $f'(t_0) < 0$, $f^{(i)}(t_0) > 0$, $i = 2, 3, \ldots k - 1$ and $f^{(k)} > 0$. Moreover:*

(i) There exists a solution s of $f'(t) = 0$ with $s > t_0$, then s is the unique minimum of $f(t)$ in $[t_0, \bar{t}_0]$ and the function f is decreasing in the interval $[t_0, s)$.

(ii) If $f(s) \leq 0$, then $f(t) = 0$ has solutions in the interval $[t_0, +\infty)$. Denote by t_* the least such solution. Then, $t_0 \leq t_* \leq s$. Moreover, assume conditions (C1) and (C2) hold and there exists a solution $s \in (t_0, \bar{t}_0)$ of $f'(t) = 0$. Then, $\lim_{n \longrightarrow +\infty} t_n = t_*$ and $t_n \leq t_{n+1} \leq t_*$ for each $n = 0, 1, 2, \ldots$.

The main semi-local result follows for method (38.1).

Theorem 21. Suppose: conditions (C1) and (C2) holds:

(i) There exists a solution $s \in (t_0, \bar{t}_0)$ of $f'(t) = 0$ and such that $f(s) < 0$.

(ii) $S[x_0, t_* - t_0] \subseteq \Omega$. Then, the following assertions hold:

(a) $\|F'(x_n)^{-1}F'(x_0)\| \leq -\dfrac{1}{f'(t_n)}$.

(b) $\|F'(x_0)^{-1}F^{(i)}(x_n)\| \leq f^{(i)}(t_n),\ i = 2, 3, \ldots k - 1$.

(c) $\|F'(x_0)^{-1}F(x_n)\| \leq f(t_n)$.

(d) $\|x_{n+1} - x_n\| \leq t_{n+1} - t_n$.

(e) There exists $x_* \in S[x_0, t_* - t_0]$ solving $F(x) = 0$; $x_n, x_* \in S[x_0, t_* - t_0]$ and $\|x_* - x_n\| \leq t_* - t_n$.

Moreover, if the real function f has solutions t_* and t_{**} such that $t_0 < t_* \leq t_{**}$, then x_* is unique in $\Omega_0 = S(x_0, t_{**} - t_0) \cap \Omega$ if $t_* < t_{**}$ or in $\Omega_1 = S[x_0, t_{**} - t_0] \cap \Omega$ if $t_* = t_{**}$. Furthermore, the apriori estimates are available

(iii) if $t_* < t_{**}$, then

$$\frac{(t_{**} - t_*)q^{2^n}}{\sqrt{p_1} - q^{2^n}} < t_* - t_{**} < \frac{(t_{**} - t_*)\Delta^{2^n}}{\sqrt{p_1} - \Delta^{2^n}},$$

where $q = \frac{t_*}{t_{**}}\sqrt{p_1}$, $\Delta = \frac{t_*}{t_{**}}\sqrt{p_2}$, $p1 = \min\{h_1(t); t \in [0, t_*]\}$, $p_2 = \max\{h_1(t); t \in [0, t_*]\}$, $h_1(t) = \frac{(t_{**} - t)g'(t) - g(t)}{(t_* - t)g'(t) - g(t)}$ provided $f(t) = (t_* - t)(t_{**} - t)g(t)$, $q < 1$ and $\Delta < 1$.

(iv) If $t_* = t_{**}$, then $p_3^n t_* \leq t_* - t_n \leq p_4^n t_*$, where $p_3 = \min\{h_2(t) : t \in [0, t_*]\}, p_4 = \max\{h_2(t) : t \in [0, t_*]\}$ and $h_2(t) = \frac{(t_* - t)g'(t) - g(t)}{(t_* - t)f'(t) - 2g(t)}$ provided that $p_3 < 1$ and $p_4 < 1$.

Proof. Simply replace F by $F'(x_0)^{-1}F$ in the corresponding proofs of Section 3 in [7].

□

Remark. Notice that if $k = 2$ and f is a quadratic polynomial of the form $f(t) = \frac{\ell}{2}t^2 - t + \eta$, then Theorem 21 gives the Newton-Kantorovich result for Newton's method in affine invariant form [9]. So far we have rewritten the result in [9] in affine instead of noaffine invariant form. In the next Section, some ways are presented to improve the semi-local convergence analysis of this Section under the same or weaker convergence conditions.

3. Extended Semi-local Analysis

It is convenient to define real sequences in turn by

$$\begin{aligned} u_m &= u^{(i)}(t_{m-1,m}) \\ &= \sum_{j=1}^{k-1} \frac{1}{(j-1)!} f^{(j+i-1)}(t_{m-1})(t_m - t_{m-1})^{j-1} \\ &\quad + \frac{1}{(k-1-i)!}\int_{t_{m-1}}^{t_m} f^{(k)}(t_m - t)^{k-i-1} dt, \\ v_m &= v_{m-1,m} = \sum_{j=2}^{k-1} \frac{1}{j!} u^{(i)}(t_{m-1})(t_m - t_{m-1})^j \\ &\quad + \frac{1}{(k-1)!}\int_0^1 f^{(k)}(t_{m-1} + t(t_m - t_{m-1}))(1-t)^{k-1}(t_m - t_{m-1})^k dt, \\ r_{n-1} &= r_n - \frac{v_n}{f'(t_n)} \end{aligned} \tag{38.4}$$

and

$$
\begin{aligned}
w_m &= \frac{1}{1-\psi_0(s_m)}\left(\sum_{j=2}^{k-1}\frac{1}{(i-1)!}f^{(i)}(s_{m-1})(s_m-s_{m-1})^{i-1}\right.\\
&\quad \left.+\frac{1}{(k-2)!}\int_{s_{m-1}}^{s_m} f^{(k)}(t)(s_m-t)^{k-2}d\theta\right)\\
\mu_m &= \mu^{(i)}(s_{m-1},s_m)=\sum_{j=1}^{k-1}\frac{1}{(j-1)!}f^{(k+i-1)}(s_{m-1})(s_m-s_{m-1})^{j-1}\\
&\quad +\frac{1}{(k-1-i)!}\int_{s_{m-1}}^{s_m} f^{(k)}(t)(s_m-t)^{k-i-1}dt,\\
\lambda_m &= \lambda_{m-1,m}=\sum_{j=2}^{k-1}\frac{1}{j!}\mu^{(j)}(s_{m-1})(s_m-s_{m-1})^{j}\\
&\quad +\frac{1}{(n-1)!}\int_0^1 f^{(k)}(s_{m-1}+t(s_m-s_{m-1}))(1-t)^{k-1}(s_m-s_{m-1})^k dt,\\
s_{n+1} &= s_n+\frac{\lambda_n}{1-w_n}, \qquad (38.5)
\end{aligned}
$$

where we also imposed the condition

$$\|F'(x_0)(F'(x)-F'(x_0))\| \leq \psi(\|x-x_0\|) \tag{38.6}$$

for each $x \in S(x_0, t-t_0), t \in [t_0, \bar{t_0}]$ and some nondecreasing continuous function on the interval $[t_0, \bar{t_0}]$.

It follows by the proof of Lemma 7 in [7] that tighter sequences $\{r_n\}, \{s_n\}$ can replace $\{t_n\}$ in Theorem 21. In particular, we have

$$0 \leq s_n \leq r_n \leq t_n \tag{38.7}$$

and

$$0 \leq s_{n+1}-s_n \leq r_{n+1}-r_n \leq t_{n+1}-t_n \tag{38.8}$$

and

$$\lim_{n\to+\infty} s_n \leq \lim_{n\to+\infty} r_n \leq \lim_{n\to+\infty} t_n = t_*. \tag{38.9}$$

Estimates (38.7)-(38.9) show that the results in Theorem 21 are improved under the same conditions. Notice also that the assumption on the solution s of $f'(t)$ can be dropped if replaced by weaker convergence criterion for $f'(t)=0$ as follows. Assume:

$$\bar{w}_n < 1 \quad \text{and} \quad s_n \leq \gamma \tag{38.10}$$

for some $\gamma > 0$ and each $n = 1, 2, \ldots$

It follows by this condition that

$$0 \leq s_n \leq s_{n+1} \leq \gamma \tag{38.11}$$

and there exists $\gamma_* \in [0, \gamma]$ such that

$$\lim_{n\to+\infty} s_n = \gamma. \tag{38.12}$$

Thus, the majorizing sequence $\{s_n\}$ can replace $\{t_n\}$.

It is worth noticing that the condition (38.9) is weaker than the one involving the existence of s.

In view of the condition (38.6) there are two more versions of the sequence $\{w_n\}$ in (38.5).

We have

$$\|F'(x_0)^{-1}(F'(x_n)-F'(x_0))\| \leq \psi(\|x_n-x_0\|)$$

so if the additional condition $\psi(t)<1$ is imposed, then

$$\|F'(x_n)^{-1}F'(x_0)\| \leq \frac{1}{1-\psi(\|x_n-x_0\|)}.$$

Thus, we can set

$$w_n = \psi(s_n).$$

Denote the resulting sequence by $\bar{s}_n$.

Moreover, by the calculation

$$\begin{aligned} I-F'(x_0)^{-1}F'(x_n) &= I-F'(x_0)^{-1}F(x_{n-1}) \\ &-\sum_{i=2}^{k-1} F'(x_0)^{-1}F^{(i)}(x_{n-1})(x_n-x_{n-1})^{i-1} \\ &-\frac{1}{(k-2)!}\int_0^1 F'(x_0)^{-1}F^{(k)}(x_{n-1}+\theta(x_n-x_{n-1})) \\ &(1-\theta)^{k-2}(x_n-x_{n-1})^{k-1}d\theta. \end{aligned}$$

So, we get

$$\begin{aligned} \|I-F'(x_0)^{-1}F(x_n)\| &\leq \psi(\|x_{n-1}-x_0\|) \\ &+\sum_{i=2}^{k-1} f^{(i)}(\|x_{n-1}-x_0\|)\|x_n-x_{n-1}\|^{i-1} \\ &+\frac{1}{(k-2)!}\int_0^1 f^{(k)}((1-\theta)\|x_{n-1}-x_0\|+\theta\|x_n-x_0\|) \\ &(1-\theta)^{k-2}\|x_n-x_{n-1}\|^{k-1}d\theta. \end{aligned}$$

Therefore another choice for w_n is given by

$$\begin{aligned} w_n &= \psi(s_{n-1})+\sum_{i=2}^{k-1} f^{(i)}(s_{n-1})(s_n-s_{n-1})^{i-1} \\ &+\frac{1}{(k-2)!}\int_0^1 f^{(k)}((1-\theta)s_{n-1}+\theta s_n)(1-\theta)^{k-2}(s_n-s_{n-1})^{k-1}d\theta \end{aligned}$$

provided that $w_n<1$.

Therefore, we have developed five majorizing sequences. In practice we shall use the tighter one.

4. References

[1] Andronow, A. A., Chaikin, C. E., *Theory of Oscillations*, Princeton University Press, New Jersey, (1949).

[2] Argyros, I. K., *Convergence and Applications of Newton-type Iterations*, Springer Verlag, Berlin, Germany, (2008).

[3] Argyros, I. K., *On the convergence of Newtons method for polynomial equations and applications in radiative transfer*, Monatsh. Math., 127 (4), (1999), 265–276.

[4] Argyros, I. K., *On the NewtonKantorovich hypothesis for solving equations*, J. Comput. Appl. Math., 169 (2), (2004), 315–332.

[5] Argyros, I. K., *The theory and applications of iteration methods*, 2nd Edition, CRC Press, Engineering Series, Taylor and Francis Group, (2022).

[6] Argyros, I. K., *Unified Convergence Criteria for Iterative Banach Space Valued Methods with Applications*, Mathematics 2021, 9(16), 1942; https://doi.org/10.3390/math9161942.

[7] Ezquerro, J. A., Gonzle, D., Hernndez-Vern, M. A., *A semilocal convergence result for Newtons method under generalized conditions of Kantorovich*, Journal of Complexity, 30, 3, (2014), 309–324.

[8] Gutirrez, J. M., *A new semilocal convergence theorem for Newtons method*, J. Comput. Appl. Math., 79 (1997), 131–145.

[9] Kantorovich, L. V., Akilov, G. P., *Functional Analysis*, Pergamon Press, Oxford (1982).

[10] Yamamoto, T., *Convergence theorem for Newton-like methods in Banach spaces*, Numer. Math., 51 (1987), 545–557.

Chapter 39

Fourth-Order Family of Methods with One Parameter

1. Introduction

The method is defined for $x_0 \in \Omega$ and each $n = 0, 1, 2, \cdots$ by

$$\begin{aligned} y_n &= x_n - \frac{2}{3} F'(x_n)^{-1} F(x_n) \\ x_{n+1} &= x_n - A_n F'(x_n)^{-1} F(x_n), \end{aligned} \tag{39.1}$$

where $a \in \mathbb{R}$ is a free parameter and

$$A_n = \left(\frac{5}{3} - a\right) I + aF'(y_n)^{-1} F'(x_n) + \frac{a}{3} F'(x_n)^{-1} F'(y_n) + \left(\frac{3}{a} - \frac{a}{3}\right)\left(F'(y_n)^{-1} F'(x_n)\right)^2.$$

The local convergence order four is established under the Taylor expansion series approach in [15] by assuming the existence of the fifth derivative not on the method and when $B_1 = B_2 = \mathbb{R}^k$. Other limitations of this approach have been given in previous chapters. That is why we present a local convergence analysis based only on the operators on the method and the concept of generalized continuity. Moreover, the more interesting semi-local convergence (not given in [15]) is also developed using majorizing sequences. Therefore, the applicability of the method is extended and in the more general setting of a Banach space. The technique can be used on other methods [1]-[14], [16,17] analogously.

2. Local Analysis

The majorant functions $g_i, i = 1, 2$ are defined as

$$g_1(t) = \frac{\int_0^1 \omega((1-\theta)t)d\theta + \frac{1}{3}(1 + \int_0^1 \omega_0(\theta t))d\theta}{1 - \omega_0(t)},$$

$$\overline{\omega}(t) = \begin{cases} \omega((1 + g_1(t))t) \\ \text{or} \\ \omega_0(t) + \omega_0(g_1(t)t) \end{cases}$$

and

$$g_2(t) = \frac{\int_0^1 \omega((1-\theta)t)d\theta}{1-\omega_0(t)}$$
$$+\left(\left|\frac{3}{4}+\frac{a}{3}\right| \frac{\overline{\omega}(t)}{1-\omega_0(g_1(t)t)} + \left|\frac{3}{8}-\frac{a}{3}\right| \left(\frac{\overline{\omega}(t)}{1-\omega_0(g_1(t)t)}\right)^2\right.$$
$$\left. + \left|\frac{a}{3}\right| \frac{\overline{\omega(t)}}{1-\omega_0(t)}\right) \frac{1+\int_0^1 \omega_0(\theta t)d\theta}{1-\omega_0(t)}.$$

The development of these functions follows from the calculations:

$$y_n - x^* = x_n - x^* - F'(x_n)^{-1}F(x_n) + \frac{1}{3}F'(x_n)^{-1}F(x_n),$$
$$\|y_n - x^*\| \le \frac{\left[\int_0^1 \omega((1-\theta)\|x_n - x^*\|)d\theta + \frac{1}{3}(1+\int_0^1 \omega_0(\theta)\|x_n - x^*\|)d\theta\right]\|x_n - x^*\|}{1-\omega_0(\|x_n - x^*\|)}$$
$$\le g_1(\|x_n - x^*\|)\|x_n - x^*\|$$
$$\le \|x_n - x^*\| < r,$$
$$x_{n+1} - x^* = x_n - x^* - F'(x_n)^{-1}F(x_n) - (A_n - I)F'(x_n)^{-1}F(x_n)$$
$$= x_{n+1} - x^* - F'(x_n)^{-1}F(x_n)$$
$$-\left[\left(\frac{3}{4}+\frac{a}{3}\right)\left(F'(y_n)^{-1}F'(x_n) - I\right) + \left(\frac{3}{8}-\frac{a}{3}\right)\left(F'(y_n)^{-1}F'(x_n) - I\right)^2\right.$$
$$\left. + \frac{a}{3}(F'(x_n)^{-1}F^{-1}(y_n) - I)\right]F'(x_n)^{-1}F(x_n)$$

and

$$\|x_{n+1} - x^*\| \le \left[\frac{\int_0^1 \omega((1-\theta)\|x_n - x^*\|)d\theta}{1-\omega_0(\|x_n - x^*\|)}\right.$$
$$+\left(\left|\frac{3}{4}+\frac{a}{3}\right| \frac{\overline{\omega}_n}{1-\omega_0(\|y_n - x^*\|)} + \left|\frac{3}{8}-\frac{a}{3}\right| \left(\frac{\overline{\omega}_n}{1-\omega_0(\|y_n - x^*\|)}\right)^2\right.$$
$$\left.\left. + \left|\frac{a}{3}\right| \frac{\overline{\omega}_n}{1-\omega_0(\|x_n - x^*\|)}\right) \frac{1+\int_0^1 \omega_0(\theta\|x_n - x^*\|)d\theta}{1-\omega_0(\|x_n - x^*\|)}\right]\|x_n - x^*\|$$
$$\le g_2(\|x_n - x^*\|)\|x_n - x^*\|$$
$$\le \|x_n - x^*\|.$$

3. Semi-local Analysis

The majorant sequence $\{\alpha_n\}$ is defined for $\alpha_0 = 0, \beta_0 \ge \frac{2}{3}\|F'(x_0)^{-1}F(x_0)\|$ and for each $n = 0, 1, 2, \cdots$ by

$$\alpha_{n+1} = \beta_n + \frac{3}{2}\left[\frac{1}{3} + \left|\frac{a}{3}+\frac{3}{4}\right| \frac{\overline{v_n}}{1-v_0(\beta_n)} + \left|\frac{a}{3}\right| \frac{\overline{v_n}}{1-v_0(\alpha_n)} + \left|\frac{3}{8}-\frac{a}{3}\right| \left(\frac{\overline{v_n}}{1-v_0(\beta_n)}\right)^2\right](\beta_n - \alpha_n)$$

where

$$\overline{\omega}(t) = \begin{cases} v_0(\beta_n - \alpha_n) \\ \text{or} \\ v_0(\alpha_n) + v_0(\beta_n) \end{cases},$$

$$\delta_{n+1} = \left(1 + \int_0^1 v_0(\alpha_n + \theta(\alpha_{n+1} - \alpha_n))d\theta\right)(\alpha_{n+1} - \alpha_n) + \frac{3}{2}(1+v_0(\alpha_n))(\beta_n - \alpha_n)$$

and $\beta_{n+1} = \alpha_{n+1} + \frac{2}{3}\frac{\delta_{n+1}}{1-v_0(\alpha_{n+1})}$.

The calculations are :

$$\begin{aligned}
x_{n+1}-y_n &= \Big[\frac{2}{3}I-\Big(\frac{5}{8}-a\Big)I-aF'(y_n)^{-1}F'(x_n)\\
&\quad -\frac{a}{3}F'(x_n)^{-1}F'(y_n)-\Big(\frac{3}{8}-\frac{a}{3}\Big)F'(y_n)^{-1}F^{-1}(x_n)\Big]\Big(-\frac{3}{2}(y_n-x_n)\Big)\\
&= \frac{3}{2}\Big[\frac{1}{3}+\Big(\frac{a}{3}+\frac{3}{4}\Big)(F'(y_n)^{-1}F'(x_n)-I)\\
&\quad +\frac{a}{3}(F'(x_n)^{-1}F^{-1}(y_n)-I)+\Big(\frac{3}{8}-\frac{a}{3}\Big)(F'(y_n)^{-1}F'(x_n)-I)^2\Big](y_n-x_n),\\
\|x_{n+1}-y_n\| &\le \frac{3}{2}\left[\frac{1}{3}+\left|\frac{a}{3}+\frac{3}{4}\right|\frac{\overline{v_n}}{1-v_0(\beta_n)}+\left|\frac{a}{3}\right|\frac{\overline{v_n}}{1-v_0(\alpha_n)}+\left|\frac{3}{8}-\frac{a}{3}\right|\Big(\frac{\overline{v_n}}{1-v_0(\beta_n)}\Big)^2\right](\beta_n-\alpha_n)\\
&= \alpha_{n+1}-\beta_n,\\
F(x_{n+1}) &= F(x_{n+1})-F(x_n)-\frac{3}{2}F'(x_n)(y_n-x_n),\\
\|L^{-1}F(x_{n+1})\| &\le \Big(1+\int_0^1 v_0(\alpha_n+\theta(\alpha_{n+1}-\alpha_n))d\theta\Big)(\alpha_{n+1}-\alpha_n)\\
&\quad +\frac{3}{2}\Big(1+v_0(\alpha_n)\Big)(\beta_n-\alpha_n)=\delta_{n+1},\\
\|y_{n+1}-x_{n+1}\| &\le \frac{2}{3}\|F'(x_{n+1})^{-1}L\|\|L^{-1}F(x_{n+1})\|\le\frac{2}{3}\frac{\delta_{n+1}}{1-v_0(\alpha_{n+1})}=\beta_{n+1}-\alpha_{n+1}.
\end{aligned}$$

The rest of the analysis follows as in the previous chapters.

4. References

[1] Artidiello, S., Cordero, A., Torregrosa, J. R., Vassileva, M. P., *Optimal High-Order Methods for Solving Nonlinear Equations*, J. Appl. Math., 2014, (2014), 1-9, Article ID 591638.

[2] Artidiello, S., Cordero, A., Torregrosa, J. R., Vassileva, M. P., *Two weighted-order classes of iterative root-finding methods*, Int. J. Comput. Math., 92-9, (2015), 1790-1805.

[3] Argyros, I. K., *Convergence and Applications of Newton-type Iterations*, Springer Verlag, Berlin, Germany, (2008).

[4] Argyros, I. K., *On the convergence of Newtons method for polynomial equations and applications in radiative transfer*, Monatsh. Math., 127 (4), (1999), 265–276.

[5] Argyros, I. K., *On the NewtonKantorovich hypothesis for solving equations*, J. Comput. Appl. Math., 169 (2), (2004), 315–332.

[6] Argyros, I. K., *The theory and applications of iteration methods*, 2nd Edition, CRC Press, Engineering Series, Taylor and Francis Group, (2022).

[7] Argyros, I. K., *Unified Convergence Criteria for Iterative Banach Space Valued Methods with Applications*, Mathematics 2021, 9(16), 1942; https://doi.org/10.3390/math9161942.

[8] Behl, R., Cordero, A., Motsa, S. S., Torregrosa, J. R., *A new efficient and optimal sixteenth-order scheme for simple roots of nonlinear equations*, Bull. Math. de la Socit des Sciences Mathmatiques de Roumanie, Nouvelle Srie, 60-108(2), (2017), 127–140.

[9] Chand, P., Chicharro, F. I., Garrido, N., Jain, P., *Design and Complex Dynamics of Potra-Ptk-Type Optimal Methods for Solving Nonlinear Equations and Its Applications*, Mathematics, (2019), 7–10.

[10] Chicharro, F. I., Cordero, A., Garrido, N., Torregrosa, J. R., *On the choice of the best members of the Kim family and the improvement of its convergence*, Mathematical Methods in the Applied Sciences, 43-14, (2020), 8051–8066.

[11] Chicharro, F. I., Cordero, A., Torregrosa, J. R., *Drawing Dynamical Parameters Planes of Iterative Families and Methods*, Sci. World J., 2013, Article ID 780153, 1–11.

[12] Chun, C., *Construction of Newton-like iterative methods for solving nonlinear equations*, Numer. Math., 104, (2006), 297–315.

[13] Chun, C., Neta, B., Kozdon, J., Scott, M., *Choosing weight functions in iterative methods for simple roots*, Appl. Math. Comput., 227, (2014), 788–800.

[14] Cordero, A., Ledesma, A., Maimo, J. G., Torregrosa, J. R., *Design and Dynamical behavior of a fourth-order family of iterative methods for solving nonlinear equations*, Mathematics, 92-2, (2015), 1790–1805.

[15] Cordero, A., Torregrosa, J. R., *On the design of optimal iterative methods for solving nonlinear equations*, Advances in Iterative Methods for Nonlinear Equations, (2016), 79–112.

[16] Petkovi, M., Neta, B., Petkovi, L., Dzuni, J., *Multipoint methods for solving nonlinear equations: A survey*, Appl. Math. Comput., 226, (2014), 635–660.

[17] Qureshi, S., Soomro, A., Shaikh, A. A., Hincal, E., Gokbulut, N., *A novel Multistep Iterative Technique for Models in Medical Sciences with Complex Dynamics*, Comput. Math. Methods Med., (2022), 1–10, ID7656451.

Chapter 40

Extended Steffensen Hybrid Method

1. Introduction

The Steffensen Hybrid Method is defined for $a \in \mathbb{R}, x_0 \in \Omega$ and each $n = 0, 1, 2, \cdots$ by $u_n = x_n + F(x_n), v_n = x_n - F(x_n)$ and

$$x_{n+1} = x_n - 2a[u_n, x_n; F]^{-1}F(x_n) - 2(1-a)[x_n, v_n; F]^{-1}F(x_n). \tag{40.1}$$

The Taylor series expansion technique is given in [6] to show the local quadratic convergence of this method provided that the $F^{(3)}$ exists, $a \in (0,1)$ and $B_1 = B_2 = \mathbb{R}$. It is also shown that for some values of $a \in (0,1)$ method (40.1) has better convergence properties than Steffensen's method [1]-[5]. Other limitations of this approach are given in earlier chapters. The new local convergence analysis is based only on operators on the method and the concept of generalized continuity in the Banach space setting. Hence, the applicability of the method is extended. The new technique can be used on other methods [1]-[5], [7].

2. Local Analysis

The majorant function g is :

$$g(t) = \frac{\phi(z, f(t))}{1 - \phi_0(t, f(t))} + \frac{|1-2a|(1+\phi_1(t))}{1-\phi_0(t, f(t))} + \frac{2|1-a|(1+\phi_1(t))}{1-\phi_0(t, g(t))},$$

where the conditions are:

$$||L^{-1}([z,x;F]-L)|| \leq \phi_0(||z-x^*||, ||x-x^*||)$$
$$||u-x^*|| \leq f(||x-x^*||), ||v-x^*|| \leq g(||x-x^*||),$$
$$||L^{-1}\Big([u,x;F]-[x,x^*;F]\Big)|| \leq \phi(||u-x^*||, ||x-x^*||),$$
$$||L^{-1}\Big([x,x^*;F]-L\Big)|| \leq \phi_1(||x-x^*||),$$

and $x, z, u, \phi_0, \phi, \phi_1$ and L are as in previous chapters. The calculations leading to the development of the function g are:

$$\begin{aligned}
x_{n+1} - x^* &= x_n - x^* - [u_n, x_n; F]^{-1} F(x_n) \\
&+ (1-2a)[u_n, x_n; F]^{-1} F(x_n) - 2(1-a)[x_n, v_n; F]^{-1} F(x_n) \\
&= [u_n, x_n; F]^{-1} \Big([u_n, x_n; F] - [x_n, x^*; F] \Big) (x_n - x^*) \\
&+ (1-2a)[u_n, x_n; F]^{-1} - 2(1-a)[x_n, v_n; F]^{-1} F(x_n), \\
||x_{n+1} - x^*|| &\le \Big[\frac{\phi(||x_n - x^*||, ||u_n - x^*||)}{1 - \phi_0(||x_n - x^*||, ||u_n - x^*||)} \\
&+ \frac{|1-2a|(1+\phi_1(||x_n - x^*||))}{1 - \phi_0(||x_n - x^*||, ||u_n - x^*||)} \\
&+ \frac{2|1-a|(1+\phi_1(||x_n - x^*||))}{1 - \phi_0(||x_n - x^*||, ||v_n - x^*||)} \Big] ||x_n - x^*|| \\
&\le g(||x_n - x^*)||x_n - x^*|| \le ||x_n - x^*|| < r,
\end{aligned}$$

where we also need the condition $U[x^*, R] \subset \Omega$, where $R = \max\{r, f(r), g(r)\}$. The rest of the analysis is as given in earlier chapters.

3. References

[1] Argyros, I. K., *Convergence and Applications of Newton-type Iterations*, Springer Verlag, Berlin, Germany, (2008).

[2] Argyros, I. K., *On the convergence of Newtons method for polynomial equations and applications in radiative transfer*, Monatsh. Math., 127 (4), (1999), 265–276.

[3] Argyros, I. K., *On the NewtonKantorovich hypothesis for solving equations*, J. Comput. Appl. Math., 169 (2), (2004), 315–332.

[4] Argyros, I. K., *The theory and applications of iteration methods*, 2nd Edition, CRC Press, Engineering Series, Taylor and Francis Group, (2022).

[5] Argyros, I. K., *Unified Convergence Criteria for Iterative Banach Space Valued Methods with Applications*, Mathematics 2021, 9(16), 1942; https://doi.org/10.3390/math9161942.

[6] Eskandari, H., *Comparing Stephensen Hybrid Method with Other Techniques for Solving Nonlinear Equations*, International Journal of Mathematical and Computational Methods, 8, (2023), 25–28.

[7] Qureshi, U. K., Shaikhi, A. A., Shaikh, F. K., Hazarewal, S. K., Laghari, T. A., *New Simpson type method for solving nonlinear equations*, Open J. Math. Sci., 5, (2021), 94–100.

Chapter 41

Local Convergence of Newton's Method for k-Differentiable Operators

1. Introduction

Newton's method is defined for each $n = 0,1,2,..$ by $x_0 \in \Omega$,

$$x_{n+1} = x_n - F'(x_n)^{-1}F(x_n).$$

There is a plethora of local and semi-local convergence results for Newton's method [7]. In particular, the semi-local convergence is given in [7], provided that the operator is k-times differentiable for $k = 2,3,\ldots$. These results extend the celebrated Newton-Kantorovich theorem, which is given in [8] for $k = 2$ and provided that the majorant function is a quadratic polynomial. In a previous chapter, we extended the semi-local convergence in this setting. In the present chapter we provide the local convergence of Newton's method not considered before. Our technique can be applied on other methods where F is k-times differentiable along the same lines [1]-[6],[8, 9].

2. Convergence

The conditions are:
There exists a simple solution $x^* \in \Omega$ of the equation $F(x) = 0$ and a function $f : [0,+\infty) \longrightarrow \mathbb{R}$ which is k-times continuously differentiable such that for each $x \in \Omega$.

$$\|F'(x^*)^{-1}F^{(i)}(x^*)\| \leq f^{(i)}(0), \ \ i = 1,2,...,k-1$$

and

$$\|F'(x^*)^{-1}F^{(k)}(x)\| \leq f^{(k)}(\|x - x^*\|)$$

for $x \in U(x^*, r) \subset \Omega$, where $r > 0$ is determined later.
Define the functions

$$\begin{aligned} q(t) &= \sum_{i=2}^{k-1} \frac{1}{(i-1)!} f^{(i)}(0) t^{i-1} \\ &+ \frac{1}{(k-2)!} \int_0^1 f^{(k)}(\theta t)(1-\theta)^{k-2} t^{k-1} d\theta, \\ p(t) &= \left(1 - \frac{1}{2}\right) f''(0) t + \left(\frac{1}{2!} - \frac{1}{3!}\right) f'''(0) t^2 \\ &+ \ldots + \left(\frac{1}{(k-2)!} - \frac{1}{(k-1)!}\right) f^{(k-1)}(0) t^{k-2} \\ &+ \frac{1}{(k-2)!} \int_0^1 f(\theta t)\theta(1-\theta)^{k-2} t^{k-1} d\theta \end{aligned}$$

and

$$g(t) = \frac{p(t)}{1 - q(t)}.$$

Moreover, suppose $q(t) - 1 = 0$ has a smallest positive solution denoted by ρ, and $g(t) - 1 = 0$ has a smallest solution $r \in (0, \rho)$. Finally, suppose $x_0 \in U(x^*, \rho) - \{x^*\}$ and $U[x^*, r] \subset \Omega$.

$$\|F'(x_n)^{-1} F'(x^*)\| \leq \frac{1}{1 - q_n}$$

by the Banach lemma on linear invertible operators [1]-[6], [8].
Let $A_n = F'(x_n)(x_n - x^*) - F(x_n)$. We need an upper bound on $\|F'(x^*)^{-1} A_n\|$. We obtain in turn

$$\begin{aligned} A_n &= \sum_{i=1}^{k-1} \frac{1}{(i-1)!} F^{(i)}(x^*)(x_n - x^*)^i - \sum_{j=1}^{k-1} \frac{1}{j!} F^{(j)}(x^*)(x_n - x^*)^j \\ &+ \frac{1}{(k-2)!} \int_0^1 F^{(k)}(x^* + \theta(x_n - x^*))(1-\theta)^{k-2}(x_n - x^*)^k d\theta \\ &+ \frac{1}{(k-2)!} \int_0^1 F^{(k)}(x^* + \theta(x_n - x^*))(1-\theta)^{k-1}(x_n - x^*)^k d\theta \\ &= \left(1 - \frac{1}{2}\right) F''(x^*)(x_n - x^*)^2 \\ &+ \left(\frac{1}{2!} - \frac{1}{3!}\right) F''(x^*)(x_n - x^*)^3 \\ &+ \ldots + \left(\frac{1}{(k-2)!} - \frac{1}{(k-1)!}\right) F^{(k-1)}(x^*)(x_n - x^*)^{k-1} \\ &+ \frac{1}{(k-2)!} \int_0^1 F^{(k)}(x^* + \theta(x_n - x^*))\theta(1-\theta)^{k-2} d\theta (x_n - x^*)^k. \end{aligned}$$

By taking norms and using the conditions we have in turn.

$$\begin{aligned}\|F'(x^*)^{-1}A_n\| &\leq \left(1-\frac{1}{2}\right)f''(0)\|x_n-x^*\|^2 \\ &+\left(\frac{1}{2!}-\frac{1}{3!}\right)f'''(0)\|x_n-x^*\|^3 \\ &+\ldots+\left(\frac{1}{(k-2)!}-\frac{1}{(k-1)!}\right)f^{(k-1)}(0)\|x_n-x^*\|^{k-1} \\ &+\frac{1}{(k-2)!}\int_0^1 f^{(k)}(\theta\|x_n-x^*\|)\theta(1-\theta)^{k-2}d\theta\|x_n-x^*\|^k \\ &= p_n\|x_n-x^*\|.\end{aligned}$$

Then, from the identity

$$x_{n+1}-x^* = x_n-x^*-F'(x_n)^{-1}F(x_n)=F'(x_n)^{-1}A_n,$$

we get

$$\begin{aligned}\|x_{n+1}-x^*\| &\leq \|F'(x_n)^{-1}F'(x^*)\|\|F'(x^*)^{-1}A_n\| \\ &\leq \frac{p_n\|x_n-x^*\|}{1-q_n}\leq g(\|x_n-x^*\|)\|x_n-x^*\| \\ &\leq \|x_n-x^*\|<r.\end{aligned}$$

3. References

[1] Argyros, I. K., *Convergence and Applications of Newton-type Iterations*, Springer Verlag, Berlin, Germany, (2008).

[2] Argyros, I. K., *On the convergence of Newtons method for polynomial equations and applications in radiative transfer*, Monatsh. Math., 127 (4), (1999), 265–276.

[3] Argyros, I. K., *On the NewtonKantorovich hypothesis for solving equations*, J. Comput. Appl. Math., 169 (2), (2004), 315–332.

[4] Argyros, I. K., *The theory and applications of iteration methods*, 2nd Edition, CRC Press, Engineering Series, Taylor and Francis Group, (2022).

[5] Argyros, I. K., *Unified Convergence Criteria for Iterative Banach Space Valued Methods with Applications*, Mathematics 2021, 9(16), 1942; https://doi.org/10.3390/math9161942.

[6] Ezquerro, J. A., Gonzalez, D., Hernndez-Vern, M. A., *A new semilocal convergence result for Newtons method, under generalized conditions of Kantorovich*, J. Complexity, 30, (2014), 309–324.

[7] Kantorovich, L. V., Akilov, G. P., *Functional Analysis*, Pergamon Press, Oxford, (1982).

[8] Yamamoto, T., *Convergence theorem for Newton-like methods in Banach spaces*, Numer. Math., 51, (1987), 545–557.

Chapter 42

Jarratt-Type Method of Order Six with Two Parameters

1. Introduction

The three step Jarratt-type method is defined for each $x_0 \in \Omega$ and $n = 0, 1, 2, \ldots$ by

$$\begin{aligned} y_n &= x_n - \frac{2}{3}F'(x_n)^{-1}F(x_n), \\ z_n &= x_n - \frac{1}{2}(3F'(y_n) - F'(x_n))^{-1}(3F'(y_n) + F'(x_n))F'(x_n)^{-1}F(x_n) \end{aligned} \tag{42.1}$$

and

$$x_{n+1} = z_n - \frac{1}{2}(aF'(x_n) + bF'(y_n))^{-1}[(5a+3b)F'(x_n) - (a+b)F'(y_n)]F'(x_n)^{-1}F(z_n),$$

where $a, b \in \mathbb{R}$ with $a + b \neq 0$.
The local convergence order six is shown in [10] under the Taylor series approach when $B_1 = B_2 = \mathbb{R}$ and provided that $F^{(7)}$ exists and is bounded. The limitations of this approach are given in previous chapters. In this chapter the local convergence is based only on the operators on the method. Moreover, the semi-local convergence of the method (not given in [10]) is developed using majorizing sequence and in the more general setting of a Banach space. The same technique can be applied to extend the applicability of other methods [1]-[9].

2. Local Convergence

The majorant functions $g_i,\ i = 1, 2, 3$ are:

$$\begin{aligned} g_1(t) &= \frac{\int_0^1 w((1-\theta)t)d\theta + \frac{1}{3}(1 + \int_0^1 w_0(\theta t)d\theta)}{1 - w_0(t)} \\ p(t) &= \frac{1}{2}(3w_0(g_1(t)t) + w_0(t)), \end{aligned}$$

$$\bar{w} = \begin{cases} w((1+g_1(t)t) \\ or \\ w_0(t)+w_0(g_1(t)t), \end{cases}$$

$$g_2(t) = \frac{\int_0^1 w((1-\theta))t)d\theta}{1-w_0(t)} + \frac{3}{4}\frac{\bar{w}(t)(1+\int_0^1 w_0(\theta t)d\theta)}{(1-p(t))(1-w_0(t))},$$

$$q(t) = \frac{1}{|a+b|}(|a|w_0(t)+bw_0(g_1(t)t))$$

$$\bar{\bar{w}} = \begin{cases} w((1+g_2(t)t) \\ or \\ w_0(t)+w_0(g_2(t)t), \end{cases}$$

and

$$g_3(t) = \left[\frac{\int_0^1 w((1-\theta)g_2(t)t)d\theta}{1-w_0(g_2(t)t)} + \frac{\bar{\bar{w}}(t)(1+\int_0^1 w_0(\theta g_2(t)t)d\theta)}{(1-w_0(t))(1-w_0(g_2(t)t))} + \frac{3}{2}\frac{\bar{w}(t)(1+\int_0^1 w_0(\theta g_2(t)t)d\theta)}{(1-q(t))(1-w_0(t))}\right] \times g_2(t).$$

The calculations leading to the development of the majorant functions are in turn:

$$\begin{aligned}
y_n - x^* &= x_n - x^* - F'(x_n)^{-1}F(x_n) + \frac{1}{3}F'(x_n)^{-1}F(x_n), \\
\|y_n - x^*\| &\leq \frac{[\int_0^1 w((1-\theta)\|x_n-x^*\|)d\theta + \frac{1}{3}(1+\int_0^1 w_0(\theta\|x_n-x^*\|)d\theta]\|x_n-x^*\|}{1-w_0(\|x_n-x^*\|)} \\
&\leq g_1(\|x_n-x^*\|)\|x_n-x^*\| \leq \|x_n-x^*\| < r, \\
&\quad \|(2L)^{-1}(3(F'(y_n)-L)+(L-F'(x_n)))\| \\
&\leq \frac{1}{2}(3w_0(\|y_n-x^*\|)+w_0(\|x_n-x^*\|)) = p_n < 1, \\
\|(3F'(y_n)-F'(x_n))^{-1}L\| &\leq \frac{1}{2(1-p_n)}, \\
z_n - x^* &= x_n - x^* - F'(x_n)^{-1}F(x_n) \\
&\quad + \frac{3}{2}(3F'(y_n)-F'(x_n))^{-1}(F'(y_n)-F'(x_n))F'(x_n)^{-1}F(x_n), \\
\|z_n - x^*\| &\leq \left[\frac{\int_0^1 w((1-\theta)\|x_n-x^*\|)d\theta}{1-w_0(\|x_n-x^*\|)} + \frac{3\bar{w}_n(1+\int_0^1 w_0(\theta\|x_n-x^*\|)d\theta)}{4(1-p_n)(1-w_0(\|x_n-x^*\|))}\right]\|x_n-x^*\| \\
&\leq g_2(\|x_n-x^*\|)\|x_n-x^*\| \leq \|x_n-x^*\|, \\
&\quad \|((a+b)L)^{-1}(aF'(x_n)+bF'(y_n)-(a+b)L)\| \\
&\leq \frac{1}{|a+b|}(|a|w_0(\|x_n-x^*\|+|b|w_0(\|y_n-x^*\|)) = q_n, \\
\|(aF'(x_n)+bF'(y_n))^{-1}L\| &\leq \frac{1}{|a+b|(1-q_n)}, \\
x_{n+1} - x^* &= z_n - x^* - F'(z_n)^{-1}F(z_n) + (F'(z_n)^{-1} - F'(x_n)^{-1})F(z_n) \\
&\quad + [I - \frac{1}{2}(aF'(x_n)+bF'(y_n))^{-1}((5a+3b)F'(x_n)-(3a+b)F'(y_n)] \\
&\quad \times F'(x_n)^{-1}F(z_n) \\
&= z_n - x^* - F'(z_n)^{-1}F(z_n) + (F'(z_n)^{-1} - F'(x_n)^{-1})F(z_n) \\
&\quad + \frac{3}{2}(a+b)(aF'(x_n)+bF'(y_n))^{-1}(F'(y_n)-F'(x_n))F'(x_n)^{-1}F(z_n)
\end{aligned}$$

and

$$\begin{aligned}
\|x_{n+1}-x^*\| &\leq \left[\frac{\int_0^1 w((1-\theta)\|z_n-x^*\|)d\theta}{1-w_0(\|z_n-x^*\|)}\right. \\
&+\frac{\bar{\bar{w}}(1+\int_0^1 w_0(\theta\|z_n-x^*\|)d\theta)}{(1-w_0(\|x_n-x^*\|))(1-w_0(\|z_n-x^*\|))} \\
&\left.+\frac{3}{2}\frac{\bar{w}_n(1+\int_0^1 w_0(\theta\|z_n-x^*\|)d\theta)}{(1-q_n)(1-w_0(\|x_n-x^*\|)}\right]\|z_n-x^*\| \\
&\leq g_3(\|x_n-x^*\|)\|x_n-x^*\| \leq \|x_n-x^*\|.
\end{aligned}$$

3. Semi-local Convergence

The majorant sequence $\{\alpha_n\}$ is defined for $\alpha_0=0, \beta_0 \geq \frac{2}{3}\|F'(x_0)^{-1}F(x_0)\|$ and each $n = 0,1,2,...$

$$\begin{aligned}
\bar{p}_n &= \begin{cases} v(\beta_n-\alpha_n) \\ or \\ v_0(\alpha_n)+v_0(\beta_n), \end{cases} \\
\gamma_n &= \beta_n+\frac{(3\bar{v}_n+4(1+v_0(\alpha_n))(\beta_n-\alpha_n)}{8(1-\bar{p}_n)},
\end{aligned}$$

$$\begin{aligned}
\lambda_n &= (1+\int_0^1 v_0(\alpha_n+\theta(\gamma_n-\alpha_n))d\theta)(\gamma_n-\alpha_n) \\
&+\frac{3}{2}(1+v_0(\alpha_n))(\beta_n-\alpha_n), \\
\bar{q}_n &= \begin{cases} v(\gamma_n-\alpha_n) \\ or \\ v_0(\alpha_n)+v_0(\gamma_n), \end{cases} \\
\alpha_{n+1} &= \gamma_n+\frac{(|3a+b|\bar{v}_n+2|a+b|(1+v_0(\alpha_n))\lambda_n}{2|a+b|(1-\bar{q}_n)(1-v_0(\alpha_n))} \\
\delta_{n+1} &= (1+\int_0^1 v_0(\alpha_n+\theta(\alpha_{n+1}-\alpha_n))d\theta)(\alpha_{n+1}-\alpha_n) \\
&+\frac{3}{2}(1+v_0(\alpha_n))(\beta_n-\alpha_n)
\end{aligned}$$

and

$$\beta_{n+1} = \alpha_{n+1}+\frac{2}{3}\frac{\delta_{n+1}}{1-v_0(\alpha_{n+1})}.$$

The calculations are:

$$\begin{aligned}
z_n - y_n &= \left[\frac{2}{3}I - \frac{1}{2}(3F'(y_n) - F'(x_n))^{-1}(3F'(y_n) + F'(x_n))\right]\left(-\frac{3}{2}(y_n - x_n)\right)\\
&= -\frac{1}{4}(3F'(y_n) - F'(x_n))^{-1}[3(F'(y_n) - F'(x_n)) - 4F'(x_n)](y_n - x_n),\\
\|z_n - y_n\| &\leq \frac{(3\bar{v}_n + 4(1 + v_0(\alpha_n)))(\beta_n - \alpha_n)}{8(1 - \bar{p}_n)} = \gamma_n - \beta_n,\\
x_{n+1} - z_n &= -\frac{1}{2}(aF'(x_n) + bF'(y_n))^{-1}[(3a + b)(F'(x_n) - F'(y_n))\\
&\quad + 2(a + b)F'(x_n)]F'(x_n)^{-1}F(z_n),\\
F(z_n) &= F(z_n) - F(x_n) + F(x_n),\\
\|L^{-1}F(z_n)\| &\leq (1 + \int_0^1 v_0(\alpha_n + \theta(\gamma_n - \alpha_n))d\theta)(\gamma_n - \alpha_n)\\
&\quad + \frac{3}{2}(1 + v_0(\alpha_n))(\beta_n - \alpha_n) = \lambda_n,\\
\|x_{n+1} - z_n\| &\leq \frac{(|3a + b|\bar{v}_n + 2|a + b|(1 + v_0(\alpha_n)))\lambda_n}{2|a + b|(1 - \bar{q}_n)(1 - v_0(\alpha_n))}\\
&= \alpha_{n+1} - \gamma_n,
\end{aligned}$$

$$\begin{aligned}
F(x_{n+1}) &= F(x_{n+1}) - F(x_n) - \frac{3}{2}F'(x_n)(y_n - x_n),\\
\|L^{-1}(x_{n+1})\| &\leq (1 + \int_0^1 v_0(\theta(\alpha_{n+1} - \alpha_n))d\theta)(\alpha_{n+1} - \alpha_n)\\
&\quad + \frac{3}{2}(1 + v_0(\alpha_n))(\beta_n - \alpha_n) = \delta_{n+1}
\end{aligned}$$

and

$$\begin{aligned}
\|y_{n+1} - x_{n+1}\| &\leq \frac{2}{3}\|F'(x_{n+1})^{-1}L\|\|L^{-1}F(x_{n+1})\|\\
&\leq \frac{2}{3}\frac{\delta_{n+1}}{1 - v_0(\alpha_{n+1})} = \beta_{n+1} - \alpha_{n+1}.
\end{aligned}$$

The rest of the analysis follows as in the previous Chapters.

4. References

[1] Argyros, I. K., *Convergence and Applications of Newton-type Iterations*, Springer Verlag, Berlin, Germany, (2008).

[2] Argyros, I. K., *On the convergence of Newtons method for polynomial equations and applications in radiative transfer*, Monatsh. Math., 127 (4), (1999), 265–276.

[3] Argyros, I. K., *On the NewtonKantorovich hypothesis for solving equations*, J. Comput. Appl. Math., 169 (2), (2004), 315–332.

[4] Argyros, I. K., *The theory and applications of iteration methods*, 2nd Edition, CRC Press, Engineering Series, Taylor and Francis Group, (2022).

[5] Argyros, I. K., *Unified Convergence Criteria for Iterative Banach Space Valued Methods with Applications*, Mathematics 2021, 9(16), 1942; https://doi.org/10.3390/math9161942.

[6] Cordero, A., Hueso, J. L., Martinez, E., Torregrosa, J. R., *Efficient three-step iterative methods with sixth order convergence for nonlinear equations*, Numer. Algor., 53, (2010), 485–495.

[7] Gupta, D. K., Parhi, S. K., *An improved class of regula falsi methods of third order for solving nonlinear equations in R*, J. Appl. Math. Comput., 33, (2010), 35–45.

[8] Kung, H. T., Traub, J. F., *Optimal order of one-point and multipoint iteration*, J. Assoc. Comput. Math., 21, (1974), 634–651.

[9] Sharma, J. R., Guha, R. K., *A family of modified Ostrowski methods with accelerated sixth-order convergence*, Appl. Math. Comput., 190, (2007), 111–115.

[10] Wang, X., Kou, J., Li, Y., *A variant of Jarratt method with sixth-order convergence*, Appl. Math. Comput., 204, (2008), 14–19.

Chapter 43

An Efficient Fourth-Order Method

1. Introduction

The method is defined for $x_0 \in \Omega$ and each $n = 0, 1, 2, ...$ by

$$y_n = x_n - \frac{2}{3} F'(x_n)^{-1} F(x_n)$$

and

$$x_{n+1} = x_n - \frac{1}{2}(-I + \frac{9}{4} F'(y_n)^{-1} F'(x_n) + \frac{3}{4} F'(x_n)^{-1} F'(y_n)) F'(x_n)^{-1} F(x_n) \tag{43.1}$$

The local convergence order four is determined in [13] under the Taylor series approach when $B_1 = B_2 = \mathbb{R}^k$ and provided that $F^{(5)}$ exists and is bounded limiting the applicability of the method. That is why the local convergence in this chapter relies only on the operators on the method. Moreover, the more interesting semi-local convergence analysis (not given in [13]) is also studied using majorizing sequences. Hence, the applicability of the method is extended. The same methodology can be applied on other methods [1]-[12], [14,15].

2. Local Convergence

The majorant functions $g_i, i = 1, 2$ are:

$$g_1(t) = \frac{\int_0^1 w((1-\theta)t)d\theta + \frac{1}{3}(1 + \int_0^1 w_0(\theta t)d\theta)}{1 - w_0(t)}$$

and

$$\begin{aligned} g_2(t) &= \frac{\int_0^1 w((1-\theta)t)d\theta}{1 - w_0(t)} \\ &+ \frac{3}{8}\left(\frac{3\bar{w}(t)}{1 - w_0(g_1(t)t)} + \frac{\bar{w}(t)}{1 - w_0(t)}\right)(1 + \int_0^1 w_0(\theta t)d\theta). \end{aligned}$$

The motivation for the development of these functions follows from the calculations:

$$
\begin{aligned}
y_n - x^* &= x_n - x^* - F'(x_n)^{-1}F(x_n) + \frac{1}{3}F'(x_n)^{-1}F(x_n),\\
\|y_n - x^*\| &\leq \frac{[\int_0^1 w((1-\theta)\|x_n - x^*\|)d\theta + \frac{1}{3}(1 + \int_0^1 w_0(\theta\|x_n - x^*\|)d\theta)]}{1 - w_0(\|x_n - x^*\|)}\\
&\quad \times \|x_n - x^*\|\\
&\leq g_1(\|x_n - x^*\|)\|x_n - x^*\| \leq \|x_n - x^*\| < r,\\
x_{n+1} - x^* &= x_n - x^* - F'(x_n)^{-1}F(x_n) - \frac{1}{2}(-I - 2I + \frac{9}{4}F'(y_n)^{-1}F'(x_n)\\
&\quad + \frac{3}{4}F'(x_n)^{-1}F'(y_n))F'(x_n)^{-1}F(x_n)\\
&= x_n - x^* - F'(x_n)^{-1}F(x_n) - \frac{3}{8}(3(F'(y_n)^{-1}F'(x_n) - I)\\
&\quad + (F'(x_n)^{-1}F'(y_n) - I))F'(x_n)^{-1}F(x_n),
\end{aligned}
$$

and

$$
\begin{aligned}
\|x_{n+1} - x^*\| &\leq \left[\frac{\int_0^1 w((1-\theta)\|x_n - x^*\|)d\theta}{1 - w_0(\|x_n - x^*\|)}\right.\\
&\quad + \frac{3}{8}\left(\frac{3\bar{w}_n}{1 - w_0(\|y_n - x^*\|)} + \frac{\bar{w}_n}{1 - w_0(\|x_n - x^*\|)}\right)(1 + \int_0^1 w_0(\theta\|x_n - x^*\|)d\theta)]\\
&\quad \times \|x_n - x^*\|\\
&\leq g_2(\|x_n - x^*\|)\|x_n - x^*\| \leq \|x_n - x^*\|,
\end{aligned}
$$

where

$$
\bar{w}(t) = \begin{cases} w((1 + g_1(t)t) \\ or \\ w_0(t) + w_0(g_1(t)t), \end{cases}
$$

and

$$
\bar{w}_n = \begin{cases} w(\|x_n - x^*\| + \|y_n - x^*\|) \\ or \\ w_0(\|x_n - x^*\|) + w_0(\|y_n - x^*\|), \end{cases}
$$

3. Semi-local Convergence

The majorant sequence $\{\alpha_n\}$ is defined for $\alpha_0 = 0, \beta_0 \geq \frac{2}{3}\|F'(x_0)^{-1}F(x_0)\|$ and each $n = 0, 1, 2, \ldots$ by

$$
\begin{aligned}
\bar{v}_n &= \begin{cases} v(\beta_n - \alpha_n) \\ or \\ v_0(\alpha_n) + v_0(\beta_n), \end{cases}\\
\alpha_{n+1} &= \beta_n \frac{3}{2}\left(\frac{1}{3} + \frac{9}{8(1 - v_0(\beta_n))}\bar{v}_n + \frac{3}{8(1 - v_0(\alpha_n))}\right)(\beta_n - \alpha_n),\\
\delta_{n+1} &= (1 + \int_0^1 v_0(\alpha_n + \theta(\alpha_{n+1} - \alpha_n))d\theta)(\alpha_{n+1} - \alpha_n)\\
&\quad + \frac{3}{2}(1 + v_0(\alpha_n))(\beta_n - \alpha_n)
\end{aligned}
$$

and

$$\beta_{n+1} = \alpha_{n+1} + \frac{2}{3}\frac{\delta_{n+1}}{1-v_0(\alpha_{n+1})}.$$

The calculations are:

$$\begin{aligned}
x_{n+1}-y_n &= \left[\frac{2}{3}I+\frac{1}{2}I-\frac{9}{8}F'(y_n)^{-1}F'(x_n)-\frac{3}{8}F'(x_n)^{-1}F'(y_n)\right] \\
&\quad \times\left(-\frac{3}{2}(y_n-x_n)\right), \\
\|x_{n+1}-y_n\| &= \left\|\frac{3}{2}\left(\frac{1}{3}I+\frac{9}{8}(F'(y_n)^{-1}F'(x_n)-I\right)\right. \\
&\quad \left.+\frac{3}{8}(F'(x_n)^{-1}F'(y_n)-I))(y_n-x_n)\right\| \\
&\leq \frac{3}{2}\left(\frac{1}{3}+\frac{9}{8}\frac{\bar{v}_n}{1-v_0(\beta_n)}+\frac{3}{8}\frac{\bar{v}_n}{1-v_0(\alpha_n)}\right)(\beta_n-\alpha_n)=\alpha_{n+1}-\beta_n, \\
F(x_{n+1}) &= F(x_{n+1})-F(x)-\frac{3}{2}F'(x_n)(y_n-x_n), \\
\|L^{-1}F(x_{n+1})\| &\leq (1+\int_0^1 v_0(\alpha_n+\theta(\alpha_{n+1}-\alpha_n))d\theta)(\alpha_{n+1}-\alpha_n) \\
&\quad +\frac{3}{2}(1+v_0(\alpha_n))(\beta_n-\alpha_n)=\delta_{n+1}
\end{aligned}$$

and

$$\begin{aligned}
\|y_{n+1}-x_{n+1}\| &\leq \frac{2}{3}\|F'(x_{n+1})^{-1}L\|\|L^{-1}F(x_{n+1})\| \\
&\leq \frac{2}{3}\frac{\delta_{n+1}}{1-v_0(\alpha_{n+1})}=\beta_{n+1}-\alpha_{n+1}.
\end{aligned}$$

The rest of the analysis follows as in the previous Chapters.

4. References

[1] Argyros, I. K., *Convergence and Applications of Newton-type Iterations*, Springer Verlag, Berlin, Germany, (2008).

[2] Argyros, I. K., *On the convergence of Newtons method for polynomial equations and applications in radiative transfer*, Monatsh. Math., 127 (4), (1999), 265–276.

[3] Argyros, I. K., *On the NewtonKantorovich hypothesis for solving equations*, J. Comput. Appl. Math., 169 (2), (2004), 315–332.

[4] Argyros, I. K., *The theory and applications of iteration methods*, 2nd Edition, CRC Press, Engineering Series, Taylor and Francis Group, (2022).

[5] Argyros, I. K., *Unified Convergence Criteria for Iterative Banach Space Valued Methods with Applications*, Mathematics 2021, 9(16), 1942; https://doi.org/10.3390/math9161942.

[6] Cordero, A., Martnez, E., Torregrosa, J. R., *Iterative methods of order four and five for systems of nonlinear equations*, Appl. Math. Comput. 231, (2009), 541–551.

[7] Darvishi, M. T., Barati, A., *A fourth-order method from quadrature formulae to solve systems of nonlinear equations*, Appl. Math. Comput., 188, (2007), 257–261.

[8] Frontini, M., Sormani, E., *Some variant of Newtons method with third-order convergence*, Appl. Math. Comput., 140, (2003), 419–426.

[9] Gautschi, W., *Numerical Analysis: An Introduction*, Birkhuser, Boston, (1997).

[10] Gutirrez, J. M., Hernndez, M. A., *A family of Chebyshev-Halley type methods in Banach spaces*, Bull. Aust. Math. Soc. 55, (1997), 113–130.

[11] Homeier, H. H. H., *A modified Newton method with cubic convergence: the multivariable case*, J. Comput. Appl. Math., 169, (2004), 161–169.

[12] Ozban, A. Y., *Some new variants of Newtons method*, Appl. Math. Lett., 17, (2004), 677–682.

[13] Sharma, J. R., Guha, R. K., Sharma, R., *An efficient fourth order weighted-Newton method for systems of nonlinear equations*, Numer. Algor., 62, (2013), 307–323. https://doi.org/10.1007/s11075-012-9585-7.

[14] Traub, J. F., *Iterative Methods for the Solution of Equations*, Prentice-Hall, Englewood Cliffs, (1964).

[15] Weerakoon, S., Fernando, T. G. I., *A variant of Newtons method with accelerated third-order convergence*, Appl. Math. Lett., 13, (2000), 87–93.

Chapter 44

Parametric Iterative Method

1. Introduction

The iteration is defined for $x_0 \in \Omega$ and each $n = 0, 1, 2, ...$ by

$$\begin{aligned} y_n &= x_n - F'(x_n)^{-1}F(x_n), \\ v_n &= x_n + F'(x_n)^{-1}F(x_n), \\ u_n &= x_n - \frac{2}{3}F'(x_n)^{-1}F(x_n), \\ A_n &= aF'(x_n) + bF'(u_n) + cF'(y_n), \\ B_n &= a_1F(y_n) + b_1F(x_n) + c_1F(v_n) \end{aligned} \tag{44.1}$$

and

$$x_{n+1} = x_n - A_n^{-1}B_n,$$

where $a, b, c, a_1, b_1, c_1 \in \mathbb{R}$ with $a + b + c \neq 0$. The method is shown to be of order four if and only if

$$\begin{aligned} b &= \frac{3}{2}\frac{(1-2c_1)(1-4c_1)}{1+2c_1},\ a = 1-2c_1-b,\ c = 2c_1, \\ b_1 &= 1-2c_1,\ a_1 = 1-5c_1-b \end{aligned}$$

and $c_1 \neq \frac{1}{2}$, provided that $B_1 = B_2 = \mathbb{R}^k$ and $F^{(5)}$ exists [4]. Moreover, higher order methods are developed by introducing more steps and also considering component by component multiplication and division. That is for $\alpha = (\alpha_1, \alpha_2, ..., \alpha_k)^T \in \mathbb{R}^k$ and $\beta = (\beta_1, \beta_2, ..., \beta_k)^T \in \mathbb{R}^k$.

$$\alpha.\beta = (\alpha_1\beta_1, \alpha_2\beta_2, ..., \alpha_3\beta_3)^T \in \mathbb{R}^k$$

and

$$\frac{\alpha}{\beta} = \left(\frac{\alpha_1}{\beta_1}, \frac{\alpha_2}{\beta_2}, ..., \frac{\alpha_k}{\beta_k}\right) \in \mathbb{R}^k.$$

In this chapter, the local convergence is based only on the operators on the method in the setting of a Banach space.

2. Convergence

The majorant functions g_i, $i=1,2$ are:

$$\begin{aligned}
g_1(t) &= \frac{\int_0^1 w((1-\theta)t)d\theta}{1-w_0(t)},\\
h_1(t) &= \frac{\int_0^1 w((1-\theta)t)d\theta+\frac{1}{3}(1+\int_0^1 w_0(\theta t)d\theta)}{1-w_0(t)},\\
h_2(t) &= \frac{\int_0^1 w((1-\theta)t)d\theta+2(1+\int_0^1 w_0(\theta t)d\theta)}{1-w_0(t)},\\
q(t) &= \frac{1}{|a+b+c|}[|a|w_0(t)+|b|w_0(h_1(t)t)+|c|w_0(g_1(t)t)],\\
\lambda(t) &= |a_1|(1+\int_0^1 w_0(\theta g_1(t)t)d\theta)g_1(t)t\\
&\quad +|b_1-1|(1+\int_0^1 w_0(\theta t)d\theta)t+|c_1|(1+\int_0^1 w_0(\theta h_2(t)td\theta)h_2(t)t,\\
\mu(t) &= |a-1|w_0(t)+|b|w_0(h_1(t)t)+|c|w_0(g_1(t)t)\\
&\quad +|a-1|+|b|+|c|,
\end{aligned}$$

and

$$\begin{aligned}
g_2(t) &= \frac{\int_0^1 w((1-\theta)t)d\theta t}{1-w_0(t)}+\frac{\lambda(t)}{1-w_0(t)}\\
&\quad +\frac{\mu(t)}{|a+b+c|(1-q(t))(1-w_0(t))}.
\end{aligned}$$

The motivational calculations are:

$$\begin{aligned}
y_n-x^* &= x_n-x^*-F'(x_n)^{-1}F(x_n),\\
\|y_n-x^*\| &\leq \frac{\int_0^1 w((1-\theta)\|x_n-x^*\|)d\theta\|x_n-x^*\|}{1-w_0(\|x_n-x^*\|)}\\
&\leq g_1(\|x_n-x^*\|)\|x_n-x^*\|\leq\|x_n-x^*\|<r,\\
u_n-x^* &= x_n-x^*-F'(x_n)^{-1}F(x_n)+\frac{1}{3}F'(x_n)^{-1}F(x_n),\\
\|u_n-x^*\| &\leq \frac{[\int_0^1 w((1-\theta)\|x_n-x^*\|)d\theta+\frac{1}{3}(1+\int_0^1 w_0(\theta\|x_n-x^*\|)d\theta)]\|x_n-x^*\|}{1-w_0(\|x_n-x^*\|)}\\
&\leq h_1(\|x_n-x^*\|)\|x_n-x^*\|,\\
v_n-x^* &= x_n-x^*-F'(x_n)^{-1}F(x_n)+2F'(x_n)^{-1}F(x_n),\\
\|v_n-x^*\| &\leq \frac{[\int_0^1 w((1-\theta)\|x_n-x^*\|)d\theta+2(1+\int_0^1 w_0(\theta\|x_n-x^*\|)d\theta)]\|x_n-x^*\|}{1-w_0(\|x_n-x^*\|)}\\
&\leq h_2(\|x_n-x^*\|)\|x_n-x^*\|,
\end{aligned}$$

$$
\begin{aligned}
& \|((a+b+c)L)^{-1}(A_n-(a+b+c)L)\| \\
\leq\ & \frac{1}{|a+b+c|}[|a|\|L^{-1}(F'(x_n)-L)\|+|b|\|L^{-1}(F'(y_n)-L)\| \\
& +|c|\|L^{-1}(F'(y_n)-L)\|] \\
\leq\ & \frac{1}{|a+b+c|}[|a|w_0(\|x_n-x^*\|)+|b|w_0(\|u_n-x^*\|)+|c|w_0(\|y_n-x^*\|)] \\
=\ & q_n,
\end{aligned}
$$

$$
\begin{aligned}
x_{n+1}-x^* =\ & x_n-x^*-A_n^{-1}B_n=x_n-x^*-F'(x_n)^{-1}F(x_n) \\
& +F'(x_n)^{-1}(F(x_n)-B_n)+(F'(x_n)^{-1}-A_n^{-1})B_n
\end{aligned}
$$

and

$$
\begin{aligned}
\|x_{n+1}-x^*\| \leq\ & \frac{\int_0^1 w((1-\theta)\|x_n-x^*\|)d\theta\|x_n-x^*\|}{1-w_0(\|x_n-x^*\|)} \\
& +\frac{\lambda_n}{1-w_0(\|x_n-x^*\|)}+\frac{\mu_n}{|a+b+c|(1-q_n)(1-w_0(\|x_n-x^*\|))} \\
\leq\ & g_2(\|x_n-x^*\|)\|x_n-x^*\| \leq \|x_n-x^*\|,
\end{aligned}
$$

provided that $q_n < 1$. The rest of the analysis is omitted as similar to previous chapters.

3. References

[1] Ahmad, F., Soleymani, F., Haghani, F. K., Serra-Capizzano, S., *Higher order derivative-free iterative methods with and without memory for systems of nonlinear equations*, Appl. Math. Comput., 314, (2017), 199–211.

[2] Argyros, I. K., *Convergence and Applications of Newton-type Iterations*, Springer Verlag, Berlin, Germany, (2008).

[3] Argyros, I. K., *The theory and applications of iteration methods*, 2nd Edition, CRC Press, Engineering Series, Taylor and Francis Group, (2022).

[4] Argyros, I. K., George, S., *Mathematical Modeling For The Solution Of Equations And Systems Of Equations With Applications*, Volume-III, Nova Publishes, US, (2019).

[5] Behl, R., Arora, H., *A novel scheme having seventh-order convergence for nonlinear systems*, J. Comput. Appl. Math., (2020), 113301.

[6] Behl, R., Cordero, A., Motsa, S. S., Torregrosa, J. R., *Construction of fourth-order optimal families of iterative methods and their dynamics*, Appl. Math. Computation, 271, (2015), 89–101.

[7] Changbum Chun, B., Neta, *An efficient derivative-Free method for the solution of systems of equations*, Numer. Func. Anal. Optimiz., 42, (2021), 834–848.

[8] Changbum Chun, B., Neta, *Developing high order methods for the solution of systems of nonlinear equations*, Appl. Math. Comput., 342, (2019), 178–190.

[9] Cuyt, A. A., Van der Cruyssen, P., *Abstract Pad-approximants for the solution of systems of nonlinear equations*, Comput. Math. Appl., 9, (1983), 617–624.

[10] Dehghan, M., Shirilord, A., *Three-step iterative methods for numerical solution of systems of nonlinear equations*, Engineering with Computers, 38, (2022), 1015–1028.

[11] Ham, Y., Chun, C., *A fifth-order iterative method for solving nonlinear equations*, Appl. Math. Comput., 194, (2007), 287–290.

[12] Singh, H., Sharma, J. R., *Simple and efficient fifth order solvers for systems of nonlinear problems*, Mathematical Modelling and Analysis, 28, (2023), 1–22.

[13] Su, Q., *A unified model for solving a system of nonlinear equations*, Appl. Math. Comput., 290, (2016), 46–55.

[14] Wang, X., *Fixed-point iterative method with eighth-order constructed by undetermined parameter technique for solving nonlinear systems*, Symmetry, 13(5), (2021), 863.

[15] Zhanlav, T., Mijiddorj, R., *Development of higher order iterative methods in $\mathbb{R}^n$ with specific rules*, International Journal of Computer Mathematics (To appear).

[16] Zhanlav, T., Mijiddorj, R., Otgondorj, K., *A family of Newton-type methods with seventh and eighth-order of convergence for solving systems of nonlinear equations*, (accepted) Hacet. J. Math. Stat. (2023).

Chapter 45

A Comparison Between Four Second-Order Methods

1. Introduction

The four single step quadratically convergent methods are defined for $x_0 \in \Omega$ and each $n = 0, 1, 2...$ by

Newton's Method [1,3]

$$x_{n+1} = x_n - F'(x_n)^{-1}F(x_n). \tag{45.1}$$

Steffensen Method [1, 3]

$$\begin{aligned} A_n &= [u_n, x_n; F],\ u_n = x_n + F(x_n), \\ x_{n+1} &= x_n - A_n^{-1}F(x_n) \end{aligned} \tag{45.2}$$

Trapezoidal Method [4]

$$\begin{aligned} B_n &= F'(x_n) + F'(u_n),\ u_n = x_n + F(x_n), \\ x_{n+1} &= x_n - 2B_n^{-1}F(x_n) \end{aligned} \tag{45.3}$$

Quadrature Method [4]

$$\begin{aligned} v_n &= x_n + 2F(x_n),\ C_n = F'(x_n) + 4F'(u_n) + F'(v_n) \\ x_{n+1} &= x_n - 6C_n^{-1}F(x_n). \end{aligned}$$

The quadratic convergence of these methods is established in [4] when $B_1 = B_2 = R$. The convergence conditions require that $F^{(2)}$ or $F^{(3)}$ exists and is bounded.
The local as well as the semi-local convergence involves weaker conditions. Hence, the applicability of these methods is extended. Similar extensions are obtained when the methodology of this chapter is extended to other methods [1,2,3].

(I) **Local Convergence**
The majorant function g is:

$$g(t) = \frac{\int_0^1 w((1-\theta)t)d\theta}{1 - w_0(t)}.$$

The calculations leading to the development of this function are:

$$x_{n+1} - x^* = x_n - x^* - F'(x_n)^{-1}F(x_n)$$

and

$$\begin{aligned}\|x_{n+1} - x^*\| &\leq \frac{\int_0^1 w((1-\theta)\|x_n - x^*\|)d\theta\|x_n - x^*\|}{1 - w_0(\|x_n - x^*\|)}\\ &\leq g(\|x_n - x^*\|)\|x_n - x^*\| \leq \|x_n - x^*\| < r.\end{aligned}$$

(II) **Semi-local Convergence**

The majorant sequence $\{\alpha_n\}$ is defined for $\alpha_0 = 0, \alpha_1 \geq \|F'(x_0)^{-1}F(x_0)\|$ and each $n = 0, 1, 2, ...$ by

$$\begin{aligned}\delta_{n+1} &= \int_0^1 v((1-\theta)(\alpha_{n+1} - \alpha_n))d\theta(\alpha_{n+1} - \alpha_n),\\ \alpha_{n+2} &= \alpha_{n+1} + \frac{\delta_{n+1}}{1 - v_0(\alpha_{n+1})}.\end{aligned}$$

The motivational calculations are:

$$\begin{aligned}F(x_{n+1}) &= F(x_{n+1}) - F(x_n) - F'(x_n)(x_{n+1} - x_n)\\ &= \int_0^1 (F'(x_n + \theta(x_{n+1} - x_n)) - F'(x_n))(x_{n+1} - x_n),\\ \|L^{-1}F(x_{n+1})\| &\leq \int_0^1 v(\theta\|x_{n+1} - x_n\|)d\theta\|x_{n+1} - x_n\|\\ &\leq \int_0^1 v(\theta(\alpha_{n+1} - \alpha_n))d\theta(\alpha_{n+1} - \alpha_n) = \delta_{n+1},\\ x_{n+2} - x_{n+1} &= -F'(x_{n+1})^{-1}F(x_{n+1}),\end{aligned}$$

and

$$\begin{aligned}\|x_{n+2} - x_{n+1}\| &\leq \|F'(x_{n+1})^{-1}L\|\|L^{-1}F(x_{n+1})\|\\ &\leq \frac{\delta_{n+1}}{1 - v_0(\|x_{n+1} - x_0\|)} \leq \frac{\delta_{n+1}}{1 - v_0(\alpha_{n+1})} = \alpha_{n+2} - \alpha_{n+1}.\end{aligned}$$

2. Steffensen Method

(I) **Local Convergence**

The majorant function g is:

$$g(t) = \frac{\psi(f(t), t)}{1 - \psi_0(f(t), t)}.$$

The calculations are

$$\begin{aligned}x_{n+1} - x^* &= x_n - x^* - A_n^{-1}F(x_n)\\ &= A_n^{-1}(A_n - [x_n, x^*; F])(x_n - x^*)\end{aligned}$$

and

$$\begin{aligned}\|x_{n+1}-x^*\| &\leq \frac{\phi(f(\|x_n-x^*\|),\|x_n-x^*\|)\|x_n-x^*\|}{1-\phi_0(f\|x_n-x^*\|,\|x_n-x^*\|)}\\ &\leq g(\|x_n-x^*\|)\|x_n-x^*\| \leq \|x_n-x^*\| < r,\end{aligned}$$

where we also used for each $x \in \Omega$

$$\begin{aligned}\|u-x^*\| &\leq f(\|x-x^*\|),\ u = x+F(x)\\ \|L^{-1}([u,x;F]-L)\| &\leq \phi_0(\|u-x^*\|,\|x-x^*\|)\end{aligned}$$

and

$$\|L^{-1}([u,x;F]-[x,x^*;F])\| \leq \phi(\|u-x^*\|,\|x-x^*\|).$$

(II) **Semi-local Convergence**

The majorant sequence $\{\alpha_n\}$ is defined for $\alpha_0 = 0, \alpha_1 \geq \|A_0^{-1}F(x_0)\|$ and each $n = 0,1,2,...$ by

$$\begin{aligned}\delta_{n+1} &= (1+\psi_0(\alpha_n,\alpha_{n+1}))(\alpha_{n+1}-\alpha_n)\\ &\quad +(1+\psi_0(f_1(\alpha_n),\alpha_n))(\alpha_{n+1}-\alpha_n)\end{aligned}$$

and

$$\alpha_{n+2} = \alpha_{n+1} + \frac{\delta_{n+1}}{1-\psi_0(f_1(\alpha_{n+1}),\alpha_{n+1})}.$$

The motivational calculations are:

$$\begin{aligned}F(x_{n+1}) &= F(x_{n+1})-F(x_n)-A_n(x_{n+1}-x_n),\\ \|L^{-1}F(x_{n+1})\| &\leq \|L^{-1}([x_{n+1},x_n;F]-L+L)(x_{n+1}-x_n)\|\\ &\quad +\|L^{-1}(A_n-L+L)(x_{n+1}-x_n)\|\\ &\leq (1+\psi_0(\|x_{n+1}-x_0\|,\|x_n-x_0\|))\|x_{n+1}-x_n\|\\ &\quad +1+\psi_0(\|u_n-x_0\|,\|x_n-x_0\|))\|x_{n+1}-x_n\|\\ &\leq (1+\psi_0(\alpha_{n+1},\alpha_n))(\alpha_{n+1}-\alpha_n)\\ &\quad +(1+\psi_0(f_1(\alpha_n),\alpha_n))(\alpha_{n+1}-\alpha_n) = \delta_{n+1},\end{aligned}$$

and

$$\begin{aligned}\|x_{n+2}-x_{n+1}\| &\leq \|A_{n+1}^{-1}L\|\|L^{-1}F(x_{n+1})\|\\ &\leq \frac{\delta_{n+1}}{1-\psi_0(f_1(\|x_{n+1}-x_0\|),\|x_{n+1}-x_0\|)}\\ &\leq \frac{\delta_{n+1}}{1-\psi_0(f_1(\alpha_{n+1}),\alpha_{n+1})}\\ &= \alpha_{n+2}-\alpha_{n+1},\end{aligned}$$

where we also used for each $x \in \Omega$

$$\|u-x_0\| \leq f_1(\|x-x_0\|),\ u = x+F(x),$$

and

$$\|L^{-1}([u,x;F]-L)\| \leq \psi_0(\|u-x_0\|,\|x-x_0\|).$$

3. Trapezoidal Method

(I) **Local Convergence**

The majorant functions q and g are:

$$q(t) = \frac{1}{2}(w_0(t)+w_0(f(t)))$$
$$g(t) = \frac{\int_0^1 w((1-\theta)t)d\theta+\int_0^1 w(f(t)+\theta t)d\theta}{2(1-q(t))}.$$

The calculations are:

$$\begin{aligned}\|(2L)^{-1}(B_n-2L)\| &\leq \frac{1}{2}[\|L^{-1}(F'(x_n)-L)\|+\|L^{-1}(F'(u_n)-L)\|] \\ &\leq \frac{1}{2}(w_0(\|x_n-x^*\|)+w_0(f(\|x_n-x^*\|))) \\ &= q_n<1, \\ \|B_n^{-1}L\| &\leq \frac{1}{2(1-q_n)}\end{aligned}$$

$$\begin{aligned}x_{n+1}-x^* &= x_n-x^*-2B_n^{-1}F(x_n) \\ &= B_n^{-1}(B_n-2[x_n,x^*;F])(x_n-x^*) \\ &= B_n^{-1}[F'(x_n)+F'(u_n)-2\int_0^1 F'(x^*+\theta(x_n-x^*))d\theta](x_n-x^*)\end{aligned}$$

and

$$\begin{aligned}&\|x_{n+1}-x^*\| \\ \leq\ & \frac{[\int_0^1 w((1-\theta)\|x_n-x^*\|)d\theta+\int_0^1 w(\|u_n-x^*\|+\theta\|x_n-x^*\|)d\theta]}{2(1-q_n)} \\ & \times\|x_n-x^*\| \\ \leq\ & g(\|x_n-x^*\|)\|x_n-x^*\|\leq\|x_n-x^*\|<r.\end{aligned}$$

(II) **Semi-local Convergence**

The majorant sequence $\{\alpha_n\}$ is defined for $\alpha_0,\alpha_1\geq 2\|B_0^{-1}F(x_0)\|$ and each $n=0,1,2,...$ by

$$\begin{aligned}\delta_{n+1} &= (1+\int_0^1 v_0(\alpha_n+\theta(\alpha_{n+1}-\alpha_n))d\theta(\alpha_{n+1}-\alpha_n) \\ &\quad +\frac{1}{2}(2+v_0(\alpha_n)+v_0(f_1(\alpha_n))(\alpha_{n+1}-\alpha_n)\end{aligned}$$

and

$$\alpha_{n+2} = \alpha_{n+1}+\frac{\delta_{n+1}}{1-v_0(\alpha_{n+1})-v_0(f_1(\alpha_{n+1}))}.$$

The calculations are:

$$\begin{aligned}F(x_{n+1}) &= F(x_{n+1})-F(x_n)-\frac{1}{2}B_n(x_{n+1}-x_n), \\ \|L^{-1}F(x_{n+1})\| &\leq (1+\int_0^1 v_0(\|x_n-x_0\|+\theta\|x_{n+1}-x_n\|)d\theta)\|x_{n+1}-x_n\| \\ &\quad +\frac{1}{2}(2+v_0(\|x_n-x_0\|)+v_0(f_1(\|x_n-x_0\|))) \\ &= \bar{\delta}_{n+1}\leq\delta_{n+1}\end{aligned}$$

and

$$\begin{aligned}\|x_{n+2}-x_{n+1}\| &\leq \frac{2\bar{\delta}_{n+1}}{2(1-v_0(\|x_{n+1}-x_0\|)-v_0(f_1(\|x_{n+1}-x_0\|))}\\ &\leq \frac{\delta_{n+1}}{1-v_0(\alpha_{n+1})-v_0(f_1(\alpha_{n+1}))}=\alpha_{n+2}-\alpha_{n+1}.\end{aligned}$$

4. Quadrature Method

(I) **Local Convergence**

The majorant function g is:

$$\begin{aligned}p(t) &= \int_0^1 w((1-\theta)t)d\theta+4\int_0^1 w(f(t)+\theta t)d\theta\\ &\quad +\int_0^1 w(f(t)+\theta t)d\theta,\\ q(t) &= \frac{1}{6}(w_0(t)4w_0(f(t))+w_0(f_2(t)))\end{aligned}$$

and

$$g(t) = \frac{p(t)}{6(1-q(t))}.$$

The motivational calculations are:

$$\begin{aligned} &\|(6L)^{-1}(C_n-6L)\|\\ \leq\ &\frac{1}{6}[\|L^{-1}(F'(x_n)-L)\|\\ &+4\|L^{-1}(F'(u_n)-L)\|+\|L^{-1}(F'(v_n)-L)\|]\\ \leq\ &\frac{1}{6}[w_0(\|x_n-x^*\|)+4w_0(f(\|x_n-x^*\|))+w_0(f_2(\|x_n-x^*\|))]\\ =\ &q_n<1,\end{aligned}$$

$$\|C_n^{-1}L\| \leq \frac{1}{6(1-q_n)}$$

$$\begin{aligned} &\|L^{-1}(C_n-6\int_0^1 F'(x^*+\theta(x_n-x^*))d\theta)\|\\ \leq\ &\int_0^1 w((1-\theta)\|x_n-x^*\|)d\theta+4\int_0^1 w(\|u_n-x^*\|+\theta\|x_n-x^*\|)d\theta\\ &+\int_0^1 w(\|v_n-x^*\|+\theta\|x_n-x^*\|)d\theta=p_n,\end{aligned}$$

$$\begin{aligned}x_{n+1}-x^* &= x_n-x^*-6C_n^{-1}F(x_n)\\ &= C_n^{-1}(C_n-6\int_0^1 F'(x^*+\theta(x_n-x^*))d\theta)(x_n-x^*)\end{aligned}$$

and

$$\begin{aligned}\|x_{n+1}-x^*\| &\leq \|C_n^{-1}L\|\|L^{-1}(C_n-6\int_0^1 F'(x^*+\theta(x_n-x^*))d\theta)\|\|x_n-x^*\|\\ &\leq \frac{P_n\|x_n-x^*\|}{6(1-q_n)}\leq g(\|x_n-x^*\|)\|x_n-x^*\|\leq\|x_n-x^*\|<r,\end{aligned}$$

where we also used

$$\|v-x^*\|\leq f_2(\|x-x^*\|),\ x\in\Omega.$$

(II) **Semi-local Convergence**

The majorant sequence $\{\alpha_n\}$ is defined for $\alpha_0 = 0, \alpha_1 \geq 6\|C_0^{-1}F(x_0)\|$ and each $n = 0,1,2,...$

$$\begin{aligned} \delta_{n+1} &= (1+\int_0^1 v_0(\alpha_n+\theta(\alpha_{n+1}-\alpha_n))d\theta)(\alpha_{n+1}-\alpha_n) \\ &+\frac{1}{6}(6+v_0(\alpha_n)+v_0(f_1(\alpha_n))+v_0(f_3(\alpha_n))) \end{aligned}$$

and

$$\alpha_{n+2} = \alpha_{n+1}\frac{\delta_{n+1}}{1-v_0(\alpha_{n+1})-v_0(f_1(\alpha_{n+1}))-v_0(f_3(\alpha_{n+1}))}.$$

The calculations are:

$$\begin{aligned} F(x_{n+1}) &= F(x_{n+1})-F(x_n)-\frac{1}{6}C_n(x_{n+1}-x_n), \\ \|L^{-1}F(x_{n+1})\| &\leq (1+\int_0^1 v_0(\|x_n-x_0\|+\theta\|x_n-x_n\|)d\theta)\|x_{n+1}-x_n\| \\ &+\frac{1}{6}(6+v_0(\|x_n-x_0\|)+v_0(\|u_n-x_0\|)+v_0(\|v_n-x_0\|) \\ &= \bar{\delta}_{n+1} \leq \delta_{n+1} \end{aligned}$$

and

$$\begin{aligned} \|x_{n+2}-x_{n+1}\| &\leq 6\|C_{n+1}^{-1}L\|\|L^{-1}F(x_{n+1})\| \\ &\leq \alpha_{n+2}-\alpha_{n+1}, \end{aligned}$$

where we also used

$$\|v_n-x_0\| \leq f_3(\|x_n-x_0\|).$$

The rest of the analysis is given as in previous chapters.

5. References

[1] Argyros, I. K., *Convergence and Applications of Newton-type Iterations*, Springer Verlag, Berlin, Germany, (2008).

[2] Argyros, I. K., *The theory and applications of iteration methods*, 2nd Edition, CRC Press, Engineering Series, Taylor and Francis Group, (2022).

[3] Argyros, I. K., George, S., *Mathematical Modeling For The Solution Of Equations And Systems Of Equations With Applications*, Volume-III, Nova Publishes, US, (2019).

[4] Qureshi, U. K., Shaikhi, A. A., Shaikh, E. K., Hazarewal, S. K., Laghari, T. A., *New Simpson type method for solving nonlinear equations*, Open Journal of Mathematical Sciences, 5(1), (2021), 94–100, DOI: 10.30538/oms2021.0148.

Chapter 46

A Unified Model for Equations

1. Introduction

The two step unified Model is defined for $x_0 \in \Omega$ and each $n = 0, 1, 2, ...$ by

$$
\begin{aligned}
y_n &= x_n - F'(x_n)^{-1}F(x_n), \\
A_n &= \sum_{i=1}^{p} c_i F'(a_i x_n + b_i y_n), \\
B_n &= \sum_{j=1}^{q} F'(h_j x_n + c_j y_n), \\
a_i + b_i &= 1,\ h_j + c_j = 1,\ c = \sum_{i=1}^{p} c_i \\
x_{n+1} &= x_n - A_n^{-1} B_n,
\end{aligned}
\tag{46.1}
$$

where p, q are natural numbers and a_i, b_i, h_j and c_j are real numbers and $c \neq 0$. The local fourth order is determined in [11] using Taylor series expansions. If $B_1 = B_2 = \mathbb{R}^k$ for some additional choices of the parameters and by assuming the existence and boundedness of $F^{(5)}$. The new local analysis is based only on the operators on the method. Moreover the semi-local analysis (not given in [11]) is also provided in this chapter based on majorizing sequences. The technique can be used to extend the applicability of other methods along the same lines [1]-[10], [12].

2. Local Convergence

The majorant functions $g_i, i = 1, 2$, are:

$$
\begin{aligned}
g_1 &= \frac{\int_0^1 w((1-\theta)t)d\theta}{1 - w_0(t)}, \\
\lambda(t) &= \frac{1}{|c|}\sum_{i=1}^{p} |c_i| w_0(\|(|a_i| + |b_i| g_1(t))t\|), \\
\mu(t) &= \sum_{i=1}^{p} |c_i| w((|a_i| + |b_i| g_1(t)t) + \sum_{i=1}^{p} |1 - c_i|(1 + w_0(t)),
\end{aligned}
$$

$$\begin{aligned}\zeta(t) &= \sum_{i=1}^{q}|d_i|(1+\int_0^1 w_0(\theta(|h_j|t+|c_j|g_i(t)t)d\theta)(|h_j|+|c_j|g_i(t))\\ &+(1+\int_0^1 w_0(\theta t)d\theta),\end{aligned}$$

and

$$\begin{aligned}g_2(t) &= \frac{\int_0^1 w((1-\theta)t)d\theta}{1-w_0(t)}+\frac{\mu(t)(1+\int_0^1 w_0(\theta t)d\theta)}{|c|(1-\lambda(t))(1-w_0(t))}\\ &+\frac{\zeta(t)}{|c|(1-\lambda(t))}.\end{aligned}$$

The motivation follows from the calculations

$$\begin{aligned}x_{n+1}-x^* &= x_n-x^*-F'(x_n)^{-1}F(x_n),\\ \|x_{n+1}-x^*\| &\leq \frac{\int_0^1 w((1-\theta)\|x_n-x^*\|)d\theta\|x_n-x^*\|}{1-w_0(\|x_n-x^*\|)}\\ &\leq g_1(\|x_n-x^*\|)\|x_n-x^*\|\leq\|x_n-x^*\|<r,\end{aligned}$$

$$\begin{aligned}\|(cL)^{-1}(A_n-cL)\| &\leq \frac{1}{|c|}\sum_{i=1}^{p}|c_i|w_0(\|a_ix_n+b_iy_n-x^*\|)\\ &\leq \frac{1}{|c|}\sum_{i=1}^{p}|c_i|w_0((|a_i|\|x_n-x^*\|+|b_i|\|y_n-x^*\|)\\ &= \lambda_n<1,\\ \|A_n^{-1}L\| &\leq \frac{1}{|c|(1-\lambda_n)},\\ A_n-F'(x_n) &= \sum_{i=1}^{p}c_i(F'(a_ix_n+b_iy_n)-F'(x_n))-\sum_{i=1}^{p}(1-c_i)F'(x_n),\end{aligned}$$

$$\begin{aligned}\|L^{-1}(A_n-F'(x_n))\| &\leq \sum_{i=1}^{p}|c_i|w(|a_i-1|\|x_n-x^*\|+|b_i|\|y_n-x^*\|)\\ &+\sum_{i=1}^{p}|1-c_i|(1+w_0(\|x_n-x^*\|))=\mu_n,\\ B_n-F(x_n) &= \sum_{j=1}^{q}d_j(F(h_jx_n+c_jy_n)-F'(x^*))+(F(x^*)-F(x_n)),\\ \|L^{-1}(B_n-F(x_n))\| &\leq \sum_{j=1}^{q}|d_j|(1+\int_0^1 w_0(\theta(|h_j|\|x_n-x^*\|+|c_j|\|y_n-x^*\|)d\theta)\\ &\times(|h_j|+|c_j|g_1(\|x_n-x^*\|))\|x_n-x^*\|\\ &+(1+\int_0^1 w(\theta\|x_n-x^*\|)d\theta)\|x_n-x^*\|=\zeta_n\|x_n-x^*\|,\end{aligned}$$

$$\begin{aligned}x_{n+1}-x^* &= x_n-x^*-F'(x_n)^{-1}F(x_n)+F'(x_n)^{-1}F(x_n)-A_n^{-1}B_n\\ &= x_n-x^*-F'(x_n)^{-1}F(x_n)+(F'(x_n)^{-1}-A_n^{-1})F(x_n)+A_n^{-1}(F(x_n)-B_n)\end{aligned}$$

and

$$\|x_{n+1}-x^*\| \leq \left[\frac{\int_0^1 w((1-\theta)\|x_n-x^*\|)d\theta}{1-w_0(\|x_n-x^*\|)} + \frac{\mu_n(1+\int_0^1 w_0(\theta\|x_n-x^*\|)d\theta}{|c|(1-\lambda_n)(1-w_0(\|x_n-x^*\|))} + \frac{\zeta_n}{|c|(1-\lambda_n)}\right]\|x_n-x^*\|.$$

3. Semi-local Analysis

The majorant sequence $\{\alpha_n\}$ is defined for $\alpha_0=0, \beta_0 \geq \|F'(x_0)^{-1}F(x_0)\|$ and each $n=0,1,2,...$ by

$$\begin{aligned}
\bar{\mu}_n &= \sum_{i=1}^{p}|c_i|v(|a_i-1|\alpha_n+|b_i|\beta_n)+\sum_{i=1}^{p}|1-c_i|(1+v_0(\alpha_n)),\\
\bar{\lambda}_n &= \frac{1}{|c|}\sum_{i=1}^{p}v_0(|a_i|\alpha_n+|b_i|\beta_n),\\
\bar{\zeta}_n &= (1+\int_0^1 v_0(\theta(|h_j|\alpha_n+|c_j|\beta_n)d\theta)(|h_j|\alpha_n+|c_j|\beta_n)\\
&\quad +(1+v_0(\alpha_n))\alpha_n,\\
\alpha_{n+1} &= \beta_n+\frac{\bar{\mu}_n(\beta_n-\alpha_n)}{|c|(1-\bar{\lambda}_n)}+\frac{\bar{\zeta}_n}{|c|(1-\bar{\lambda}_n)},\\
\delta_{n+1} &= \int_0^1 v((1-\theta)(\alpha_{n+1}-\alpha_n))d\theta(\alpha_{n+1}-\alpha_n)\\
&\quad +(1+v_0(\alpha_n))d\theta(\alpha_{n+1}-\alpha_n)
\end{aligned}$$

and

$$\beta_{n+1} = \alpha_{n+1}+\frac{\delta_{n+1}}{1-v_0(\alpha_{n+1})}.$$

The calculations are similar to the ones in the local case:

$$\begin{aligned}
x_{n+1}-y_n &= F'(x_n)^{-1}F(x_n)-A_n^{-1}B_n\\
&= (F'(x_n)^{-1}-A_n^{-1})F(x_n)+A_n^{-1}(F(x_n)-B_n)\\
&= -A_n^{-1}(F'(x_n)-A_n)F'(x_n)^{-1}F(x_n)+A_n^{-1}(F(x_n)-B_n)\\
&= A_n^{-1}(F'(x_n)-A_n)(y_n-x_n)+A_n^{-1}(F(x_n)-B_n),\\
\|x_{n+1}-y_n\| &\leq \|A_n^{-1}L\|\|L^{-1}(F'(x_n)-A_n)\|\|y_n-x_n\|\\
&\quad +\|A_n^{-1}L\|\|L^{-1}(F(x_n)-B_n)\|\\
&\leq \frac{\bar{\mu}_n}{|c|(1-\bar{\lambda}_n)}+\frac{\bar{\mu}_n}{|c|(1-\lambda_n)}=\alpha_{n+1}-\beta_n,\\
F(x_{n+1}) &= F(x_{n+1})-F(x_n)-F'(x_n)(x_{n+1}-x_n)+F(x_n)+F'(x_n)(x_{n+1}-y_n),\\
\|L^{-1}F(x_{n+1})\| &\leq \int_0^1 w((1-\theta)(\alpha_{n+1}-\alpha_n))d\theta(\alpha_{n+1}-\alpha_n)\\
&\quad +(1+v_0(\alpha_n))(\alpha_{n+1}-\beta_n)=\delta_{n+1},
\end{aligned}$$

and

$$\begin{aligned}\|y_{n+1}-x_{n+1}\| &\leq \|F'(x_{n+1})^{-1}L\|\|L^{-1}F(x_{n+1})\| \\ &\leq \frac{\delta_{n+1}}{1-v_0(\alpha_{n+1})}=\beta_{n+1}-\alpha_{n+1}.\end{aligned}$$

The rest of the analysis follows as in the previous chapters.

4. References

[1] Argyros, I. K., *Convergence and Applications of Newton-type Iterations*, Springer Verlag, Berlin, Germany, (2008).

[2] Argyros, I. K., *The theory and applications of iteration methods*, 2nd Edition, CRC Press, Engineering Series, Taylor and Francis Group, (2022).

[3] Argyros, I. K., George, S., *Mathematical Modeling For The Solution Of Equations And Systems Of Equations With Applications*, Volume-III, Nova Publishes, US, (2019).

[4] Babolian, E., Biazar, J., Vahidi, A. R., *Solution of a system of nonlinear equations by Adomian decomposition method*, Appl. Math. Comput., 150, (2004), 847–854.

[5] Cordero, A., Torregrosa, J. R., *Variants of Newtons method for functions of several variables*, Appl. Math. Comput., 183, (2006), 199–208.

[6] Darvishi, M. T., Barati, A., *A third-order Newton-type method to solve systems of nonlinear equations*, Appl. Math. Comput., 187, (2007), 630–635.

[7] Grau-Sanchez, M., Peris, J. M., Gutierrez, J. M., *Accelerated iterative methods for finding solutions of a system of nonlinear equations*, Appl. Math. Comput., 190, (2007), 1815–1823.

[8] Kou, J., *A third-order modification of Newton method for systems of nonlinear equations*, Appl. Math. Comput., 191, (2007), 117–121.

[9] Montazeri, H., Soleymani, F., Shateyi, S., Motsa, S. S., *On a new method for computing the numerical solution of systems of nonlinear equations*, J. Appl. Math., 2012(12), (2012), 919–938.

[10] Noor, M. A., Waseem, M., *Some iterative methods for solving a system of nonlinear equations*, Comput. Math. Appl., 57, (2009), 101–106.

[11] Sharma, J. R., Guha, R. K., Sharma, R., *An efficient fourth order weighted-Newton method for systems of nonlinear equations*, Numer. Algorithms, 62, (2013), 307–323.

[12] Su, Q., *A unified model for solving a system of nonlinear equations*, Appl. Math. Comput., 290, (2016), 46–55.

Chapter 47

Efficient Derivative-Free Methods of Order Six

1. Introduction

The methods are defined for $x_0 \in \Omega$ and each $n = 0, 1, 2, \ldots$ by

$$\begin{aligned}
u_n &= x_n + F(x_n), v_n = x_n - F(x_n), \\
y_n &= x_n - [u_n, v_n; F]^{-1} F(x_n), \qquad (47.1) \\
A_n &= 2[y_n, x_n; F] - [u_n, v_n; F], \\
z_n &= y_n - A_n^{-1} F(y_n), \\
x_{n+1} &= z_n - A_n^{-1} F(z_n), \\
y_n &= x_n - [u_n, v_n; F]^{-1} F(x_n), \qquad (47.2) \\
z_n &= y_n - (3I - 2[u_n, v_n; F]^{-1} [y_n, x_n; F])[u_n, v_n; F]^{-1} F(y_n)
\end{aligned}$$

and

$$x_{n+1} = z_n - (3I - 2[u_n, v_n; F]^{-1} [y_n, x_n; F])[u_n, v_n; F]^{-1} F(z_n),$$

where $[\cdot, \cdot; F] : \Omega \times \Omega \to \mathcal{L}(B)$ is a divided difference of order one. The local convergence order six is established using the Taylor series expansion technique when $B = \mathbb{R}^k$ in [10] and assuming $F^{(6)}$ exists and is bounded. In this chapter only conditions on the operators on the methods are used to show the local convergence. Moreover, the more interesting semi-local convergence (not provided in [10]) is also presented. The technique can be applied to extend the applicability of other methods along the same lines [1]-[9], [11,12,13].

2. Local Analysis

Method (47.1):

The majorant functions $g_i, i = 1,2,3$ are:

$$\begin{aligned} g_1(t) &= \frac{\psi_1(t, f_1(t), f_2(t))}{1-\psi_0(f_1(t), f_2(t))}, \\ q(t) &= \psi_2(t, g_1(t)t, f_1(t), f_2(t)) + \psi(t, g_1(t)t), \\ g_2(t) &= \frac{q(t)g_1(t)}{1-q(t)} \end{aligned}$$

and

$$g_3(t) = \frac{(\psi_1(t, g_1(t)t, g_2(t)t) + \psi_2(t, g_1(t)t, f_1(t), f_2(t)))g_2(t)}{1-q(t)}.$$

The calculations are:

$$\begin{aligned} y_n - x^* &= x_n - x^* - [u_n, v_n; F]^{-1}F(x_n) \\ &= [u_n, v_n; F]^{-1}([u_n, v_n; F] - [x_n, x^*; F])(x_n - x^*), \\ \|y_n - x^*\| &\le \frac{\psi_1(\|x_n - x^*\|, \|u_n - x^*\|, \|v_n - x^*\|)\|x_n - x^*\|}{1-\psi_0(\|u_n - x^*\|, \|v_n - x^*\|)} \\ &\le g_1(\|x_n - x^*\|)\|x_n - x^*\| \le \|x_n - x^*\| < r, \\ \|L^{-1}(A_n - L)\| &\le \psi_2(\|x_n - x^*\|, \|y_n - x^*\|, \|u_n - x^*\|, \|v_n - x^*\|) \\ &\quad + \psi(\|x_n - x^*\|, \|y_n - x^*\|) = q_n < 1, \\ \|A_n^{-1}L\| &\le \frac{1}{1-q_n}, \\ z_n - x^* &= y_n - x^* - A_n^{-1}F(y_n) \\ &= A_n^{-1}(A_n - [y_n, x^*; F])(y_n - x^*), \\ \|z_n - x^*\| &\le \frac{1}{1-q_n}\Big[\psi_2(\|x_n - x^*\|, \|y_n - x^*\|, \|u_n - x^*\|, \|v_n - x^*\|) \\ &\quad + \psi(\|x_n - x^*\|, \|y_n - x^*\|)\Big]\|y_n - x^*\| \\ &\le g_2(\|x_n - x^*\|)\|x_n - x^*\| \le \|x_n - x^*\|, \\ x_{n+1} - x^* &= A_n^{-1}(A_n - [z_n, x^*; F])(z_n - x^*) \end{aligned}$$

and

$$\begin{aligned} \|x_{n+1} - x^*\| &\le \frac{a_n\|z_n - x^*\|}{1-q_n} \le g_3(\|x_n - x^*\|)\|x_n - x^*\| \\ &\le \|x_n - x^*\|, \end{aligned}$$

where

$$\begin{aligned} a_n &= \psi_1(\|x_n - x^*\|, \|y_n - x^*\|, \|z_n - x^*\|) \\ &\quad + \psi_2(\|x_n - x^*\|, \|y_n - x^*\|, \|u_n - x^*\|, \|v_n - x^*\|). \end{aligned}$$

Method (47.2):

The function g_1 is given in mehod (47.1),

$$\begin{aligned}
p(t) &= \psi_0(f_1(t), f_2(t)),\\
g_2(t) &= \left[\frac{\psi_1(g_1(t)t, f_1(t), f_2(t))}{1-p(t)}\right.\\
&\quad \left.+\frac{2\psi_2(t, g_1(t)t, f_1(t), f_2(t))\psi_3(t)}{(1-p(t))^2}\right] g_1(t),
\end{aligned}$$

and

$$\begin{aligned}
g_3(t) &= \left[\frac{\psi_1(g_2(t)t, f_1(t), f_2(t))}{1-p(t)}\right.\\
&\quad \left.+\frac{2\psi_2(t, g_1(t)t, f_1(t), f_2(t))\psi_3(g_2(t)t)}{(1-p(t))^2}\right] g_2(t).
\end{aligned}$$

The motivation for the functions g_2 and g_3 is:

$$\begin{aligned}
z_n - x^* &= y_n - x^* - [u_n, v_n; F]^{-1}F(y_n)\\
&\quad -2[u_n, v_n; F]^{-1}([u_n, v_n; F] - [y_n, x_n; F])[u_n, v_n; F]^{-1}F(y_n),\\
\|z_n - x^*\| &\leq \left[\frac{\psi_1(\|y_n - x^*\|, \|u_n - x^*\|, \|v_n - x^*\|)}{1-p_n}\right.\\
&\quad \left.+\frac{2\psi_2(\|x_n - x^*\|, \|y_n - x^*\|, \|u_n - x^*\|, \|v_n - x^*\|)\psi_3(\|y_n - x^*\|)}{(1-p_n)^2}\right] \|y_n - x^*\|\\
&\leq g_2(\|x_n - x^*\|)\|x_n - x^*\| \leq \|x_n - x^*\|,\\
x_{n+1} - x^* &= z_n - x^* - [u_n, v_n; F]^{-1}F(z_n)\\
&\quad -2[u_n, v_n; F]^{-1}([u_n, v_n; F] - [y_n, x_n; F])[u_n, v_n; F]^{-1F(z_n)}
\end{aligned}$$

and

$$\begin{aligned}
\|x_{n+1} - x^*\| &\leq \left[\frac{\psi_1(\|z_n - x^*\|, \|u_n - x^*\|, \|v_n - x^*\|)}{1-p_n}\right.\\
&\quad \left.+\frac{2\psi_2(\|x_n - x^*\|, \|y_n - x^*\|, \|u_n - x^*\|, \|v_n - x^*\|)\psi_3(\|z_n - x^*\|)}{(1-p_n)^2}\right] \|z_n - x^*\|\\
&\leq g_3(\|x_n - x^*\|)\|x_n - x^*\|\\
&\leq \|x_n - x^*\|.
\end{aligned}$$

3. Semi-local Convergence

Method (47.1)

The majorant sequence for the method (47.1) is defined for $\alpha_0 = 0, \beta_0 \geq \|[u_0, v_0; F]^{-1}F(x_0)\|$ and each $n = 0, 1, 2, \ldots$ by

$$\begin{aligned}
\bar{q}_n &= \Psi_2(f_3(\alpha_n), f_4(\alpha_n), \alpha_n, \beta_n) + \Psi(\alpha_n, \beta_n),\\
\gamma_n &= \beta_n + \frac{\Psi_2(\alpha_n, \beta_n, f_3(\alpha_n), f_4(\beta_n))(\beta_n - \alpha_n)}{1 - \bar{q}_n},\\
d_n &= \Psi_2(\alpha_n, \beta_n, f_3(\alpha_n), f_4(\beta_n))(\beta_n - \alpha_n),\\
h_n &= (1 + \Psi_0(\beta_n, \gamma_n))(\gamma_n - \beta_n) + d_n,\\
\alpha_{n+1} &= \gamma_n + \frac{h_n}{1 - \bar{q}_n},\\
\delta_{n+1} &= (1 + \Psi_0(\alpha_n, \alpha_{n+1}))(\alpha_{n+1} - \alpha_n)\\
&\quad + (1 + \Psi_0(f_3(\alpha_n), f_4(\alpha_n)))(\alpha_{n+1} - \beta_n),\\
\beta_{n+1} &= \alpha_{n+1} + \frac{\delta_{n+1}}{1 - \Psi_0(f_3(\alpha_{n+1}), f_4(\alpha_{n+1}))}.
\end{aligned}$$

The calculations are:

$$\begin{aligned}
F(y_n) &= F(y_n) - F(x_n) + F(x_n)\\
&= ([y_n, x_n; F] - [u_n, v_n; F])(y_n - x_n),\\
\|z_n - y_n\| &\leq \frac{\Psi_2(\|x_n - x_0\|, \|y_n - x_0\|, \|u_n - x_0\|, \|v_n - x_0\|)\|y_n - x_n\|}{1 - \bar{q}_n}\\
&\leq \gamma_n - \beta_n,\\
x_{n+1} - z_n &= -A_n^{-1}(F(z_n) - F(y_n) + F(y_n)),\\
\|x_{n+1} - z_n\| &\leq \frac{(1 + \Psi_0(\beta_n, \gamma_n))(\gamma_n - \beta_n) + d_n}{1 - \bar{q}_n}\\
&\leq \alpha_{n+1} - \gamma_n,\\
F(x_{n+1}) &= F(x_{n+1}) - F(x_n) - [x_{n+1}, x_n; F](x_{n+1} - x_n)\\
&\quad + [u_n, x_n; F](x_{n+1} - y_n),\\
\|L^{-1}F(x_{n+1})\| &\leq (1 + \Psi_0(\alpha_n, \alpha_{n+1}))(\alpha_{n+1} - \alpha_n)\\
&\quad + (1 + \Psi_0(f_3(\alpha_n), f_4(\alpha_n)))(\alpha_{n+1} - \beta_n) = \delta_{n+1},\\
\|y_{n+1} - x_{n+1}\| &\leq \|[u_{n+1}, v_{n+1}; F]^{-1}L\|\|L^{-1}F(x_{n+1})\|\\
&\leq \frac{\delta_{n+1}}{1 - \bar{q}_{n+1}} = \beta_{n+1} - \alpha_{n+1}.
\end{aligned}$$

Method (47.2)

The majorant sequence $\{\alpha_n\}$ is defined for $\alpha_0 = 0, \beta_0 \geq \|[u_0, x_0; F]^{-1}F(x_0)\|$ and each $n = 0, 1, 2, \ldots$ by

$$\begin{aligned} e_n &= \Psi_0(f_3(\alpha_n), f_4(\alpha_n)), \\ \lambda_n &= \Psi_2(\alpha_n, \beta_n, f_3(\alpha_n), f_4(\alpha_n))(\beta_n - \alpha_n), \\ \gamma_n &= \beta_n + \left[\frac{1}{1-e_n} + \frac{2\Psi_2(\alpha_n, \beta_n, f_3(\alpha_n), f_4(\alpha_n))}{(1-e_n)^2}\right]\lambda_n, \\ \alpha_{n+1} &= \gamma_n + \left[\frac{1}{1-e_n} + \frac{2\Psi_2(\alpha_n, \beta_n, f_3(\alpha_n), f_4(\alpha_n))}{(1-e_n)^2}\right]h_n \end{aligned}$$

and

$$\beta_{n+1} = \alpha_{n+1} + \frac{\delta_{n+1}}{1-e_{n+1}}.$$

The calculations are:

$$\begin{aligned} z_n - y_n &= -[u_n, v_n; F]^{-1}F(y_n) \\ &\quad -2[u_n, v_n; F]^{-1}([u_n, v_n; F] - [y_n, x_n; F])[u_n, v_n; F]^{-1}F(y_n), \\ \|z_n - y_n\| &\leq \left[\frac{1}{1-e_n} + \frac{2\Psi_2(\alpha_n, \beta_n, f_3(\alpha_n), f_4(\alpha_n))}{(1-e_n)^2}\right]\lambda_n \\ &= \gamma_n - \beta_n, \\ x_{n+1} - z_n &= -[u_n, v_n; F]^{-1}F(z_n) \\ &\quad -2[u_n, v_n; F]^{-1}([u_n, v_n; F] - [y_n, x_n; F])[u_n, v_n; F]^{-1}F(z_n), \\ \|x_{n+1} - z_n\| &\leq \left[\frac{1}{1-e_n} + \frac{2\Psi_2(\alpha_n, \beta_n, f_3(\alpha_n), f_4(\alpha_n))}{(1-e_n)^2}\right]h_n \\ &\leq \alpha_{n+1} - \gamma_n, \\ \|y_{n+1} - x_{n+1}\| &\leq \frac{\delta_{n+1}}{1-e_{n+1}} = \beta_{n+1} - \alpha_{n+1}. \end{aligned}$$

The rest of the analysis is provided in previous chapters.

4. References

[1] Argyros, I. K., *Convergence and Applications of Newton-type Iterations*, Springer Verlag, Berlin, Germany, (2008).

[2] Argyros, I. K., *The theory and applications of iteration methods*, 2nd Edition, CRC Press, Engineering Series, Taylor and Francis Group, (2022).

[3] Argyros, I. K., George, S., *Mathematical Modeling For The Solution Of Equations And Systems Of Equations With Applications*, Volume-III, Nova Publishes, US, (2019).

[4] Cordero, A., Hueso, J. L., Martnez, E., Torregrosa, J. R., *Increasing the convergence order of an iterative method for nonlinear systems*, Appl. Math. Lett., 25, (2012), 2369–2374.

[5] Darvishi, M. T., Barati, A., *A fourth-order method from quadrature formulae to solve systems of nonlinear equations*, Appl. Math. Comput., 188, (2007), 257–261.

[6] Frontini, M., Sormani, E., *Third-order methods from quadrature formulae for solving systems of nonlinear equations*, Appl. Math. Comput., 149, (2004), 771–782.

[7] Grau-Snchez, M., Noguera, M., Amat, S., *On the approximation of derivatives using divided difference operators preserving the local convergence order of iterative methods*, J. Comput. Appl. Math., 237, (2013), 363–372.

[8] Homeier, H. H. H., *A modified Newton method with cubic convergence: the multivariate case*, J. Comput. Appl. Math., 169, (2004), 161–169.

[9] Petkovi, M. S., Neta, B., Petkovi, L. D., Duni, J., *Multipoint methods for solving nonlinear equations*, Elsevier, Boston, (2013).

[10] Sharma, J. R., Arora, H., *Efficient derivative-free numerical methods for solving systems of nonlinear equations*, Comp. Appl. Math., 35, (2016), 269–284, DOI 10.1007/s40314-014-0193-0.

[11] Sharma, J. R., Arora, H., *Efficient Jarratt-like methods for solving systems of nonlinear equations*, Calcolo, 51, (2014), 193–210. doi:10.1007/s10092-013-0097-1.

[12] Sharma, J. R., Guha, R. K., Sharma, R., *An efficient fourth order weighted-Newton method for systems of nonlinear equations*, Numer. Algorithms, 62, (2013), 307–323.

[13] Steffensen, J. F., *Remarks on iteration*, Skand. Aktuar. Tidskr., 16, (1933), 64–72.

Chapter 48

Methods of Order Six and Seven Without Derivatives

1. Introduction

The methods are defined for $x_0 \in \Omega$ and each $n = 0, 1, 2, \ldots$ by

$$\begin{aligned} u_n &= x_n + F(x_n), v_n = x_n - F(x_n), \\ y_n &= x_n - [u_n, v_n; F]^{-1} F(x_n), \\ z_n &= y_n - \frac{1}{2}[y_n, x_n; F]^{-1} F(y_n), \\ x_{n+1} &= z_n - \frac{1}{2}[y_n, x_n; F]^{-1} F(z_n) \end{aligned} \tag{48.1}$$

and

$$\begin{aligned} y_n &= x_n - [u_n, x_n; F]^{-1} F(x_n), \\ z_n &= h(x_n), \\ A_n &= [z_n, x_n; F] + [z_n, y_n; F] - [y_n, x_n; F] \\ x_{n+1} &= z_n - A_n^{-1} F(z_n), \end{aligned} \tag{48.2}$$

where $[\cdot, \cdot; F] : \Omega \times \Omega \to \mathcal{L}(B)$ is a divided difference of order one and h is any iteration function of order four. The local sixth and seventh order, respectively is determined in [8] when $B = \mathbb{R}^k$ under the Taylor expansion series approach and provided that $F^{(7)}$ and $F^{(8)}$ exist and are bounded. Other limitations of this approach are reported in earlier chapters. The local convergence is extended in this chapter using only the operators on the methods and in the more general setting of a Banach Space. Moreover, the more interesting semi-local convergence not given in [8] is developed based on majorizing sequences. The new approach can be applied on other methods with the same benefits in analogous way [1]-[7], [9].

2. Local Convergence

The conditions used for both methods are:

$$\begin{aligned}
\|L^{-1}([x,y;F]-L)\| &\leq \psi_0(\|x-x^*\|,\|y-x^*\|),\\
\|u-x^*\| &\leq f_1(\|x-x^*\|)\leq \|x-x^*\|,\\
\|v-x^*\| &\leq f_2(\|x-x^*\|)\leq \|x-x^*\|,\\
\|L^{-1}([x,y;F]-[z,x^*;F])\| &\leq \psi(\|x-x^*\|,\|y-x^*\|,\|z-x^*\|),\\
\|L^{-1}([y,x;F]-[y,x^*;F])\| &\leq \psi_1(\|x-x^*\|,\|y-x^*\|)
\end{aligned}$$

and

$$\|L^{-1}([x,x^*;F]-L)\| \leq \psi_2(\|x-x^*\|),$$

where f_1, f_2, ψ, ψ_1 and ψ_2 are as given in earlier chapters.

Method (48.1)

The majorant functions $g_i, i=1,2,3$ are:

$$\begin{aligned}
g_1(t) &= \frac{\psi(t,f_1(t),f_2(t))}{1-\psi_0(f_1(t),f_2(t))},\\
g_2(t) &= \frac{[\psi_1(t,g_1(t)t)+\frac{1}{2}(1+\psi_2(g_1(t)t))]g_1(t)}{1-\psi_0(t,g_1(t)t)}.
\end{aligned}$$

ans

$$g_3(t) = \frac{[\psi(t,g_1(t)t,g_2(t)t)+\frac{1}{2}(1+\psi_2(g_2(t)t))]g_2(t)}{1-\psi_0(t,g_1(t)t)}.$$

The motivation is:

$$\begin{aligned}
y_n-x^* &= [u_n,v_n;F]^{-1}([u_n,v_n;F]-[x_n,x^*;F])(x_n-x^*),\\
\|y_n-x^*\| &\leq \frac{\psi(\|x_n-x^*\|,\|u_n-x^*\|,\|v_n-x^*\|)\|x_n-x^*\|}{1-\psi_0(\|u_n-x^*\|,\|v_n-x^*\|)}\\
&\leq g_1(\|x_n-x^*\|)\|x_n-x^*\|\leq\|x_n-x^*\|<r,\\
z_n-x^* &= [y_n,x_n;F]^{-1}([y_n,x_n;F]-[y_n,x^*;F])(y_n-x^*)\\
&\quad+\frac{1}{2}[y_n,x_n;F]^{-1}F(y_n),\\
\|z_n-x^*\| &\leq \frac{[\psi_1(\|x_n-x^*\|,\|y_n-x^*\|)+\frac{1}{2}(1+\psi_2(\|y_n-x^*\|))]\|y_n-x^*\|}{1-\psi_0(\|x_n-x^*\|,\|y_n-x^*\|)}\\
&\leq g_2(\|x_n-x^*\|)\|x_n-x^*\|\leq\|x_n-x^*\|,\\
x_{n+1}-x^* &= [y_n,x_n;F]^{-1}([y_n,x_n;F]-[z_n,x^*;F])(z_n-x^*)\\
&\quad+\frac{1}{2}[y_n,x_n;F]^{-1}F(z_n)
\end{aligned}$$

and

$$\begin{aligned}\|x_{n+1}-x^*\| &\leq \frac{1}{1-\psi_0(\|x_n-x^*\|,\|y_n-x^*\|)}\\ &\quad \left[\psi(\|x_n-x^*\|,\|y_n-x^*\|,\|z_n-x^*\|)\right.\\ &\quad \left.+\frac{1}{2}(1+\psi_2(\|z_n-x^*\|))\right]\|z_n-x^*\|\\ &\leq g_3(\|x_n-x^*\|)\|x_n-x^*\| \leq \|x_n-x^*\|.\end{aligned}$$

Method (48.2)

The additional conditions is

$$\|h(x)-x^*\| \leq g_2(\|x-x^*\|)\|x-x^*\|.$$

The function g_1 is given in Method (48.1). Then, from the last substep we have in turn:

$$x_{n+1}-x^* = A_n^{-1}(A_n-[z_n,x^*;F])(z_n-x^*)$$

and

$$\begin{aligned}\|x_{n+1}-x^*\| &\leq \frac{1}{1-q_n}[\psi_1(\|x_n-x^*\|,\|z_n-x^*\|)+\psi_0(\|y_n-x^*\|,\|z_n-x^*\|)\\ &\quad +\psi_0(\|x_n-x^*\|,\|y_n-x^*\|)]\|z_n-x^*\|\\ &\leq g_3(\|x_n-x^*\|)\|x_n-x^*\| \leq \|x_n-x^*\|,\end{aligned}$$

where the following estimate is also used:

$$\begin{aligned}\|L^{-1}(A_n-L)\| &\leq \psi_0(\|x_n-x^*\|,\|z_n-x^*\|)\\ &\quad +\psi_0(\|y_n-x^*\|,\|z_n-x^*\|)+\psi_0(\|x_n-x^*\|,\|y_n-x^*\|)\\ &= q_n<1,\end{aligned}$$

thus

$$\|A_n^{-1}L\| \leq \frac{1}{1-q_n}.$$

3. Semi-Local Convergence

The functions $f_1, f_2, \|\psi\|$ are replaced by $f_3, f_4, \|\Psi\|$ and x^* by x_0, respectively.

Method (48.1)

The majorant sequence $\{\alpha_n\}$ is defined for $\alpha_0 = 0, \beta_0 \geq \|[u_0,v_0;F]^{-1}F(x_0)\|$ and each $n = 0,1,2,\ldots$ by

$$\begin{aligned}\lambda_n &= (1+\Psi_0(\alpha_n,\beta_n))(\beta_n-\alpha_n)+2(1+\Psi_0(f_3(\alpha_n),f_4(\beta_n)))(\beta_n-\alpha_n),\\ \gamma_n &= \beta_n+\frac{\lambda_n}{2(1-\Psi_0(\alpha_n,\beta_n))},\\ \mu_n &= (1+\Psi_0(\beta_n,\gamma_n))(\gamma_n-\beta_n)+\lambda_n,\\ \alpha_{n+1} &= \gamma_n+\frac{\mu_n}{2(1-\Psi_0(\alpha_n,\beta_n))},\\ \delta_{n+1} &= (1+\Psi_0(\alpha_n,\alpha_{n+1}))(\alpha_{n+1}-\alpha_n)\\ &\quad +(1+\Psi_0(f_3(\alpha_n),f_4(\alpha_n)))(\beta_n-\alpha_n)\end{aligned}$$

and

$$\beta_{n+1} = \alpha_{n+1} + \frac{\delta_{n+1}}{1 - \Psi_0(f_3(\alpha_{n+1}, f_4(\alpha_{n+1})))}.$$

The motivation follows from the calculations

$$\begin{aligned}
F(y_n) &= F(y_n) - F(x_n) - 2[u_n, v_n; F](y_n - x_n), \\
\|z_n - y_n\| &\leq \frac{(1 + \Psi_0(\alpha_n, \beta_n))(\beta_n - \alpha_n) + 2(1 + \Psi_0(f_3(\alpha_n), f_4(\beta_n)))(\beta_n - \alpha_n)}{2(1 - \Psi_0(\alpha_n, \beta_n))} \\
&= \frac{\lambda_n}{2(1 - \Psi_0(\alpha_n, \beta_n))} = \gamma_n - \beta_n, \\
x_{n+1} - z_n &= -\frac{1}{2}[y_n, x_n; F]^{-1}(F(z_n) - F(y_n) + F(y_n)), \\
\|x_{n+1} - z_n\| &\leq \frac{\mu_n}{2(1 - \Psi_0(\alpha_n, \beta_n))} = \alpha_{n+1} - \gamma_n, \\
F(x_{n+1}) &= F(x_{n+1}) - F(x_n) - [u_n, v_n; F](y_n - x_n), \\
\|L^{-1}F(x_{n+1})\| &\leq \delta_{n+1}, \\
\|y_{n+1} - x_{n+1}\| &\leq \|[u_{n+1}, v_{n+1}; F]^{-1}L\| \|L^{-1}F(x_{n+1})\| \\
&\leq \frac{\delta_{n+1}}{1 - \Psi_0(f_3(\alpha_{n+1}), f_4(\alpha_{n+1}))} = \beta_{n+1} - \alpha_{n+1}.
\end{aligned}$$

Method (48.2)

We impose

$$\|h(x_n) - y_n\| \leq \gamma_n - \beta_n.$$

The majorant sequence $\{\alpha_n\}$ is defined for α_0 and β_0 as in method (48.1). Then, for

$$\begin{aligned}
p_n &= (1 + \Psi_0(\alpha_n, \gamma_n))(\gamma_n - \alpha_n) + (1 + \Psi_0(f_3(\alpha_n), f_4(\alpha_n)))(\beta_n - \alpha_n) \\
\bar{q}_n &= \Psi_0(\alpha_n, \gamma_n) + \Psi_0(\beta_n, \gamma_n) + \Psi_0(\alpha_n, \beta_n), \\
\alpha_{n+1} &= \gamma_n + \frac{p_n}{1 - \bar{q}_n}
\end{aligned}$$

and the iterate β_{n+1} is given in method (48.1).

Notice that

$$F(z_n) = F(z_n) - F(x_n) - [u_n, v_n; F](y_n - x_n),$$

thus

$$\|L^{-1}F(z_n)\| \leq p_n.$$

The rest of the analysis is given in previous Chapters.

4. References

[1] Argyros, I. K., *Convergence and Applications of Newton-type Iterations*, Springer Verlag, Berlin, Germany, (2008).

[2] Argyros, I. K., *The theory and applications of iteration methods*, 2nd Edition, CRC Press, Engineering Series, Taylor and Francis Group, (2022).

[3] Argyros, I. K., George, S., *Mathematical Modeling For The Solution Of Equations And Systems Of Equations With Applications*, Volume-III, Nova Publishes, US, (2019).

[4] Cordero, A., Hueso, J. L., Martnez, E., *Steffensen type methods for solving nonlinear equations*, J. Comput. Appl. Math., 236, (2012), 3058–3064.

[5] Cordero, A., Hueso, J. L., Martnez, E., Torregrosa, J. R., *Efficient high-order methods based on golden ratio for nonlinear systems*, Appl. Math. Comput., 217, (2011), 4548–4556.

[6] Grau-Snchez, M., Grau, A., Noguera, M., *Ostrowski type methods for solving systems of nonlinears*, Appl. Math. Comput., 218, (2011), 2377–2385.

[7] Liu, Z., Zheng, Q., Zhao, P., *A variant of Steffensens method of fourth-order convergence and its applications*, Appl. Math. Comput., 2146, (2010), 1978–1983.

[8] Wang, X., Zhang, T., *A family of Steffensen type methods with seventh-order convergence*, Numer. Algor., 62, (2013), 429–444, DOI 10.1007/s11075-012-9597-3.

[9] Zheng, Q., Zhao, P., Huang, F., *A family of fourth-order Steffensen-type methods with the applications on solving nonlinear ODEs*, Appl. Math. Comput., 217, (2011), 8196–8203.

Chapter 49

Jarratt-Like Methods of Order Six and Eight

1. Introduction

The two three step iterative methods are defined for $x_0 \in \Omega$ and each $n = 0, 1, 2, \ldots$ by

$$\begin{aligned} y_n &= x_n - \frac{2}{3}F'(x_n)^{-1}F(x_n), \\ A_n &= 6F'(y_n) - 2F'(x_n), \\ z_n &= x_n - A_n^{-1}(3F'(y_n) + F'(x_n))F'(x_n)^{-1}F(x_n), \\ B_n &= 2aF'(x_n) + 2bF'(y_n), \\ x_{n+1} &= z_n - B_n^{-1}[(5a+3b)F'(x_n) - (3a+b)F'(y_n)]F'(x_n)^{-1}F(z_n) \end{aligned} \tag{49.1}$$

and

$$\begin{aligned} y_n &= x_n - \frac{2}{3}F'(x_n)^{-1}F(x_n), \\ z_n &= x_n - A_n^{-1}(3F'(y_n) + F'(x_n))F'(x_n)^{-1}F(x_n) \\ x_{n+1} &= z_n - F'(z_n)^{-1}F(z_n), \end{aligned} \tag{49.2}$$

where $a, b \in \mathbb{R} - \{0\}$ and $a + b \neq 0$.

The local orders six and eight are determined in [11] under the Taylor expansion series approach when $B_1 = B_2 = \mathbb{R}^k$ and provided that the seventh and ninth order derivatives exist and are bounded, respectively. The limitations of this approach are well established in previous chapters. The new local convergence analysis uses only the operators on the methods. Moreover, the more interesting semi-local convergence of these methods (not given in [11]) is presented using majorizing sequences. The same technique can be used to expand the applicability of other methods [1]-[10].

2. Local Convergence

Method (49.1)

The majorant functions $g_i, i = 1,2,3$ are:

$$
\begin{aligned}
g_1(t) &= \frac{\int_0^1 w((1-\theta)t)d\theta + \frac{1}{3}\left(1+\int_0^1 w_0(\theta t)d\theta\right)}{1-w_0(t)},\\
q(t) &= \frac{3w_0(g_1(t)t)+w_0(t)}{2},\\
\bar{w}(t) &= \begin{cases} w((1+g_1(t))t) \\ w_0(t)+w_0(g_1(t)t), \end{cases}\\
g_2(t) &= \frac{\int_0^1 w((1-\theta)t)d\theta}{1-w_0(t)} + \frac{3\bar{w}(t)\left(1+\int_0^1 w_0(\theta t)d\theta\right)}{4(1-q(t))(1-w_0(t))},\\
\bar{\bar{w}}(t) &= \begin{cases} w((1+g_2(t))t) \\ w_0(t)+w_0(g_2(t)t), \end{cases}\\
\lambda(t) &= \frac{|a|w_0(t)+|b|w_0(g_1(t)t)}{2},
\end{aligned}
$$

and

$$
\begin{aligned}
g_3(t) &= \left[\frac{\int_0^1 w((1-\theta)g_2(t)t)d\theta}{1-w_0(g_2(t)t)} + \frac{\bar{\bar{w}}(t)\left(1+\int_0^1 w_0(\theta g_2(t)t)d\theta\right)}{(1-w_0(t))(1-w_0(g_2(t)t))}\right.\\
&\quad \left. + \frac{3|a+b|\left(1+\int_0^1 w_0(\theta g_2(t)t)d\theta\right)}{4(1-\lambda(t))(1-w_0(t))}\right] g_2(t).
\end{aligned}
$$

The calculations are:

$$
\begin{aligned}
y_n - x^* &= x_n - x^* - F'(x_n)^{-1}F(x_n) + \frac{1}{3}F'(x_n)^{-1}F(x_n),\\
\|y_n - x^*\| &\le \frac{1}{1-w_0(\|x_n-x^*\|)}\left[\int_0^1 w((1-\theta)\|x_n-x^*\|)d\theta\right.\\
&\quad \left. + \frac{1}{3}\left(1+\int_0^1 w_0(\theta\|x_n-x^*\|)d\theta\right)\right]\|x_n-x^*\|\\
&\le g_1(\|x_n-x^*\|)\|x_n-x^*\| \le \|x_n-x^*\| < r,\\
z_n - x^* &= x_n - x^* - F'(x_n)^{-1}F(x_n)\\
&\quad +[I - A_n^{-1}(3F'(y_n)+F'(x_n))]F'(x_n)^{-1}F(x_n)\\
&= x_n - x^* - F'(x_n)^{-1}F(x_n)\\
&\quad +3A_n^{-1}(F'(y_n)-F'(x_n))F'(x_n)^{-1}F(x_n),
\end{aligned}
$$

$$\begin{aligned}
\|(4L)^{-1}(A_n-4L)\| &\leq \frac{6w_0(\|y_n-x^*\|)+2w_0(\|x_n-x^*\|)}{4}=q_n<1,\\
\|A_n^{-1}L\| &\leq \frac{1}{4(1-q_n)},\\
\|z_n-x^*\| &\leq \left[\frac{\int_0^1 w((1-\theta)\|x_n-x^*\|)d\theta}{1-w_0(\|x_n-x^*\|)}\right.\\
&\left.+\frac{3\bar{w}_n\left(1+\int_0^1 w_0(\theta\|x_n-x^*\|)d\theta\right)}{4(1-q_n)(1-w_0(\|x_n-x^*\|))}\right]\|x_n-x^*\|\\
&\leq g_2(\|x_n-x^*\|)\|x_n-x^*\|\leq\|x_n-x^*\|\\
\|(4L)^{-1}(B_n-4L)\| &\leq \frac{2|a|w_0(\|x_n-x^*\|)+2|b|w_0(\|y_n-x^*\|)}{4}=\lambda_n<1,\\
\|B_n^{-1}L\| &\leq \frac{1}{4(1-\lambda_n)},\\
x_{n+1}-x^* &= z_n-x^*-F'(z_n)^{-1}F(z_n)\\
&+(F'(z_n)^{-1}-F'(x_n)^{-1})F(z_n)\\
&+(I-B_n^{-1}[2aF'(x_n)+2bF'(y_n)\\
&-(5a+3b)F'(x_n)+(3a+b)F'(y_n)])\\
&\times F'(x_n)^{-1}F(z_n)\\
&= z_n-x^*-F'(z_n)^{-1}F(z_n)\\
&+F'(z_n)^{-1}(F'(x_n)-F'(z_n))F'(x_n)^{-1}F(z_n)\\
&+3(a+b)(F'(y_n)-F'(x_n))F'(x_n)^{-1}F(z_n)
\end{aligned}$$

and

$$\begin{aligned}
\|x_n-x^*\| &\leq \left[\frac{\int_0^1 w((1-\theta)\|z_n-x^*\|)d\theta}{1-w_0(\|z_n-x^*\|)}\right.\\
&+\frac{\bar{\bar{w}}_n\left(1+\int_0^1 w_0(\theta\|z_n-x^*\|)d\theta\right)}{(1-w_0(\|y_n-x^*\|))(1-w_0(\|z_n-x^*\|))}\\
&\left.+\frac{3|a+b|\bar{w}_n\left(1+\int_0^1 w_0(\theta\|z_n-x^*\|)d\theta\right)}{4(1-\lambda_n)(1-w_0(\|x_n-x^*\|))}\right]\|z_n-x^*\|\\
&\leq g_3(\|x_n-x^*\|)\|x_n-x^*\|\leq\|x_n-x^*\|.
\end{aligned}$$

Method (49.2)

The functions g_1 and g_2 are as defined in method (49.1). Moreover, the function g_3 is given by

$$g_3(t) = \frac{\int_0^1 w((1-\theta)g_2(t)t)d\theta}{1-w_0(g_2(t)t)}g_2(t).$$

The motivational calculation is

$$x_{n+1}-x^* = z_n-x^*-F'(z_n)^{-1}F(z_n)$$

so

$$\begin{aligned}\|x_{n+1}-x^*\| &\leq \frac{\int_0^1 w((1-\theta)\|z_n-x^*\|)d\theta\|z_n-x^*\|}{1-w_0(\|z_n-x^*\|)}\\ &\leq g_3(\|x_n-x^*\|)\|x_n-x^*\| \leq \|x_n-x^*\|.\end{aligned}$$

3. Semi-Local Convergence

Method (49.1)

The majorant sequence $\{\alpha_n\}$ is defined for $\alpha_0 = 0, \beta_0 \geq \frac{2}{3}\|F'(x_0)^{-1}F(x_0)\|$ and each $n = 0, 1, 2, \ldots$ by

$$\begin{aligned}
\bar{v}_n &= \begin{cases} v(\beta_n-\alpha_n) \\ v_0(\alpha_n)+v_0(\beta_n), \end{cases},\\
\bar{q}_n &= \frac{3v_0(\beta_n)+v_0(\alpha_n)}{2},\\
\gamma_n &= \beta_n + \frac{(3\bar{v}_n+4(1+v_0(\alpha_n)))(\beta_n-\alpha_n)}{8(1-\bar{q}_n)}\\
\bar{\lambda}_n &= \frac{|a|v_0(\alpha_n)+|b|v_0(\beta_n)}{2},\\
\mu_n &= \left(1+\int_0^1 v_0(\alpha_n+\theta(\gamma_n-\alpha_n))d\theta\right)(\gamma_n-\alpha_n)+\frac{3}{2}(1+v_0(\alpha_n))(\beta_n-\alpha_n),\\
\alpha_{n+1} &= \gamma_n + \frac{[|3a+b|\bar{v}_n+2|a+b|(1+v_0(\alpha_n))]\mu_n}{4(1-\bar{\lambda}_n)(1-v_0(\alpha_n))},\\
\delta_{n+1} &= \left(1+\int_0^1 v_0(\alpha_n+\theta(\alpha_{n+1}-\alpha_n))d\theta\right)(\alpha_{n+1}-\alpha_n)\\
&\quad +\frac{3}{2}(1+v_0(\alpha_n))(\beta_n-\alpha_n)
\end{aligned}$$

and

$$\beta_{n+1} = \alpha_{n+1}+\frac{2}{3}\frac{\delta_{n+1}}{1-v_0(\alpha_{n+1})}.$$

The motivational calculations follow as in the local case and they are:

$$
\begin{aligned}
z_n - y_n &= \left[\frac{2}{3}I - A_n^{-1}(3F'(y_n) + F'(x_n))\right]\left(-\frac{3}{2}(y_n - x_n)\right) \\
&= -\frac{1}{2}A_n^{-1}[3(F'(y_n) - F'(x_n)) - 4F'(x_n)](y_n - x_n), \\
\|z_n - y_n\| &\leq \frac{(3\bar{v}_n + 4(1 + v_0(\alpha_n)))(\beta_n - \alpha_n)}{8(1 - \bar{q}_n)} = \gamma_n - \beta_n, \\
F(z_n) &= F(z_n) - F(x_n) - \frac{3}{2}F'(x_n)(y_n - x_n), \\
\|L^{-1}F(z_n)\| &\leq \left(1 + \int_0^1 v_0(\|x_n - x_0\| + \theta\|z_n - x_n\|)d\theta\right)\|z_n - x_n\| \\
&\quad + \frac{3}{2}(1 + v_0(\|x_n - x_0\|))\|y_n - x_n\| = \bar{\mu}_n \leq \mu_n, \\
&\quad (5a + 3b)F'(x_n) - (3a + b)F'(z_n) \\
&= (3a + b)(F'(x_n) - F'(y_n)) \\
&\quad + 2(a + b)F'(y_n),
\end{aligned}
$$

$$
\begin{aligned}
\|x_{n+1} - z_n\| &\leq \frac{[|3a + b|\bar{v}_n + 2|a + b|(1 + v_0(\alpha_n))]\mu_n}{4(1 - \bar{\lambda}_n)(1 - v_0(\alpha_n))} = \alpha_{n+1} - \gamma_n, \\
F(x_{n+1}) &= F(x_{n+1}) - F(x_n) - \frac{3}{2}F'(x_n)(y_n - x_n), \\
\|L^{-1}F(x_{n+1})\| &\leq \left(1 + \int_0^1 v_0(\alpha_n + \theta(\alpha_{n+1} - \alpha_n))d\theta\right)(\alpha_{n+1} - \alpha_n) \\
&\quad + \frac{3}{2}(1 + v_0(\alpha_n))(\beta_n - \alpha_n) = \delta_{n+1} \\
\|y_{n+1} - x_{n+1}\| &\leq \frac{2}{3}\|F'(x_{n+1}^{-1}L)\|\|L^{-1}F(x_{n+1})\| \\
&\leq \frac{2}{3}\frac{\delta_{n+1}}{1 - v_0(\alpha_{n+1})} = \beta_{n+1} - \alpha_{n+1}.
\end{aligned}
$$

Method (49.2)

The iterates γ_n and β_{n+1} are given in method (49.1), whereas

$$
\alpha_{n+1} = \gamma_n + \frac{\mu_n}{1 - v_0(\gamma_n)},
$$

since the motivational calculation is

$$
x_{n+1} - z_n = -F'(z_n)^{-1}F(z_n)
$$

so

$$
\begin{aligned}
\|x_{n+1} - z_n\| &\leq \|F'(z_n)^{-1}L\|\|L^{-1}F(z_n)\| \\
&\leq \frac{\mu_n}{1 - v_0(\gamma_n)} = \alpha_{n+1} - \gamma_n.
\end{aligned}
$$

The rest of the convergence analysis is as given in previous chapters.

4. References

[1] Argyros, I. K., *Convergence and Applications of Newton-type Iterations*, Springer Verlag, Berlin, Germany, (2008).

[2] Argyros, I. K., *The theory and applications of iteration methods*, 2nd Edition, CRC Press, Engineering Series, Taylor and Francis Group, 2022.

[3] Argyros, I. K., George, S., *Mathematical Modeling For The Solution Of Equations And Systems Of Equations With Applications*, Volume-III, Nova Publishes, US, (2019).

[4] Cordero, A., Hueso, J. L., Martnez, E., Torregrosa, J. R., *Efficient three-step iterative methods with sixth order convergence for nonlinear equations*, Numer. Algor., 53, (2010), 485–495.

[5] Darvishi, M. T., Barati, A., *A third-order Newton-type method to solve systems of nonlinear equations*, Appl. Math. Comput., 187, (2007), 630–635.

[6] Grau-Snchez, M., Gutierrez, J. M., *Zero-finders methods derived from Obreshkovs techniques*, Appl. Math. Comput., 215, (2009), 2992–3001.

[7] Kou, J., Li, Y., Wang, X., *A modification of Newton method with third-order convergence*, Appl. Math. Comput., 181, (2006), 1106–1111.

[8] Kung, H. T., Traub, J. F., *Optimal order of one-point and multipoint iteration*, J. Assoc. Comput. Math., 21, (1974), 634–651.

[9] Sharma, J. R., Guha, R. K., *A family of modified Ostrowski methods with accelerated sixth-order convergence*, Appl. Math. Comput., 190, (2007), 111–115.

[10] Sharma, J. R., Sharma, R., *A new family of modified Ostrowskis methods with accelerated eighth order convergence*, Numer. Algor., 54, (2010), 445–458.

[11] Soleymani, F., *Revisit of Jarratt method for solving nonlinear equations*, Numer. Algor., 57, (2011), 377–388, DOI 10.1007/s11075-010-9433-6.

Chapter 50

Extended Newtons Method for Generalized Equations Using the Aubin Property

1. Introduction

A solution $x^* \in B_1$ of the generalized equation

$$g(x) + G(x) \ni 0, \tag{50.1}$$

is approximated by the Newton's method

$$g(x_n) + g'(x_n)(x_{n+1} - x_n) + G(x_{n+1}) \ni 0,\ n = 0, 1, 2, \ldots \tag{50.2}$$

where B_1, B_2 are Banach spaces, $g : B_1 \longrightarrow B_2$ is a continuously differentiable operator, and $G : B_1 \rightrightarrows B_2$ is a set-valued operator with a closed nonempty graph. A plethora of problems can be formulated like equation (50.1)[1]-[4],[6,9,10,14,21]. In this Chapter, the local convergence analysis of Newton's method using majorant conditions and the Aubin property is investigated. This work extends our earlier results in [5,6,7]. In the special case when $G = O$ this Chapter extends the results in [5,6,7]. The same technique can be used to solve other cases of generalized equations with the same benefits [1,2,3,16,21,24,25].

2. Preliminaries

We assume familiarity with the concepts of graph (gph G), the domain (dom), the range (rge) and the inverse G^{-1} of a set-valued operator G. Moreover, we use d, e which are the standard notations for the distance and excess, respectively between two subsets of B_1. Let $R > 0$.

Define the linearization error for some continuously differentiable function $f : [0, R) \longrightarrow (-\infty), +\infty)$

$$e_f(v_1, v_2) = f(v_2) - [f(v_1) + f'(v_1)(v_2 - v_1)$$

for each $v_1, v_2 \in [0, R)$,

$$E_g(u_1, u_2) = g(u_2) - [g(u_2) + g'(u_1)(u_2 - u_1)]$$

for each $u_1, u_2 \in B_1$ and

$$E_{g+G,x^*}(v) = g(x) + f'(x)(v - x^*) + G(v)$$

for each $x, v \in B_1$.

The definition of the Aubin property and the following version of the contraction mapping principle are given in order to make the Chapter as self contained as possible. More information can be found in [10, 11, 12, 13, 16].

Definition 13. Let $\tilde{y} \in B_2$ for $\tilde{x} \in B_1$. Then, the inverse operator G^{-1} of G is said to have the Aubin property at $(\tilde{y}, \tilde{x})$ with modulus c, provided that $\tilde{x} \in G^{-1}(\tilde{y})$, if $gphG^{-1}$ is locally closed at $(\tilde{y}, \tilde{x})$, for $\tilde{x} \in G^{-1}(\tilde{y}$, and there exist constants $a, b > 0$ so that

$$e^{-1}(G^{-1}(y) \cap U[\tilde{x}, a], G^{-1}(y_1)) \leq c\|y - y_1\|$$

for each $y, y_1 \in U[\tilde{y}, b]$.

Theorem 22. [2,3,10,13] Let $\Psi : B_1 \rightrightarrows B_1$ be a set-valued operator and $z \in B_1$. Suppose: There exist scalars $\beta > 0$ and $p \in (0,1)$ so that $gph\Psi \cap (U[z,\beta] \times U[z,\beta])$ is closed and

(i) $d(z, \Psi(z)) \leq \beta(1-p)$

(ii) $e(\Psi(x) \cap U[z,\beta], \Psi(z_1)) \leq p\|x - z_1\|$ for each $x, z_1 \in U[z,\beta]$.

Then, there exists $z^* \in U[z,\beta]$ so that $z^* \in \Psi(z^*)$, i.e. Ψ admits a fixed point in $U[z,\beta]$.

It is worth noticing that the Aubin property is related to the metric regularity [14]. Consequently, the results are provided in terms of metric regularity. But first, we need a relationship between different types of majorant conditions. Let $\bar{\rho} = \sup\{s \in [0,R) : U(x^*, s) \subset B_1\}$.

Definition 14. A function $h_0 : [0,R) \longrightarrow (-\infty, +\infty)$ which is twice continuously differentiable is said to be a center-majorant function for g on $U(x^*, \bar{\rho})$ with modulus $\bar{c}$ if for each $y \in U(x^*, \bar{\rho})$

$$\bar{c}\|g'(y) - g'(x^*)\| \leq h_0'(\|y - x^*\|) - h_0'(0).$$

(A1) The function h_0' is convex and strictly increasing.

(A2) $h_0(0) = 0$ and $h_0'(0) = -1$.

Let $\rho \in [0, \bar{\rho}]$ be such that $\rho = \sup\{s \in [0, \bar{\rho}) : h_0'(s) < 0\}$.

Definition 15. A function $h = h(h_0) : [0,\rho) \longrightarrow (-\infty, +\infty)$ which is twice continuously differentiable is said to be a restricted -majorant function for g on $U(x^*, \rho)$ with modulus $\bar{c}$ if for each $\theta \in [0,1]$, $y \in U(x^*, \rho)$

$$\bar{c}\|g'(y) - g'(x^* + \theta(y - x^*))\| \leq h'(\|y - x^*\|) - h'(\theta\|y - x^*\|).$$

(A3) The function h' is convex and strictly increasing.

(A4) $h(0) = 0$ and $h'(0) = -1$.

Notice that the function h_0 depends on x^* and $\bar{\rho}$, whereas the function h on x^*, ρ and h_0.

Definition 16. A function $h_1 : [0,R) \longrightarrow (-\infty, +\infty)$ which is twice continuously differentiable is said to be a majorant function for g on $U(x^*, \rho)$ with modulus $c > 0$ if for each $\theta \in [0,1]$, $y \in U(x^*, \bar{\rho})$

$$c\|g'(y) - g'(x^* + \theta(y - x^*))\| \leq h_1'(\|y - x^*\|) - h_1'(\theta\|y - x^*\|).$$

(A3)' The function h_1' is convex and strictly increasing.

(A4)' $h_1(0) = 0$ and $h_1'(0) = -1$.

It follows by these definitions that

$$h_0'(s) \leq h_1'(s) \text{ and } h'(s) \leq h_1'(s) \text{ for each } s \in [0, \rho). \tag{50.3}$$

Assume $\bar{c} \leq c$. Thus, the results in the literature using only h_1 (see e.g.[11,15,16] for $G = 0$) can be replaced by the pair (h_0, h) resulting to finer error distances, a larger convergence radius and a more precise and larger uniqueness radius for the solution x^*. These advantages are obtained under the same computational cost, since in practice the computation of the function h_1 requires that of h_0 and h as special cases.

We need the parameters:

$$c_1 = \sup\{s \in [0, \rho) : h_0'(s) < 0\},$$

$$c_2 = \sup\{s \in [0, c_1) : \frac{h(s) - sh'(s)}{sh_0'(s)} < 1\}$$

and

$$c_3 = \sup\{s \in (0, c_2) : h_0(s) < 0\}.$$

Moreover, define the Newton iteration for solving the equation $h(s) = 0$ given by

$$s_0 = 0,\ s_1 = |\frac{s_0 h_0'(s_0) - h_0(s_0)}{h_0'(s_0)}|,\ s_{n+1} = |\frac{s_n h'(s_n) - h(s_n)}{h_0'(s_n)}| \tag{50.4}$$

and for each $n = 1, 2, \ldots$.

Furthermore, define the set-valued operator $\Psi_x : B_1 \rightrightarrows B_1$ by

$$\Psi_x(v) = E_{g+G,x^*}^{-1}(E_g(x, v) - E_g(x^*, v)),$$

where

$$E_{g+G,x^*}^{-1}(w) = \{w_1 \in B_1 : w \in E_{g+G,x^*}(w_1)\}.$$

(A5) The set valued operator $p \longrightarrow E_{g+H,x^*}^{-1}(p)$ possess the Aubin property at zero for x^*, with modulus $\bar{c}$ and related parameters $\bar{a}, \bar{b} > 0$.

Define the radius

$$\rho^* = \min\{\rho, c_2, \bar{a}, \frac{\bar{b}\bar{c}}{2}\} \tag{50.5}$$

and the parameter

$$c_4 = \min\{\rho, c_3, \bar{c}\bar{b}\}.$$

Then, by simply applying Theorem 22 for $\Psi = \Psi_x$, $\beta = c_2$ and $p = |h_0'(\|x^* - x\|)$, where $q(t) = t - \frac{h(t)}{h'(t)}$ and $c_5 = \frac{|q(s)|}{s^2}\|x^* - x\|^2$ for $0 < s < c_1$ and $0 < \|x^* - x\| < s < \rho^*$ and using standard majorization techniques, we arrive at the main local convergence result for the Newton's method (50.2).

Theorem 23. Assume that the conditions (A1)- (A5) are valid with the radius ρ^* as defined in (50.5). Then, for the starter $x_0 \in U(x^*,\rho^*) - \{x^*\}$, the sequence $\{x_n\} \subset U(x^*,\rho^*)$ and converges $Q-$quadratically, to x^* so that

$$\|x^* - x_{n+1}\| \leq \frac{s_{n+1}}{s_n^2}\|x_n - x^*\|^2,$$

where the sequence $\{s_n\}$ is strictly decreasing and convergent to zero, and $\{\frac{s_{n+1}}{s_n^2}\}$ is strictly decreasing. Additionally, if for $\alpha \in (0,\rho)$, $\dfrac{h(\alpha)}{(\alpha h_0'(\alpha))-1} = 1$, then $\rho^* = \alpha$ is the largest convergence radius. Moreover, if the operator defined in (A3) is single valued in $U(0,\bar{b})$, then the sequence $\{x_n\}$ is unique and x^* is the unique solution of the equation (50.2) in the ball $U(x^*,c_4)$.

Special Cases (Lipschitz) Suppose $G = 0$, and define:
$h_1(t) = \frac{\ell_1}{2}t^2 - t$, $h_0(t) = \frac{\ell_0}{2}t^2 - t$ and $h(t) = \frac{\ell}{2}t^2 - t$. Notice that

$$\ell_0 \leq \ell_1 \tag{50.6}$$

and

$$\ell \leq \ell_1. \tag{50.7}$$

Then, using only h_1 the radius is

$$r_1 = \frac{2}{3\ell_1} \tag{50.8}$$

attributed to Rheinboldt and Traub for $G = 0$ [21,24]. In the case h_0 and h_1 are used the radius is

$$r_2 = \frac{2}{2\ell_0 + \ell_1} \tag{50.9}$$

and if h_0 and h or only h are utilized the radius is

$$r_3 = \frac{2}{2\ell_0 + \ell} \text{ or } r = \frac{2}{3\ell}, \tag{50.10}$$

respectively. Notice that in view of (50.6) and (50.7)

$$r_1 \leq r \leq r_3 \tag{50.11}$$

and

$$r_1 \leq r_2. \tag{50.12}$$

The corresponding iteration to (50.8) is

$$s_{n+1,1} = \frac{\ell_1}{2|1 - \ell_1 s_{n,1}|}|s_{n,1}|^2. \tag{50.13}$$

Moreover, under our approach we have

$$s_2 = \frac{\ell_0|s_1|^2}{2|1 - \ell_0 s_1|}, \tag{50.14}$$

and

$$s_{n+1} = \frac{\ell}{2|1-\ell s_n|}|s_n|^2, n = 2,3,\ldots. \tag{50.15}$$

Notice also that $\lim_{n\longrightarrow\infty} s_n \leq \lim_{n\longrightarrow\infty} s_{n+1,1} = 0$ and

$$s_{n+1} \leq s_{n+1,1}.$$

Moreover, the uniqueness ball under h_0 and h or only h is the largest, since $\frac{2}{\ell_1} \leq \frac{2}{\ell_0}$. Similar advantages can be obtained if the Lipschitz are replaced by Smale or Nesterov- Nemirovskii or generalized continuity conditions [5,6,7,26]. Examples where (50.6) and (50.7) hold as strict inequalities can be found in [5,6,7].

3. References

[1] Adly, S., Ngai, H. V., Nguyen, V. V., *Newtons method for solving generalized equations: Kantorovichs and Smales approaches*, J. Math. Anal. Appl., 439(1), 2016, 396–418.

[2] Aragn Artacho, F. J., Belyakov, A., Dontchev, A. L., Lpez, M., *Local convergence of quasi-Newton methods under metric regularity*, Comput. Optim. Appl., 58(1), (2014), 225–247.

[3] Aragn Artacho, F. J., Dontchev, A. L., Gaydu, M., Geoffroy, M. H., Veliov, V. M., *Metric regularity of Newtons iteration*, SIAM J. Control Optim., 49(2), 2011, 339–362.

[4] Argyros, I. K., *Convergence and Applications of Newton-Type Iterations*, Springer-Verlag, New York, (2008).

[5] Argyros, I. K., *The theory and application of iteration methods*, second edition, Engineering series, Boca Raton, Florida, USA, (2022).

[6] Argyros, I. K., George, S., *Mathematical modeling for the solution of equations and systems of equations with applications*, Volume-IV, Nova Publishes, NY, (2020).

[7] Argyros, I. K., George, S., *On the complexity of extending the convergence region for Traubs method*, Journal of Complexity 56, (2022), 101423.

[8] Argyros, I. K., Magrean, A. A., *Iterative methods and their dynamics with applications*, CRC Press, New York, Taylor & Francis, (2017).

[9] Cibulka, R., Dontchev, A., Geoffroy, M. H., *Inexact Newton methods and Dennis-Mo theorems for nonsmooth generalized equations*, SIAM Journal on Control and Optimization, 53(2), (2015), 1003–1019.

[10] Dokov, S. P., Dontchev, A. L., *Robinsons strong regularity implies robust local convergence of Newtons method. In Optimal control (Gainesville, FL, 1997)*, volume 15 of Appl. Optim., pages 116–129. Kluwer Acad. Publ., Dordrecht, (1998).

[11] Dontchev, A. L., *Local convergence of the Newton method for generalized equations*, C. R. Acad. Sci. Paris Sr. I Math., 322(4), (1996), 327–331.

[12] Dontchev, A. L., Rockafellar, R. T., *Convergence of inexact Newton methods for generalized equations*, Math. Program., 139(1–2, Ser. B), (2013), 115–137.

[13] Dontchev, A. L., Rockafellar, R. T., *Implicit functions and solution mappings. Springer Series in Operations Research and Financial Engineering*, Springer, New York, second edition, (2014).

[14] Dontchev, A. L., Rockafellar, R. T., *Newtons method for generalized equations: a sequential implicit function theorem*, Math. Program., 123(1, Ser. B), (2010), 139–159.

[15] Ferreira, O. P., *Local convergence of Newtons method under majorant condition*, J. Comput. Appl. Math., 235(5), (2011), 1515–1522.

[16] Ferreira, O. P., Silva, G. N., *Local convergence analysis of Newtons method for solving strongly regular generalized equations*, J. Math. Anal. Appl., 458(1), (2018), 481–496.

[17] Magrean, A. A., Argyros, I. K., *A contemporary study of iterative methods: Convergence dynamics and applications*, Academic Press, Elsevier, (2018).

[18] Nesterov, Y., Nemirovskii, A., *Interior-point polynomial algorithms in convex programming*, Society for Industrial and Applied Mathematics, (1994).

[19] Ortega, J. M., Rheinboldt, W. C., *Iterative solution of nonlinear equations in several variables*, Academic Press, New York, (1970).

[20] Proinov, P. D., *A new convergence theorem for the Weierstrass method from data at one point*, C. R. Acad. Bulgare Sci., 59, 2 (2006), 131–136.

[21] Proinov, P. D., *Semi-local convergence of two iterative methods for simultaneous computation of polynomial zeros*, C. R. Acad. Bulgare Sci., 59, 7 (2006), 705–712.

[22] Rheinboldt, W. C., *An adaptive continuation process for solving systems of nonlinear equations*, In: Mathematical models and numerical methods (A. N. Tikhonov et al. eds.) pub.3, (1977), 129–142 Banach Center, Warsaw Poland.

[23] Sharma, J. R., Arora, A., *An efficient derivative-free numerical method for solving systems of nonlinear equations*, Appl. Anal. and Disc. Math., 7(2), (2013), 390–403.

[24] Sharma, J. R., Guha, R. K., Sharma, R., *An efficient fourth order weighted Newton method for systems of nonlinear equations*, Numer. Algor., 62, (2013), 307–323.

[25] Traub, J., *Iterative methods for solution of equations*, Prentice-Hall, Englewood Cliffs, New Jersey, USA, 1964.

[26] Wang, J., Ouyang, W., *Newtons method for solving generalized equations without Lipschitz condition*, J. Optim. Theory Appl., 192(2), (2022), 510–532.

Chapter 51

Novel Comparable Sixth- and Seventh-Order Methods

1. Introduction

The three step order methods are defined for $x_0 \in \Omega$ and each $n = 0, 1, 2, \ldots$ by

$$\begin{aligned} y_n &= x_n - F'(x_n)^{-1}F(x_n), \\ z_n &= y_n - (2I - F'(x_n)^{-1}F'(y_n))F'(x_n)^{-1}F(y_n), \\ x_{n+1} &= z_n - \left(\frac{7}{2}I - F'(x_n)^{-1}F'(y_n)(4I - \frac{3}{2}F'(x_n)^{-1}F'(y_n)\right)F'(x_n)^{-1}F(z_n) \end{aligned} \tag{51.1}$$

$$\begin{aligned} y_n &= x_n - F'(x_n)^{-1}F(x_n), \\ z_n &= x_n - 2(F'(x_n) + F'(y_n))^{-1}F(x_n), \\ x_{n+1} &= z_n\left[\frac{7}{2}I - 4F'(x_n)^{-1}F'(y_n) + \frac{3}{2}(F'(x_n)^{-1}F'(y_n))^2\right]F'(x_n)^{-1}F(z_n), \end{aligned} \tag{51.2}$$

and

$$\begin{aligned} y_n &= x_n - F'(x_n)^{-1}F(x_n), \\ z_n &= y_n - F'(x_n)^{-1}(2I - F'(x_n)^{-1}F'(y_n))F(y_n) \\ x_{n+1} &= z_n - F'(y_n)^{-1}F(z_n). \end{aligned} \tag{51.3}$$

The local convergence order seven, six and six, are determined in [3,6,8], respectively using Taylor series expansion approach for $B_1 = B_2 = \mathbb{R}^K$. The limitations of this approach are provided in previous chapters. The local convergence is using only the operators on these methods in a Banach space setting. Moreover, the more interesting semi-local convergence analysis (not given in [3,6,8]) is presented in this chapter based on majorizing sequences. The methodology can be used to extend the applicability of other methods [1,2,4,5,7,9,10,11,12].

2. Local Convergence

Method (51.1)

The majorant sequences g_i, $i = 1,2,3$ are:

$$g_1(t) = \frac{\int_0^1 w((1-\theta)t)d\theta}{1-w_0(t)},$$

$$\bar{w}(t) = \begin{cases} w((1+g_1(t))t) \\ or \\ w_0(t)+w_0(g_1(t)t), \end{cases}$$

$$g_2(t) = \left[\frac{\int_0^1 w((1-\theta)g_1(t)t)d\theta}{1-w_0(g_1(t)t)} + \frac{\bar{w}(t)(1+\int_0^1 w_0(\theta g_1(t)t)d\theta)}{(1-w_0(t))(1-w_0(g_1(t)t)} + \frac{\bar{w}(t)(1+\int_0^1 w_0(\theta g_2(t)t)d\theta)}{(1-w_0(t))^2}\right] g_1(t)$$

$$\bar{\bar{w}}(t) = \begin{cases} w((1+g_2(t))t) \\ or \\ w_0(t)+w_0(g_1(t)t), \end{cases}$$

and

$$g_3(t) = \left[\frac{\int_0^1 w((1-\theta)g_2(t)t)d\theta}{1-w_0(g_2(t)t)} + \frac{\bar{\bar{w}}(t)(1+\int_0^1 w_0(\theta g_2(t)t)d\theta)}{(1-w_0(t))(1-w_0(g_2(t)t)} + \frac{1}{2}\frac{\bar{w}(t)(1+\int_0^1 w_0(\theta g_2(t)t)d\theta)}{(1-w_0(t))^2}\left(2+\frac{3\bar{w}(t)}{1-w_0(t)}\right)\right] g_2(t).$$

The calculations motivating the functions g_i are in turn:

$$\begin{aligned}
y_n - x^* &= x_n - x^* - F'(x_n)^{-1}F(x_n),\\
\|y_n - x^*\| &\leq \frac{\int_0^1 w((1-\theta)\|x_n-x^*\|)d\theta\|x_n-x^*\|}{1-w_0(\|x_n-x^*\|)}\\
&\leq g_1(\|x_n-x^*\|)\|x_n-x^*\| \leq \|x_n-x^*\| < r,\\
z_n - x^* &= y_n - x^* - F'(y_n)^{-1}F(y_n) + (F'(y_n)^{-1} - F'(x_n)^{-1})F(y_n)\\
&\quad -F'(x_n)^{-1}(F'(x_n)-F'(y_n))F'(x_n)^{-1}F(y_n),\\
\|z_n - x^*\| &\leq \left[\frac{\int_0^1 w((1-\theta)\|y_n-x^*\|)d\theta}{1-w_0(\|y_n-x^*\|)} + \frac{\bar{w}_n(1+\int_0^1 w_0(\theta\|y_n-x^*\|)d\theta)}{(1-w_0(\|x_n-x^*\|))(1-w_0(\|y_n-x^*\|))}\right.\\
&\quad \left. + \frac{\bar{w}_n(1+\int_0^1 w_0(\theta\|y_n-x^*\|)d\theta)}{(1-w_0(\|x_n-x^*\|))^2}\right]\|y_n-x^*\| \leq g_2(\|x_n-x^*\|)\|x_n-x^*\|\\
&\leq \|x_n-x^*\|,
\end{aligned}$$

$$\begin{aligned}
x_{n+1} - x^* &= z_n - x^* - F'(z_n)^{-1}F(z_n) + (F'(z_n)^{-1} - F'(x_n)^{-1})F(z_n)\\
&\quad -\frac{1}{2}(5I - 8F'(x_n)^{-1}F'(y_n) + 3(F'(x_n)^{-1}F'(y_n))^2)F'(x_n)^{-1}F(z_n)\\
&= z_n - x^* - F'(z_n)^{-1}F(z_n) + (F'(z_n)^{-1} - F'(x_n)^{-1})F(z_n)\\
&\quad \frac{1}{2}(F'(x_n)^{-1}F'(y_n) - I)[3(F'(x_n)^{-1}F'(y_n) - I) - 2I]F'(x_n)^{-1}F(z_n)
\end{aligned}$$

and

$$\begin{aligned}\|x_{n+1}-x^*\| &\le \left[\frac{\int_0^1 w((1-\theta)\|z_n-x^*\|)d\theta}{1-w_0(\|z_n-x^*\|)}+\frac{\bar{\bar{w}}_n(1+\int_0^1 w_0(\theta\|z_n-x^*\|)d\theta)}{(1-w_0(\|x_n-x^*\|))(1-w_0(\|z_n-x^*\|))}\right.\\ &\left.+\frac{1}{2}\frac{\bar{w}_n(1+\int_0^1 w_0(\theta\|z_n-x^*\|)d\theta)}{(1-w_0(\|x_n-x^*\|))^2}\left(2+\frac{3\bar{w}_n}{1-w_0(\|x_n-x^*\|)}\right)\right]\|z_n-x^*\|\\ &\le g_3(\|x_n-x^*\|)\|x_n-x^*\|\le\|x_n-x^*\|.\end{aligned}$$

Method (51.2)
The calculations for the functions g_i and g_3 are given in method (51.1). Moreover, for the function g_2 we have in turn:

$$\begin{aligned}&z_n-x^*\\ =\ & x_n-x^*-F'(x_n)^{-1}F(x_n)+[F'(x_n)^{-1}-2(F'(x_n)+F'(y_n))^{-1}]F(x_n)\\ =\ & x_n-x^*-F'(x_n)^{-1}F(x_n)+F'(x_n)^{-1}(F'(y_n)-F'(x_n))(F'(x_n)+F'(y_n))^{-1}F(x_n)\end{aligned}$$

and

$$\|z_n-x^*\| \le \left[\frac{\int_0^1 w((1-\theta)\|x_n-x^*\|)d\theta}{1-w_0(\|x_n-x^*\|)}+\frac{\bar{w}_n(1+\int_0^1 w_0(\theta\|x_n-x^*\|)d\theta)}{2(1-q_n)(1-w_0(\|x_n-x^*\|))}\right]\|x_n-x^*\|,$$

where we also used

$$\begin{aligned}&\|(2L)^{-1}(F'(x_n)+F'(y_n)-2L)\|\\ \le\ & \frac{p}{2}(w_0(\|x_n-x^*\|)+w_0(\|y_n-x^*\|))=q_n<1,\end{aligned}$$

$$\|(F'(x_n)+F'(y_n))^{-1}L\|\le\frac{1}{2(1-q_n)}.$$

Thus, we choose

$$g_2(t) = \frac{\int_0^1 w((1-\theta)t)d\theta}{1-w_0(t)}+\frac{\bar{w}(t)(1+\int_0^1 w_0(\theta t)d\theta)}{2(1-w_0(t))(1-q(t))},$$

where

$$q(t)=\frac{1}{2}(w_0(t)+w_0(g_1(t)t)).$$

Method (51.3)
The functions g_1, g_2 are given as in method (51.1), whereas

$$\tilde{w}(t) = \begin{cases} w((g_2(t)+g_3(t))t)\\ or\\ w_0(g_2(t)t)+w_0(g_3(t)t),\end{cases}$$

and

$$g_3(t) = \left[\frac{\int_0^1 w((1-\theta)g_2(t)t)d\theta}{1-w_0(g_2(t)t)}+\frac{\tilde{w}(t)(1+\int_0^1 w_0(\theta g_2(t)t)d\theta)}{(1-w_0(g_1(t)t))(1-w_0(g_2(t)t))}\right]g_2(t).$$

The calculations for the function g_3 are:

$$x_{n+1}-x^* = z_n-x^*-F'(z_n)^{-1}F(z_n)+(F'(z_n)^{-1}-F'(y_n)^{-1})F(z_n)$$

and

$$\begin{aligned}\|x_{n+1}-x^*\| &\leq \Bigg[\frac{\int_0^1 w((1-\theta)\|z_n-x^*\|)d\theta}{1-w_0(\|z_n-x^*\|)} \\ &+\frac{\tilde{w}_n(1+\int_0^1 w_0(\theta\|z_n-x^*\|)d\theta)}{(1-w_0(\|y_n-x^*}))(1-w_0(\|z_n-x^*\|))\Bigg]\|z_n-x^*\| \\ &\leq g_3(\|x_n-x^*\|)\|x_n-x^*\| \leq \|x_n-x^*\|,\end{aligned}$$

where

$$\tilde{w}(t) = \begin{cases} w(\|z_n-y_n\|) \\ or \\ w_0(\|y_n-x^*\|)+w_0(\|z_n-x^*\|).\end{cases}$$

3. Semi-local Convergence

The computations are similar to the ones in the local case but x^*, w are replaced by x_0, v respectively.

Method (51.3)

The majorant sequence $\{\alpha_n\}$ is defined for $\alpha_0 = 0, \beta_0 \geq \|F'(x_0)^{-1}F(x_0)\|$ and each $n = 0,1,2,..$ by

$$\begin{aligned}\bar{v}(t) &= \begin{cases} v(\beta_n-\alpha_n) \\ or \\ v_0(\alpha_n)+v_0(\beta_n).\end{cases} \\ \lambda_n &= \int_0^1 v((1-\theta)(\beta_n-\alpha_n))(\beta_n-\alpha_n) \\ \gamma_n &= \beta_n+\frac{1}{1-v_0(\alpha_n)}\left(1+\frac{\bar{v}_n}{1-v_0(\alpha_n)}\right)\lambda_n, \\ \mu_n &= (1+\int_0^1 v_0(\beta_n+\theta(\gamma_n-\beta_n))d\theta)(\gamma_n-\beta_n)+\lambda_n, \\ \alpha_{n+1} &= \gamma_n+\frac{1}{2}\left(1+\frac{2\bar{v}_n}{1-v_0(\alpha_n)}+3\left(\frac{\bar{v}_n}{1-v_0(\alpha_n)}\right)^2\right)\frac{\mu_n}{1-v_0(\alpha_n)} \\ \delta_{n+1} &= (1+\int_0^1 v_0(\alpha_n+\theta(\alpha_{n+1}-\alpha_n))d\theta)(\alpha_{n+1}-\alpha_n) \\ &+(1+v_0(\alpha_n))(\beta_n-\alpha_n)\end{aligned}$$

and

$$\beta_{n+1} = \alpha_{n+1}+\frac{\delta_{n+1}}{1-v_0(\alpha_{n+1})}.$$

The motivational calculations are in turn:

$$\begin{aligned}
F(y_n) &= F(y_n)-F(x_n)-F'(x_n)(y_n-x_n),\\
\|L^{-1}F(y_n)\| &\leq \int_0^1 v((1-\theta)(\beta_n-\alpha_n))d\theta(\beta_n-\alpha_n)=\lambda_n\\
z_n-y_n &= -F'(x_n)^{-1}F(y_n)-F'(x_n)^{-1}(F'(x_n)-F'(y_n))F'(x_n)^{-1}F(y_n),\\
\|z_n-y_n\| &\leq \left[\frac{1}{1-v_0(\alpha_n)}+\frac{\bar{v}_n}{(1-v_0(\alpha_n))^2}\right]\lambda_n=\gamma_n-\beta_n,\\
F(z_n) &= F(z_n)-F(y_n)+F(y_n),\\
\|L^{-1}F(z_n)\| &\leq (1+\int_0^1 v_0(\beta_n+\theta)\gamma_n-\beta_n))d\theta)(\gamma_n-\beta_n)\\
&\quad +\lambda_n=\mu_n,
\end{aligned}$$

$$\begin{aligned}
\|x_{n+1}-z_n\| &\leq \frac{1}{2}[1+2\|F'(x_n)^{-1}F'(x_n)-I\|+3\|F'(x_n)^{-1}F'(y_n)-I\|^2]\|F'(x_n)^{-1}F(z_n)\|\\
&\leq \frac{1}{2}\left[1+\frac{2\bar{v}_n}{1-v_0(\alpha_n)}+3\left(\frac{\bar{v}_n}{1-v_0(\alpha_n)}\right)^2\right]\frac{\mu_n}{1-v_0(\alpha_n)}=\alpha_{n+1}-\alpha_n,
\end{aligned}$$

$$\begin{aligned}
F(x_{n+1}) &= F(x_{n+1})-F(x_n)-F'(x_n)(x_{n+1}-x_n)+F'(x_n)(y_n-x_n),\\
\|L^{-1}F(x_{n+1})\| &\leq \int_0^1 v((1-\theta)\|x_{n+1}-x_n\|)d\theta\|x_{n+1}-x_n\|\\
&\quad +(1+v_0(\|x_n-x_0\|))\|y_n-x_n\|=\bar{\delta}_{n+1}\leq\delta_{n+1}
\end{aligned}$$

and

$$\begin{aligned}
\|y_{n+1}-x_{n+1}\| &\leq \|F'(x_{n+1})^{-1}L\|\|L^{-1}F(x_{n+1})\|\\
&\leq \frac{\bar{\delta}_{n+1}}{1-v_0(\|x_{n+1}-x_0\|)}\leq\frac{\delta_{n+1}}{1-v_0(\alpha_{n+1})}=\beta_{n+1}-\alpha_{n+1}.
\end{aligned}$$

Method (51.2)

The iterates β_{n+1} and α_{n+1} are given in method (51.1) whereas

$$\gamma_n = \beta_n+\frac{\bar{v}_n(\beta_n-\alpha_n)}{2(1-\bar{q}_n)}.$$

The calculations are in turn

$$\begin{aligned}
z_n-y_n &= [F'(x_n)^{-1}-2(F'(x_n)+F'(y_n))^{-1}]F(x_n)\\
&= -(F'(x_n)+F'(y_n))^{-1}(F'(x_n)-F'(y_n)))F'(x_n)^{-1}F(x_n)\\
&= (F'(x_n)+F'(y_n))^{-1}(F'(x_n)-F'(y_n)))(y_n-x_n)
\end{aligned}$$

and

$$\|z_n-y_n\| \leq \frac{\bar{v}_n(\beta_n-\alpha_n)}{2(1-q_n)}=\gamma_n-\beta_n,$$

where

$$\bar{q}_n=\frac{1}{2}(v_0(\alpha_n)+v_0(\beta_n)).$$

Method (51.3)

The iterates γ_n and β_{n+1} are given in method (51.1), whereas

$$\alpha_{n+1} = \gamma_n + \frac{\mu_n}{1-\nu_0(\beta_n)}.$$

The calculations are

$$\|x_{n+1}-z_n\| \leq \|F'(y_n)^{-1}L\|\|L^{-1}F(z_n)\| \leq \frac{\mu_n}{1-\nu_0(\beta_n)} = \alpha_{n+1}-\gamma_n.$$

The rest of the analysis is as given in previous chapters.

4. References

[1] Argyros, I. K., George, S., *High convergence order Q-step methods for solving equations and systems of equations*, Contemp. Math., 1(3), (2020), 119–126, http://dx.doi.org/10.37256/cm.132020403.

[2] Artidiello, S., Cordero, A., Torregrosa, J. R., Vassileva, M. P., *Multidimensional generalization of iterative methods for solving nonlinear problems by means of weight-function procedure*, Appl. Math. Comput., 268, (2015), 1064–1071.

[3] Behl, R., Arora, H., *A novel scheme having seventh order convergence for nonlinear systems*, J. Comput. Appl. Math., 404, (2022), 13301–113301.

[4] Behl, R., Sarria, I., Gonzlez, R., Magrean, A. A., *Highly efficient family of iterative methods for solving nonlinear models*, J. Comput. Appl. Math., 346, (2019), 110–132.

[5] Chicharro, F. I., Cordero, A., Garrido, N., Torregrosa, J. R., *On the improvement of the order of convergence of iterative methods for solving nonlinear systems by means of memory*, Appl. Math. Lett., 104, (2020), 106277, http://dx.doi.org/10.1016/j.aml.2020.106277.

[6] Cordero, A., Hueso, J. L., Martnez, E., Torregrosa, J. R., *Increasing the convergence order of an iterative method for nonlinear systems*, Appl. Math. Lett., 25, (2012), 2369–2374.

[7] Grau-Snchez, M., Grau, ., Noguera, M., *On the computational efficiency index and some iterative methods for solving systems of nonlinear equations*, J. Comput. Appl. Math., 236, (2011), 1259–1266.

[8] Lotfi, T., Bakhtiari, P., Cordero, A., Mahdiani, K., Torregrosa, J. R., *Some new efficient multipoint iterative methods for solving nonlinear systems of equations*, Int. J. Comput. Math., 92, (2015), 1921–1934.

[9] Sharma, J. R., Guna, R. K., Sharma, R., *An efficient fourth order weighted-Newton method for systems of nonlinear equations*, Numer. Algorithms, 2, (2013), 307–323.

[10] Sharma, J. R., Kumar, D., *A fast and efficient composite Newton-Chebyshev method for systems of nonlinear equations*, J. Complexity, 49, (2018), 56–73.

[11] Sharma, J. R., Kumar, D., Jntschi, L., *On a reduced cost higher order Traub-Steffensen-Like method for nonlinear systems*, Symmetry, 11(7), (2019), 891, http://dx.doi.org/10.3390/sym11070891.

[12] Wang, X., Zhang, T., *A family of Steffensen type methods with seventh-order convergence*, Numer. Algorithms, 62, (2013), 429–444.

Chapter 52

Comparable Eighth-Order Methods

1. Introduction

The studied methods are defined for $x_0 \in \Omega$ and each $n = 0, 1, 2, \ldots$ by

$$\begin{aligned} y_n &= x_n - F'(x_n)^{-1}F(x_n), \\ z_n &= y_n - \left[\frac{1}{4}I + \frac{1}{2}F'(y_n)^{-1}F'(x_n) + \frac{1}{4}(F'(y_n)^{-1}F'(x_n))^2\right]F'(x_n)^{-1}F(y_n) \\ x_{n+1} &= z_n - \frac{1}{2}(I + (F'(y_n)^{-1}F'(x_n))^2)F'(x_n)^{-1}F(z_n), \end{aligned} \tag{52.1}$$

$$\begin{aligned} y_n &= x_n - F'(x_n)^{-1}F(x_n), \\ z_n &= y_n - \left[\frac{5}{4}I - \frac{1}{4}F'(y_n)^{-1}F'(x_n) + \frac{1}{4}(F'(y_n)^{-1}F'(x_n))^2\right]F'(x_n)^{-1}F(y_n), \\ x_{n+1} &= z_n - \left[\frac{3}{2}I - F'(y_n)^{-1}F'(x_n) + \frac{1}{2}(F'(y_n)^{-1}F'(x_n))^2\right]F'(y_n)^{-1}F(z_n) \end{aligned} \tag{52.2}$$

and

$$\begin{aligned} y_n &= x_n - F'(x_n)^{-1}F(x_n),\ A_n = F'(x_n)^{-1}(F(x_n) - F'(y_n)), \\ z_n &= y_n - \left[I + \left(I + \frac{5}{4}A_n\right)A_n\right]F'(x_n)^{-1}F(y_n), \\ x_{n+1} &= z_n - \left[I + \left(I + \frac{3}{2}A_n\right)A_n\right]F'(x_n)^{-1}F(z_n). \end{aligned} \tag{52.3}$$

The local convergence order eight is shown in [6,9,21], respectively if $B_1 = B_2 = \mathbb{R}^k$ under the Taylor series expansion approach and by assuming $F^{(9)}$ (not on the methods) exists and is bounded. Other limitations of this approach are reported in previous chapters. The new local convergence is given in a Banach space setting and uses only the operator on the methods. Moreover, the more interesting semi-local convergence (not given in [6,9,21]) is provided using majorizing sequences. The new technique can be used to extend the applicability of other methods [1]-[5], [7, 8],[10]-[20].

2. Local Convergence

Method (52.1)
The majorant sequences g_i, $i = 1,2,3$ are:

$$\begin{aligned}
g_1(t) &= \frac{\int_0^1 w((1-\theta)t)d\theta}{1-w_0(t)},\\
\bar{w}(t) &= \begin{cases} w((1+g_1(t))t) \\ w_0(t)+w_0(g_1(t)t), \end{cases}\\
g_2(t) &= \Bigg[\frac{\int_0^1 w((1-\theta)g_1(t)t)d\theta}{1-w_0(g_1(t)t)} + \frac{\bar{w}(t)(1+\int_0^1 w_0(\theta g_1(t)t)d\theta)}{(1-w_0(t))(1-w_0(g_1(t)t)}\\
&\quad +\frac{1}{4}\frac{\bar{w}(t)}{(1-w_0(g_1(t)t))}\left(4+\frac{\bar{w}(t)}{(1-w_0(g_1(t)t))}\right)\frac{(1+\int_0^1 w_0(\theta g_1(t)t)d\theta)}{1-w_0(t)}\Bigg] g_1(t)\\
\bar{\bar{w}}(t) &= \begin{cases} w((1+g_2(t))t) \\ w_0(t)+w_0(g_2(t)t), \end{cases}
\end{aligned}$$

and

$$\begin{aligned}
g_3(t) &= \Bigg[\frac{\int_0^1 w((1-\theta)g_2(t)t)d\theta}{1-w_0(g_2(t)t)} + \frac{\bar{\bar{w}}(t)(1+\int_0^1 w_0(\theta g_2(t)t)d\theta)}{(1-w_0(t))(1-w_0(g_2(t)t)}\\
&\quad +\frac{1}{2}\frac{\bar{w}(t)}{(1-w_0(g_1(t)t))}\left(2+\frac{\bar{w}(t)}{(1-w_0(g_1(t)t))}\right)\frac{(1+\int_0^1 w_0(\theta g_2(t)t)d\theta)}{1-w_0(t)}\Bigg] g_2(t)
\end{aligned}$$

The calculations motivating the functions g_i are in turn:

$$\begin{aligned}
y_n - x^* &= x_n - x^* - F'(x_n)^{-1}F(x_n),\\
\|y_n - x^*\| &\leq \frac{\int_0^1 w((1-\theta)\|x_n-x^*\|)d\theta}{1-w_0(\|x_n-x^*\|)}\\
&\leq g_1(\|x_n-x^*\|)\|x_n-x^*\| \leq \|x_n-x^*\| < r,\\
z_n - x^* &= y_n - x^* - F'(y_n)^{-1}F(y_n) + (F'(y_n)^{-1} - F'(x_n)^{-1})F(y_n)\\
&\quad -\frac{1}{4}[-3I + 2F'(y_n)^{-1}F'(x_n) + (F'(y_n)^{-1}F'(x_n))^2]F'(x_n)^{-1}F(y_n)\\
&= y_n - x^* - F'(y_n)^{-1}F(y_n) + (F'(y_n)^{-1} - F'(x_n)^{-1})F(y_n)\\
&\quad -\frac{1}{4}(F'(y_n)^{-1}F'(x_n) - I)[4I + (F'(y_n)^{-1}F'(x_n) - I)]F'(x_n)^{-1}F(y_n)
\end{aligned} \tag{52.4}$$

$$\begin{aligned}
\|z_n - x^*\| &\leq \Bigg[\frac{\int_0^1 w((1-\theta)\|y_n-x^*\|)d\theta}{1-w_0(\|y_n-x^*\|)}\\
&\quad +\frac{\bar{w}_n(1+\int_0^1 w_0(\theta\|y_n-x^*\|)d\theta)}{(1-w_0(\|x_n-x^*\|))(1-w_0(\|y_n-x^*\|))}\\
&\quad +\frac{1}{4}\frac{\bar{w}_n}{(1-w_0(\|y_n-x^*\|))}\left(4+\frac{\bar{w}_n}{(1-w_0(\|y_n-x^*\|))}\right)\\
&\quad \times\frac{(1+\int_0^1 w_0(\theta\|y_n-x^*\|)d\theta}{1-w_0(\|x_n-x^*\|)}\Bigg]\|y_n-x^*\|\\
&\leq g_2(\|x_n-x^*\|)\|x_n-x^*\| \leq \|x_n-x^*\|,
\end{aligned}$$

$$\begin{aligned}
x_{n+1} - x^* &= z_n - x^* - F'(z_n)^{-1}F(z_n) + (F'(z_n)^{-1} - F'(x_n)^{-1})F(z_n)\\
&\quad +\frac{1}{2}(I - (F'(y_n)^{-1}F'(x_n))^2)F'(x_n)^{-1}F(z_n)
\end{aligned}$$

and

$$\begin{aligned}\|x_{n+1}-x^*\| &\le \left[\frac{\int_0^1 w((1-\theta)\|z_n-x^*\|)d\theta}{1-w_0(\|z_n-x^*\|)}\right.\\ &+\frac{\bar{\bar{w}}_n(1+\int_0^1 w_0(\theta\|z_n-x^*\|)d\theta)}{(1-w_0(\|x_n-x^*\|))(1-w_0(\|z_n-x^*\|))}\\ &+\frac{1}{2}\frac{\bar{w}_n}{(1-w_0(\|y_n-x^*\|))}\left(2+\frac{\bar{w}_n}{1-w_0(\|y_n-x^*\|)}\right)\\ &\left.\times\frac{(1+\int_0^1 w_0(\theta\|z_n-x^*\|)d\theta)}{1-w_0(\|x_n-x^*\|)}\right]\|z_n-x^*\|\\ &\le g_3(\|x_n-x^*\|)\|x_n-x^*\|\le\|x_n-x^*\|.\end{aligned}$$

Method(52.2)

The function g_1 is as given in method (52.1). Moreover, we have for the other two functions in turn

$$\begin{aligned}z_n-x^* &= y_n-x^*-F'(y_n)^{-1}F(y_n)\\ &-\frac{1}{4}(I-2F'(y_n)^{-1}F'(x_n)+(F'(y_n)^{-1}F'(x_n))^2)F'(y_n)^{-1}F(y_n),\\ \|z_n-x^*\| &\le \left[\frac{\int_0^1 w((1-\theta)\|y_n-x^*\|)d\theta}{1-w_0(\|y_n-x^*\|)}\right.\\ &\left.+\frac{1}{4}\left(\frac{\bar{w}_n}{1-w_0(\|y_n-x^*\|)}\right)^2\frac{(1+\int_0^1 w_0(\theta\|y_n-x^*\|)d\theta)}{1-w_0(\|y_n-x^*\|)}\right]\|y_n-x^*\|\\ &\le g_2(\|x_n-x^*\|)\|x_n-x^*\|\le\|x_n-x^*\|,\\ x_{n+1}-x^* &= z_n-x^*-F'(z_n)^{-1}F(z_n)\\ &-\frac{1}{2}[I-2F'(y_n)^{-1}F'(x_n)+(F'(y_n)^{-1}F'(x_n))^2]F'(y_n)^{-1}F(z_n)\\ &+(F'(z_n)^{-1}-F'(y_n)^{-1})F(z_n),\\ \|x_{n+1}-x^*\| &\le \left[\frac{\int_0^1 w((1-\theta)\|z_n-x^*\|)d\theta}{1-w_0(\|z_n-x^*\|)}\right.\\ &+\frac{\bar{\bar{w}}_n(1+\int_0^1 w_0(\theta\|z_n-x^*\|)d\theta)}{(1-w_0(\|y_n-x^*\|))(1-w_0(\|z_n-x^*\|))}\\ &\left.+\frac{1}{2}\left(\frac{\bar{w}_n}{1-w_0(\|y_n-x^*\|)}\right)^2\frac{(1+\int_0^1 w_0(\theta\|z_n-x^*\|)d\theta)}{1-w_0(\|y_n-x^*\|)}\right]\|z_n-x^*\|\\ &\le g_3(\|x_n-x^*\|)\|x_n-x^*\|\le\|x_n-x^*\|.\end{aligned}$$

Hence we can choose

$$\begin{aligned}\bar{w}(t) &= \begin{cases} w((1+g_1(t))t)\\ w_0(t)+w_0(g_1(t)t),\end{cases},\\ g_2(t) &= \left[\frac{\int_0^1 w((1-\theta)g_1(t)t)d\theta}{1-w_0(g_1(t)t)}+\frac{1}{4}\left(\frac{\bar{w}_n(t)}{1-w_0(g_1(t)t)}\right)^2\frac{(1+\int_0^1 w_0(\theta g_1(t)t)d\theta)}{1-w_0(g_1(t)t)}\right]g_1(t),\\ \bar{\bar{w}}(t) &= \begin{cases} w((g_1(t)+g_2(t))t)\\ w_0(g_1(t)t)+w_0(g_2(t)t),\end{cases}\end{aligned}$$

and

$$\begin{aligned}g_3(t) &= \left[\frac{\int_0^1 w((1-\theta)g_2(t)t)d\theta}{1-w_0(g_2(t)t)}+\frac{\bar{\bar{w}}(t)(1+\int_0^1 w_0(\theta g_2(t)t)d\theta)}{(1-w_0(g_1(t)t))(1-w_0(g_2(t)t)}\right.\\ &\left.+\frac{1}{2}\left(\frac{\bar{w}(t)}{1-w_0(g_1(t)t)}\right)^2\frac{(1+\int_0^1 w_0(\theta g_2(t)t)d\theta)}{1-w_0(g_1(t)t)}\right]g_2(t).\end{aligned}$$

Method (52.3)

The function g_1 is as given in method (52.1). Moreover, the functions g_2 and g_3 are defined as

$$\begin{aligned} g_2(t) &= \Big[\frac{\int_0^1 w((1-\theta)g_1(t)t)d\theta}{1-w_0(g_1(t)t)} + \frac{\bar{w}(t)(1+\int_0^1 w_0(\theta g_2(t)t)d\theta)}{(1-w_0(t))(1-w_0(g_1(t)t)} \\ &+ \frac{\bar{w}(t)}{(1-w_0(t))^2}\left(1+\frac{5}{4}\frac{\bar{w}(t)}{1-w_0(t)}\right)(1+\int_0^1 w_0(\theta g_1(t)t)d\theta)\Big] g_1(t) \end{aligned}$$

and

$$\begin{aligned} g_3(t) &= \Big[\frac{\int_0^1 w((1-\theta)g_2(t)t)d\theta}{1-w_0(g_2(t)t)} + \frac{\bar{\bar{w}}(t)(1+\int_0^1 w_0(\theta g_2(t)t)d\theta)}{(1-w_0(t))(1-w_0(g_2(t)t)} \\ &+ \frac{\bar{w}(t)}{(1-w_0(t))^2}\left(1+\frac{3}{2}\left(\frac{\bar{w}(t)}{1-w_0(t)}\right)\right)(1+\int_0^1 w_0(\theta g_2(t)t)d\theta)\Big] g_2(t). \end{aligned}$$

The calculations are:

$$\begin{aligned} z_n - x^* &= y_n - x^* - F'(y_n)^{-1}F(y_n) + (F'(y_n)^{-1} - F'(x_n)^{-1})F(x_n) \\ &\quad -(I+\frac{5}{4}A_n)A_nF'(x_n)^{-1}F(y_n), \\ \|z_n - x^*\| &\leq \Big[\frac{\int_0^1 w((1-\theta)\|y_n-x^*\|)d\theta}{1-w_0(\|y_n-x^*\|)} \\ &\quad + \frac{\bar{w}_n(1+\int_0^1 w_0(\theta\|y_n-x^*\|)d\theta)}{(1-w_0(\|x_n-x^*\|))(1-w_0(\|y_n-x^*\|))} \\ &\quad + \frac{\bar{w}_n}{(1-w_0(\|x_n-x^*\|))^2}(1+\frac{5}{4}\frac{\bar{w}_n}{1-w_0(\|x_n-x^*\|)}) \\ &\quad \times(1+\int_0^1 w_0(\theta\|y_n-x^*\|)d\theta)\Big]\|y_n-x^*\| \\ &\leq g_2(\|x_n-x^*\|)\|x_n-x^*\| \leq \|x_n-x^*\|, \\ x_{n+1}-x^* &= z_n - x^* - F'(z_n)^{-1}F(z_n) + (F'(z_n)^{-1} - F'(x_n)^{-1})F(z_n) \\ &\quad -\left(I+\frac{3}{2}A_n\right)A_nF'(x_n)^{-1}F(z_n) \end{aligned}$$

and

$$\begin{aligned} \|x_{n+1}-x^*\| &\leq \Big[\frac{\int_0^1 w((1-\theta)\|z_n-x^*\|)d\theta)}{1-w_0(\|z_n-x^*\|)} \qquad (52.5) \\ &\quad + \frac{\bar{\bar{w}}_n(1+\int_0^1 w_0(\theta\|z_n-x^*\|)d\theta}{(1-w_0(\|x_n-x^*\|))(1-w_0(\|z_n-x^*\|))} \\ &\quad + \frac{\bar{w}_n}{(1-w_0(\|x_n-x^*\|))^2}\left(1+\frac{3}{2}\frac{\bar{w}_n}{1-w_0(\|x_n-x^*\|)}\right) \qquad (52.6) \\ &\quad \times(1+\int_0^1 w_0(\theta\|z_n-x^*\|)d\theta)\|z_n-x^*\| \\ &\leq g_3(\|x_n-x^*\|)\|x_n-x^*\| \leq \|x_n-x^*\|. \end{aligned}$$

3. Semi-local Convergence

Method (52.1)

The majorant sequence $\{\alpha_n\}$ is defined for $\alpha_0 = 0, \beta_0 \geq \|F'(x_0)^{-1}F(x_0)\|$ and each $n = 0,1,2,..$ by

$$\begin{aligned}
\lambda_n &= \int_0^1 v((1-\theta)(\beta_n-\alpha_n))d\theta(\beta_n-\alpha_n),\\
\gamma_n &= \beta_n+\frac{1}{4}\left(4+4\left(\frac{\bar{v}_n}{1-v_0(\beta_n)}\right)+\left(\frac{\bar{v}_n}{1-v_0(\beta_n)}\right)^2\right)\frac{\lambda_n}{1-v_0(\alpha_n)},\\
\mu_n &= (1+\int_0^1 v_0(\beta_n+\theta(\gamma_n-\beta_n))d\theta)(\gamma_n-\beta_n)+\lambda_n,\\
\alpha_{n+1} &= \gamma_n+\frac{1}{2}\left[2+\frac{2\bar{v}_n}{1-v_0(\beta_n)}+\left(\frac{\bar{v}_n}{1-v_0(\beta_n)}\right)^2\right]\frac{\mu_n}{1-v_0(\alpha_n)}\\
\delta_{n+1} &= \int_0^1 v((1-\theta)(\alpha_{n+1}-\alpha_n))d\theta(\alpha_{n+1}-\alpha_n)\\
&\quad +(1+v_0(\alpha_n))(\beta_n-\alpha_n)
\end{aligned}$$

and

$$\beta_{n+1} = \alpha_{n+1}+\frac{\delta_{n+1}}{1-v_0(\alpha_{n+1})}.$$

The calculations are in turn:

$$\begin{aligned}
z_n-y_n &= -\frac{1}{4}[4I+4(F'(y_n)^{-1}F'(x_n)-I)\\
&\quad +(F'(y_n)^{-1}F'(x_n)-I)^2]F'(x_n)^{-1}F(y_n),\\
\|z_n-y_n\| &\leq \frac{1}{4}\left(4+\left(\frac{\bar{v}_n}{1-v_0(\beta_n)}\right)^2+4\left(\frac{\bar{v}_n}{1-v_0(\beta_n)}\right)\right]\frac{\lambda_n}{1-v_0(\alpha_n)}\\
&= \gamma_n-\beta_n,\\
F(z_n) &= F(z_n)-F(y_n)+F(y_n),\\
\|L^{-1}F(z_n)\| &\leq (1+v_0(\beta_n+\theta(\gamma_n-\beta_n))d\theta)(\gamma_n-\beta_n)+\lambda_n = \mu_n,\\
\|x_{n+1}-z_n\| &\leq \frac{1}{2}\left(2+\left(\frac{\bar{v}_n}{1-v_0(\alpha_n)}\right)^2+2\left(\frac{\bar{v}_n}{1-v_0(\alpha_n)}\right)\right)\frac{\mu_n}{1-v_0(\alpha_n)}\\
&= \alpha_{n+1}-\gamma_n,\\
F(x_{n+1}) &= F(x_{n+1})-F(x_n)-F'(x_n)(x_{n+1}-x_n)+F'(x_n)(y_n-x_n),\\
\|L^{-1}F(x_{n+1})\| &\leq \bar{\delta}_{n+1} \leq \delta_{n+1}
\end{aligned}$$

and

$$\begin{aligned}
\|y_{n+1}-x_{n+1}\| &\leq \|F'(x_{n+1})^{-1}L\|\|L^{-1}F(x_{n+1})\|\\
&\leq \frac{\bar{\delta}_{n+1}}{1-v_0(\|x_{n+1}-x_0\|)} \leq \frac{\delta_{n+1}}{1-v_0(\alpha_{n+1})} = \beta_{n+1}-\alpha_{n+1}.
\end{aligned}$$

Method (52.2)

The majorant sequence $\{\alpha_n\}$ is

$$\begin{aligned}\gamma_n &= \beta_n + \frac{1}{4}\left(4 + \left(\frac{\bar{v}_n}{1 - v_0(\beta_n)}\right)^2\right)\frac{\lambda_n}{1 - v_0(\beta_n)}\\ \alpha_{n+1} &= \gamma_n + \frac{1}{2}\left(2 + \left(\frac{\bar{v}_n}{1 - v_0(\beta_n)}\right)^2\right)\frac{\mu_n}{1 - v_0(\beta_n)}\end{aligned}$$

and the iterate β_{n+1} is as given in method (52.2). The calculations are:

$$\begin{aligned}z_n - y_n &= -\frac{1}{4}(4I + (I - F'(y_n)^{-1}F'(x_n))^2)F'(y_n)^{-1}F(y_n),\\ \|z_n - y_n\| &\le \frac{1}{4}\left(4 + \left(\frac{\bar{v}_n}{1 - v_0(\beta_n)}\right)^2\right)\frac{\lambda_n}{1 - v_0(\beta_n)} = \gamma_n - \beta_n,\\ x_{n+1} - z_n &= -\frac{1}{2}(2I + (I - F'(y_n)^{-1}F'(x_n))^2)F'(x_n)^{-1}F(z_n)\end{aligned}$$

and

$$\|x_{n+1} - z_n\| \le \frac{1}{2}\left(2 + \left(\frac{\bar{v}_n}{1 - v_0(\beta_n)}\right)^2\right)\frac{\mu_n}{1 - v_0(\beta_n)} = \alpha_{n+1} - \gamma_n.$$

Method (52.3)
The majorant sequence $\{\alpha_n\}$ is:

$$\begin{aligned}\gamma_n &= \beta_n + \left(1 + \frac{\bar{v}_n}{1 - v_0(\alpha_n)} + \frac{5}{4}\left(\frac{\bar{v}_n}{1 - v_0(\alpha_n)}\right)^2\right)\frac{\lambda_n}{1 - v_0(\alpha_n)},\\ \alpha_{n+1} &= \gamma_n + \left(1 + \frac{\bar{v}_n}{1 - v_0(\alpha_n)} + \frac{3}{2}\left(\frac{\bar{v}_n}{1 - v_0(\alpha_n)}\right)^2\right)\frac{\mu_n}{1 - v_0(\alpha_n)},\end{aligned}$$

and the iterate is given in method (52.1). The corresponding calculations are:

$$\begin{aligned}\|z_n - y_n\| &\le \left[1 + \frac{\bar{v}_n}{1 - v_0(\alpha_n)} + \frac{5}{4}\left(\frac{\bar{v}_n}{1 - v_0(\alpha_n)}\right)^2\right]\frac{\lambda_n}{1 - v_0(\alpha_n)},\\ &= \gamma_n - \beta_n.\end{aligned}$$

and

$$\begin{aligned}\|x_{n+1} - z_n\| &\le \left[1 + \frac{\bar{v}_n}{1 - v_0(\alpha_n)} + \frac{3}{2}\left(\frac{\bar{v}_n}{1 - v_0(\alpha_n)}\right)^2\right]\frac{\mu_n}{1 - v_0(\alpha_n)},\\ &= \alpha_{n+1} - \gamma_n.\end{aligned}$$

The rest of the analysis is given in previous chapter.

4. References

[1] Alqahtani, H. F., Behl, R., Kansal, M., *Higher-order iteration schemes for solving nonlinear systems of equations*, Mathematics, 7(10), (2019), 937, http://dx.doi.org/10.3390/math7100937.

[2] Argyros, I. K., George, S., *High convergence order Q-step methods for solving equations and systems of equations*, Contemp. Math., 1(3), (2020), 119–126, http://dx.doi.org/10.37256/cm.132020403.

[3] Argyros, I. K., Hilout, S., *Numerical Methods in Nonlinear Analysis*, World Scientific Publ. Comp., New Jersey, 2013.

[4] Artidiello, S., Cordero, A., Torregrosa, J. R., Vassileva, M. P., *Multidimensional generalization of iterative methods for solving nonlinear problems by means of weight-function procedure*, Appl. Math. Comput., 268, (2015), 1064–1071.

[5] Behl, R., Bhalla, S., Magrean, A. A., Kumar, S., *An efficient high order iterative scheme for large nonlinear systems with dynamics*, J. Comput. Appl. Math, 404, (2020), 189–198.

[6] Behl, R., Cordero, A., Torregrosa, J. R., *A new higher-order optimal derivative free scheme for multiple roots*, J. Comput. Appl. Math., 404, (2022), 113773.

[7] Chun, C., Neta, B., *An Efficient Derivative-Free Method for the Solution of Systems of Equations*, Numer. Func. Anal. Opt., 42, (2021), 834–848.

[8] Chun, C., Neta, B., *Developing high order methods for the solution of systems of nonlinear equations*, Appl. Math. Comput., 342, (2019), 178–190.

[9] Cordero, A., Gmez, E., Torregrosa, J. R., *Efficient high-order iterative methods for solving nonlinear systems and their application on heat conduction problems*, Complexity, (2017), 6457532.

[10] Grau-Snchez, M., Noguera, M., Amat, S., *On the approximation of derivatives using divided difference operators preserving the local convergence order of iterative methods*, J. Comput. Appl. Math., 237, (2013), 363–372.

[11] Lin, Y., Bao, L., Jia, X., *Convergence analysis of a variant of the Newton method for solving nonlinear equations*, Comput. Math. Appl., 59, (2010), 2121–2127.

[12] Neta, B., *A new Derivative-Free Method to solve Nonlinear Equations*, Math., 9, (2021), 583.

[13] Sharma, J. R., Guna, R. K., Sharma, R., *An efficient fourth order weighted-Newton method for systems of nonlinear equations*, Numer. Algorithms, 2, (2013), 307–323.

[14] Sharma, J. R., Kumar, D., *A fast and efficient composite Newton-Chebyshev method for systems of nonlinear equations*, J. Complexity, 49, (2018), 56–73.

[15] Sharma, J. R., Kumar, D., Jntschi, L., *On a reduced cost higher order Traub-Steffensen-Like method for nonlinear systems*, Symmetry, 11(7), (2019), 891, http://dx.doi.org/10.3390/sym11070891.

[16] Traub, J. F., *Iterative Methods for the Solution of Equations*, Prentice-Hall, Englewood Cliffs, 1964.

[17] Wang, X., Cao, Y., *A numerically stable high-order Chebyshev-Halley type multipoint iterative method for calculating matrix sign function*, AIMS Mathematics, 8, (2023), 12456–12471.

[18] Wang, X., Li, W., *Choosing the Best Members of the Optimal Eighth-Order Petkovi's Family by Its Fractal Behavior*, Fractal and Fractional, 6, (2022), 749.

[19] Wang, X., Zhang, T., *A family of Steffensen type methods with seventh-order convergence*, Numer. Algorithms, 62, (2013), 429–444.

[20] Zhanlav, T., Otgondorj, K., *Higher order Jarratt-like iterations for solving systems of nonlinear equations*, Appl. Math. Comput., 395, (2021), 125849.

Chapter 53

Methods of Order Five with Comparable Efficiencies

1. Introduction

The methods are defined for $x_0 \in \Omega$ and each $n = 0, 1, 2, \ldots$ by

$$\begin{aligned} y_n &= x_n - F'(x_n)^{-1}F(x_n), \\ z_n &= y_n - aF'(x_n)^{-1}F(y_n), \\ x_{n+1} &= z_n - (2I - F'(x_n)^{-1}F'(y_n))F'(x_n)^{-1}F(z_n), \end{aligned} \tag{53.1}$$

$$\begin{aligned} y_n &= x_n - F'(x_n)^{-1}F(x_n), \\ z_n &= y_n - (2I - F'(x_n)^{-1}F'(y_n))F'(x_n)^{-1}F(y_n), \\ x_{n+1} &= z_n - F'(x_n)^{-1}F(z_n), \end{aligned} \tag{53.2}$$

$$\begin{aligned} y_n &= x_n - F'(x_n)F(x_n), a_n = \frac{1}{3}(2x_n + y_n), b_n = \frac{1}{3}(x_n + 2y_n), \\ A_n &= F'(x_n) + 3F'(a_n) + 3F'(b_n) + F'(y_n), \\ z_n &= x_n - 8A_n^{-1}F(x_n) \end{aligned} \tag{53.3}$$

and

$$x_{n+1} = z_n - F'(y_n)F(z_n), \tag{53.4}$$

where $a \in \mathbb{R} - \{0\}$.

The convergence order five is shown in [2], respectively for these methods using the Taylor series expansion technique provided that $B_1 = B_2 = \mathbb{R}^k$ and assuming that $F^{(6)}$ exists and is bounded limiting the applicability of these methods. Other limitations with this technique have been reported in previous chapters. That is why the new local convergence is based only on the operators on these methods. Moreover, the more interesting semi-local convergence analysis (not studied before) is also developed in this chapter utilizing majorizing sequences. The new technique can be used to extend the applicability of other methods [1, 3, 4, 5, 6].

2. Local Convergence

Method (53.1)

The majorant functions $g_i, i = 1,2,3$ defined as:

$$
\begin{aligned}
g_1(t) &= \frac{\int_0^1 w((1-\theta)t)d\theta}{1-w_0(t)},\\
\bar{w}(t) &= \begin{cases} w((1+g_1(t))t) \\ w_0(t)+w_0(g_1(t)t), \end{cases}\\
g_2(t) &= \left[\frac{\int_0^1 w((1-\theta)g_1(t)t)d\theta}{1-w_0(g_1(t)t)} + \frac{\bar{w}(t)\left(1+\int_0^1 w_0(\theta g_1(t)t)d\theta\right)}{(1-w_0(t))(1-w_0(g_1(t)t))}\right.\\
&\quad \left. + \frac{|1+a|\left(1+\int_0^1 w_0(\theta g_1(t)t)d\theta\right)}{(1-w_0(t))}\right] g_1(t),\\
\bar{\bar{w}}(t) &= \begin{cases} w((1+g_2(t))t) \\ w_0(t)+w_0(g_2(t)t) \end{cases}
\end{aligned}
$$

and

$$
\begin{aligned}
g_3(t) &= \left[\frac{\int_0^1 w((1-\theta)g_2(t)t)d\theta}{1-w_0(g_2(t)t)} + \frac{\bar{\bar{w}}(t)\left(1+\int_0^1 w_0(\theta g_2(t)t)d\theta\right)}{(1-w_0(t))(1-w_0(g_2(t)t))}\right.\\
&\quad \left. + \frac{\bar{w}(t)\left(1+\int_0^1 w_0(\theta g_2(t)t)d\theta\right)}{(1-w_0(t))^2}\right] g_2(t).
\end{aligned}
$$

The motivational calculations are in turn:

$$
\begin{aligned}
y_n - x^* &= x_n - x^* - F'(x_n)^{-1}F(x_n),\\
\|y_n - x^*\| &\leq \frac{\int_0^1 w((1-\theta)\|x_n-x^*\|)\|x_n-x^*\|}{1-w_0(\|x_n-x^*\|)}\\
&\leq g_1(\|x_n-x^*\|)\|x_n-x^*\|\\
&\leq \|x_n-x^*\| < r,\\
z_n - x^* &= y_n - x^* - F'(y_n)^{-1}F(y_n) + (F'(y_n)^{-1} - F'(x_n)^{-1})F(y_n)\\
&\quad +(1-a)F'(x_n)^{-1}F(y_n),
\end{aligned}
$$

$$\begin{aligned}
\|z_n - x^*\| &\leq \Bigg[\frac{\int_0^1 w((1-\theta)\|y_n - x^*\|)d\theta}{1 - w_0(\|y_n - x^*\|)} \\
&+ \frac{\bar{w}_n\left(1 + \int_0^1 w_0(\theta\|y_n - x^*\|)d\theta\right)}{(1 - w_0(\|x_n - x^*\|))(1 - w_0(\|y_n - x^*\|))}. \\
&+ \frac{|1-a|\left(1 + \int_0^1 w_0(\theta\|y_n - x^*\|)d\theta\right)}{1 - w_0(\|x_n - x^*\|)}\Bigg]\|y_n - x^*\| \\
&\leq g_2(\|x_n - x^*\|)\|x_n - x^*\| \\
&\leq \|x_n - x^*\|, \\
x_{n+1} - x^* &= z_n - x^* - F'(z_n)^{-1}F(z_n) + (F'(z_n)^{-1} - F'(x_n)^{-1})F(z_n) \\
&\quad - F'(x_n)^{-1}(F'(x_n) - F'(x_n))F'(x_n)^{-1}F(z_n)
\end{aligned}$$

and

$$\begin{aligned}
\|x_{n+1} - x^*\| &\leq \Bigg[\frac{\int_0^1 w((1-\theta)\|z_n - x^*\|d\theta)}{1 - w_0(\|z_n - x^*\|)} \\
&+ \frac{\bar{\bar{w}}_n\left(1 + \int_0^1 w_0(\theta\|z_n - x^*\|)d\theta\right)}{(1 - w_0(\|x_n - x^*\|))(1 - w_0(\|z_n - x^*\|))} \\
&+ \frac{\bar{w}_n\left(1 + \int_0^1 w_0(\theta\|z_n - x^*\|)d\theta\right)}{(1 - w_0(\|z_n - x^*\|))^2}\Bigg]\|z_n - x^*\| \\
&\leq g_3(\|x_n - x^*\|)\|x_n - x^*\| \\
&\leq \|x_n - x^*\|.
\end{aligned}$$

Method (53.2)

The function g_1 is as defined in method (53.1),

$$\begin{aligned}
g_2(t) &= \Bigg[\frac{\int_0^1 w((1-\theta)g_1(t)t)d\theta}{1 - w_0(g_1(t)t)} + \frac{\bar{w}(t)\left(1 + \int_0^1 w_0(\theta g_1(t)t)d\theta\right)}{(1 - w_0(t))(1 - w_0(g_1(t)t))} \\
&+ \frac{\bar{w}(t)\left(1 + \int_0^1 w_0(\theta g_1(t)t)d\theta\right)}{(1 - w_0(t))^2}\Bigg] g_1(t)
\end{aligned}$$

and

$$g_3(t) = \left[\frac{\int_0^1 w((1-\theta)g_2(t)t)d\theta}{1 - w_0(g_2(t)t)} + \frac{\bar{\bar{w}}(t)\left(1 + \int_0^1 w_0(\theta g_2(t)t)d\theta\right)}{(1 - w_0(t))(1 - w_0(g_2(t)t))}\right] g_2(t).$$

The motivational calculations are in turn:

$$
\begin{aligned}
z_n - x^* &= y_n - x^* - F'(y_n)^{-1}F(y_n) + (F'(y_n)^{-1} - F'(x_n)^{-1})F(y_n) \\
&\quad - F'(x_n)^{-1}(F'(x_n) - F'(y_n))F'(x_n)^{-1}F(y_n), \\
\|z_n - x^*\| &\leq \Bigg[\frac{\int_0^1 w((1-\theta)\|y_n - x^*\|)d\theta}{1 - w_0(\|y_n - x^*\|)} \\
&\quad + \frac{\bar{w}_n\left(1 + \int_0^1 w_0(\theta\|y_n - x^*\|)d\theta\right)}{(1 - w_0(\|x_n - x^*\|))(1 - w_0(\|y_n - x^*\|))} \\
&\quad + \frac{\bar{w}_n\left(1 + \int_0^1 w_0(\theta\|y_n - x^*\|)\right)d\theta}{(1 - w_0(\|y_n - x^*\|))^2} \Bigg] \|y_n - x^*\| \\
&\leq g_2(\|x_n - x^*\|)\|x_n - x^*\| \leq \|x_n - x^*\|, \\
x_{n+1} - x^* &\leq z_n - x^* - F'(z_n)^{-1}F(z_n) + (F'(z_n)^{-1} - F'(x_n)^{-1})F(z_n)
\end{aligned}
$$

and

$$
\begin{aligned}
\|x_{n+1} - x^*\| &\leq \Bigg[\frac{\int_0^1 w((1-\theta)\|z_n - x^*\|)d\theta}{1 - w_0(\|z_n - x^*\|)} \\
&\quad + \frac{\bar{\bar{w}}_n\left(1 + \int_0^1 w_0(\theta\|z_n - x^*\|)d\theta\right)}{(1 - w_0(\|x_n - x^*\|))(1 - w_0(\|z_n - x^*\|))} \Bigg] \|z_n - x^*\| \\
&\leq g_3(\|x_n - x^*\|)\|x_n - x^*\| \leq \|x_n - x^*\|.
\end{aligned}
$$

Method (53.3)

The majorant functions g_1 is as defined in method (53.1).

Moreover, the functions g_2 and g_3 are defined by

$$
\begin{aligned}
p(t) &= 3w\left(\frac{(1+g_1(t))t}{3}\right) + w\left(\frac{2(1+g_1(t)t)}{3}\right) + \bar{w}(t), \\
q(t) &= \frac{1}{8}\left[w_0(t) + 3w_0\left(\frac{(2+g_1(t))t}{3}\right) + 3w_0\left(\frac{(1+2g_1(t))t}{3}\right) + w_0(g_1(t)t)\right], \\
g_2(t) &= \frac{\int_0^1 w((1-\theta)t)d\theta}{1 - w_0(t)} + \frac{p(t)\left(1 + \int_0^1 w_0(\theta t)d\theta\right)}{8(1-q(t))(1-w_0(t))},
\end{aligned}
$$

$$
\tilde{w}(t) = \begin{cases} w((g_1(t) + g_2(t))t) \\ w_0(g_1(t)t) + w_0(g_2(t)t) \end{cases}
$$

and

$$
g_3(t) = \left[\frac{\int_0^1 w((1-\theta)g_2(t)t)d\theta}{1 - w_0(g_2(t)t)} + \frac{\tilde{w}(t)\left(1 + \int_0^1 w_0(\theta g_2(t)t)d\theta\right)}{(1 - w_0(g_1(t)t))} \right] g_2(t).
$$

The motivational calculations are in turn:

$$\begin{aligned}
\|(8L)^{-1}(A_n-8L)\| &\leq \frac{1}{8}\Big[\|L^{-1}(F'(x_n)-L)\|+3\|L^{-1}(F'(a_n)-L)\| \\
&\quad +3\|L^{-1}(F'(b_n)-L)\|+\|L^{-1}(F'(y_n)-L)\|\Big] \\
&\leq \frac{1}{8}\Big[w_0(\|x_n-x^*\|)+3w_0\left(\frac{2\|x_n-x^*\|+\|y_n-x^*\|}{3}\right) \\
&\quad +3w_0\left(\frac{\|x_n-x^*\|+2\|y_n-x^*\|}{3}\right)+w_0(\|y_n-x^*\|)\Big] \\
&= q_n<1, \\
\|A_n^{-1}L\| &\leq \frac{1}{8(1-q_n)}, \\
\|L^{-1}(8F'(x_n)-A_n)\| &\leq 3w\left(\frac{\|y_n-x^*\|}{3}\right)+3w\left(\frac{2\|y_n-x_n\|}{3}\right)+\bar{w}_n=p_n, \\
z_n-x^* &\leq x_n-x^*-F'(x_n)^{-1}F(x_n) \\
&\quad -A_n^{-1}(8F'(x_n)-A_n)F'(x_n)^{-1}F(x_n),
\end{aligned}$$

$$\begin{aligned}
\|z_n-x^*\| &\leq \Bigg[\frac{\int_0^1 w((1-\theta)\|x_n-x^*\|)d\theta}{1-w_0(\|x_n-x^*\|)} \\
&\quad +\frac{p_n\left(1+\int_0^1 w_0(\theta\|z_n-x^*\|)d\theta\right)}{8(1-q_n)(1-w_0(\|x_n-x^*\|))}\Bigg]\|x_n-x^*\| \\
&\leq g_2(\|x_n-x^*\|)\|x_n-x^*\| \\
&\leq \|x_n-x^*\|, \\
x_{n+1}-x^* &= z_n-x^*-F'(z_n)^{-1}F(z_n)+(F'(z_n)^{-1}-F'(y_n)^{-1})F(z_n)
\end{aligned}$$

and

$$\begin{aligned}
\|x_{n+1}-x^*\| &\leq \Bigg[\frac{\int_0^1 w((1-\theta)\|z_n-x^*\|)d\theta}{1-w_0(\|z_n-x^*\|)} \\
&\quad +\frac{\tilde{w}\left(1+\int_0^1 w_0(\theta\|z_n-x^*\|)d\theta\right)}{(1-w_0(\|y_n-x^*\|))(1-w_0(\|z_n-x^*\|))}\Bigg]\|z_n-x^*\| \\
&\leq g_3(\|x_n-x^*\|)\|x_n-x^*\|\leq\|x_n-x^*\|.
\end{aligned}$$

3. Semi-local Convergence

The role of x^*, "w" is replaced by x_0, "v", respectively in the calculations.

Method (53.1)

The majorant sequence $\{\alpha_n\}$ is defined for $\alpha_0=0$, $\beta_0 \geq \|F'(x_0)^{-1}F(x_0)\|$ and each $n=$

$0,1,2,\ldots$ by

$$\begin{aligned}
\lambda_n &= \int_0^1 v((1-\theta)(\beta_n-\alpha_n))d\theta(\beta_n-\alpha_n),\\
\gamma_n &= \beta_n+\frac{|a|\lambda_n}{1-v_0(\alpha_n)},\\
\mu_n &= \left(1+\int_0^1 v_0(\beta_n+\theta(\gamma_n-\beta_n))d\theta\right)(\gamma_n-\beta_n)+\lambda_n,
\end{aligned}$$

$$\bar{v}(t)=\begin{cases} v(\beta_n-\alpha_n)\\ v_0(\alpha_n)+v_0(\beta_n),\end{cases}$$

$$\begin{aligned}
\alpha_{n+1} &= \gamma_n+\frac{1}{1-v_0(\alpha_n)}\left(1+\frac{\bar{v}_n}{(1-v_0(\alpha_n))}\right)\lambda_n,\\
\delta_{n+1} &= \int_0^1 v((1-\theta)(\alpha_{n+1}-\alpha_n))d\theta(\alpha_{n+1}-\alpha_n)\\
&\quad +(1+v_0(\alpha_n))(\beta_n-\alpha_n)
\end{aligned}$$

and

$$\beta_{n+1} = \alpha_{n+1}+\frac{\delta_{n+1}}{1-v_0(\alpha_{n+1})}.$$

The motivational calculations are in turn:

$$\begin{aligned}
F(y_n) &= F(y_n)-F(x_n)-F'(x_n)^{-1}(y_n-x_n),\\
\|L^{-1}F(y_n)\| &\leq \int_0^1 v((1-\theta)(\beta_n-\alpha_n))d\theta(\beta_n-\alpha_n)=\lambda_n,\\
z_n-y_n &= -aF'(x_n)^{-1}F(y_n),\\
\|z_n-y_n\| &\leq |a|\|F'(x_n)^{-1}L\|\|L^{-1}F(y_n)\|\\
&\leq \frac{|a|\lambda_n}{1-v_0(\alpha_n)}=\gamma_n-\beta_n,\\
F(z_n) &= F(z_n)-F(y_n)+F(y_n),\\
\|L^{-1}F(z_n)\| &\leq \left(1+\int_0^1 v_0(\beta_n+\theta(\gamma_n-\beta_n))d\theta\right)(\gamma_n-\beta_n)+\lambda_n,\\
x_{n+1}-z_n &= -F'(x_n)^{-1}F(z_n)-F'(x_n)^{-1}(F'(x_n)-F'(y_n))F'(x_n)^{-1}F(z_n),\\
\|x_{n+1}-z_n\| &\leq \left[\frac{1}{1-v_0(\alpha_n)}+\frac{\bar{v}_n}{(1-v_0(\alpha_n))^2}\right]\mu_n=\alpha_{n+1}-\gamma_n,\\
F(x_{n+1}) &= F(x_{n+1})-F(x_n)-F'(x_n)(x_{n+1}-x_n)+F'(x_n)(y_n-x_n),\\
\|L^{-1}F(x_{n+1})\| &\leq \int_0^1 v((1-\theta)(\alpha_{n+1}-\alpha_n))d\theta(\alpha_{n+1})+(1+v_0(\alpha_n))(\beta_n-\alpha_n)\\
&= \delta_{n+1}
\end{aligned}$$

and

$$\|y_{n+1}-x_{n+1}\| \leq \|F'(x_{n+1})^{-1}L\|\|L^{-1}F(x_{n+1})\|\leq\frac{\delta_{n+1}}{1-v_0(\alpha_{n+1})}=\beta_{n+1}-\alpha_{n+1}.$$

Method (53.2)

The majorant sequence $\{\alpha_n\}$ starts as in method (53.1) and the iterates γ_n and β_{n+1} are the same. Moreover, we define

$$\alpha_{n+1} = \gamma_n + \frac{\mu_n}{1-v_0(\alpha_n)}.$$

The calculation for this iterate is:

$$\begin{aligned} x_{n+1}-z_n &= -F'(x_n)^{-1}F(z_n), \\ \|x_{n+1}-z_n\| &\leq \frac{\mu_n}{1-v_0(\alpha_n)} = \alpha_{n+1}-\gamma_n. \end{aligned}$$

Method (53.3)

The majorant sequence $\{\alpha_n\}$ starts as in method (53.1) and the iterate β_{n+1} is the same. Moreover, we define

$$\gamma_n = \beta_n + \frac{\bar{p}_n(\beta_n-\alpha_n)}{8(1-\bar{q}_n)}$$

and

$$\alpha_{n+1} = \gamma_n + \frac{\mu_n}{1-v_0(\beta_n)},$$

where

$$\bar{q}_n = \frac{1}{8}\left[v_0(\alpha_n)+3v_0\left(\frac{2\alpha_n+\beta_n}{3}\right)+3v_0\left(\frac{3\alpha_n+2\beta_n}{3}\right)+v_0(\beta_n)\right]$$

and

$$\bar{p}_n = 3v\left(\frac{\beta_n-\alpha_n}{3}\right)+3v\left(\frac{2\beta_n}{3}\right)+\bar{v}_n.$$

The motivational calculations are in turn:

$$\begin{aligned} z_n-y_n &= (F'(x_n)^{-1}-8A_n^{-1})F(x_n) = -A_n^{-1}(8F'(x_n)-A_n)F'(x_n)^{-1}F(x_n) \\ &= A_n^{-1}(8F'(x_n)-A)(y_n-x_n), \\ \|z_n-y_n\| &\leq \frac{\bar{p}_n(\beta_n-\alpha_n)}{8(1-\bar{q}_n)} = \gamma_n-\beta_n, \\ x_{n+1}-z_n &= -F'(y_n)^{-1}F(z_n) \end{aligned}$$

and

$$\|x_{n+1}-z_n\| \leq \frac{\mu_n}{1-v_0(\beta_n)} = \alpha_{n+1}-\gamma_n.$$

The rest of the convergence has appeared in previous chapters.

4. References

[1] Cordero, A., Hueso, J. L., Martnez, E., Torregrosa, J. R., *Increasing the convergence order of an iterative method for nonlinear systems*, Applied Mathematics Letters, 25(12), (2012), 2369–2374. https://doi.org/10.1016/j.aml.2012.07.005.

[2] Sharma, J. R., Gupta, P., *An efficient fifth order method for solving systems of nonlinear equations*, Computers and Mathematics with Applications, 67(3), (2014), 591–601. https://doi.org/10.1016/j.camwa.2013.12.004.

[3] Singh, H., Sharma, J. R., *Simple and efficient fifth order solvers for systems of nonlinear problems*, Mathematical Modelling and Analysis, 28(1), (2023), 1–22. https://doi.org/10.3846/mma.2023.16244.

[4] Xiao, X. Y., Yin, H. W., *A new class of methods with higher order of convergence for solving systems of nonlinear equations*, Applied Mathematics and Computation, 264, (2015), 300–309. https://doi.org/10.1016/j.amc.2015.04.094.

[5] Xu, Z., Jieqing, T., *The fifth order of three-step iterative methods for solving systems of nonlinear equations*, Mathematica Numerica Sinica, 35(3), (2013), 297–304.

[6] Zhanlav, T., Otgondorj, K., *Higher order Jarratt-like iterations for solving systems of nonlinear equations*, Applied Mathematics and Computation, 395, (2021), 125849. https://doi.org/10.1016/j.amc2020.125849.

Chapter 54

The Traub Family of Solvers

1. Introduction

The method is defined for $x_0 \in \Omega$ and each $n = 0, 1, 2, \ldots$ by

$$\begin{aligned} y_n &= x_n - F'(x_n)^{-1}F(x_n), \\ z_n &= x_n + a(y_n - x_n) \end{aligned} \tag{54.1}$$

and

$$x_{n+1} = y_n - \frac{1}{a^2}F'(x_n)^{-1}[(a-1)F(x_n) + F(z_n)],$$

where $a \neq 0$. If $a = 1$ this method coincides with the Traub method [1,5]. The local convergence order three is determined under the Taylor series expansion technique when $B_1 = B_2 = \mathbb{R}^k$ and by assuming that $F^{(4)}$ exists and is bounded [1,5]. The limitations of this approach are documented in previous chapters. The new local convergence analysis is based on generalized continuity conditions on F' and is given in a Banach space setting. Moreover, the semi-local convergence analysis (not given in [1,5]) is presented based on majorizing sequences. The technique can be used to expand the applicability of other methods [2,3,4,6].

2. Local Convergence

The majorant functions $g_i i = 1, 2, 3 \ldots$ are

$$\begin{aligned} g_1(t) &= \frac{\int_0^1 w((1-\theta)t)d\theta}{1 - w_0(t)}, \\ g_2(t) &= |1-a| + |a|g_1(t) \end{aligned}$$

and

$$g_3(t) = g_1(t) + \left|\frac{a-1}{a^2}\right| \frac{1 + \int_0^1 w_0(\theta t)d\theta}{1 - w_0(t)} + \frac{\left(1 + \int_0^1 w_0(\theta g_2(t)t)d\theta\right)}{|a^2|(1 - w_0(t))} g_2(t).$$

The motivational calculations are in turn:

$$\begin{aligned}
y_n - x^* &= x_n - x^* - F'(x_n)^{-1}F(x_n),\\
\|y_n - x^*\| &\le \frac{\int_0^1 w((1-\theta)\|x_n - x^*\|)d\theta\|x_n - x^*\|}{1 - w_0(\|x_n - x^*\|)}\\
&\le g_1(\|x_n - x^*\|)\|x_n - x^*\| \le \|x_n - x^*\| < r,\\
z_n - x^* &= x_n - x^* - a(y_n - x^*) = (1-a)(x_n - x^*) + a(y_n - x^*),\\
\|z_n - x^*\| &\le (|1-a| + |a|g_1(\|x_n - x^*\|))\|x_n - x^*\|\\
&\le g_2(\|x_n - x^*\|)\|x_n - x^*\| \le \|x_n - x^*\|,\\
x_{n+1} - x^* &= y_n - x^* - \frac{a-1}{a^2}F'(x_n)^{-1}F(z_n) - \frac{F'(x_n)^{-1}F(z_n)}{a^2}
\end{aligned}$$

and

$$\begin{aligned}
\|x_{n+1} - x^*\| &\le \left[g_1(\|x_n - x^*\|) + \left|\frac{a-1}{a^2}\right| \frac{1 + \int_0^1 w_0(\theta\|x_n - x^*\|)d\theta}{1 - w_0(\|x_n - x^*\|)}\right.\\
&\quad \left. + \frac{\left(1 + \int_0^1 w_0(\theta\|z_n - x^*\|)d\theta\right) g_2(\|x_n - x^*\|)}{|a^2|(1 - w_0(\|x_n - x^*\|))}\right] \|x_n - x^*\|\\
&\le g_3(\|x_n - x^*\|)\|x_n - x^*\| \le \|x_n - x^*\|.
\end{aligned}$$

3. Semi-local Convergence

The majorant sequence $\{\alpha_n\}$ is defined for $\alpha_0 = 0$, $\beta_0 \ge \|F'(x_0)^{-1}F(x_0)\|$ and each $n = 0, 1, 2, \ldots$ by

$$\begin{aligned}
\gamma_n &= \beta_n + |a-1|(\beta_n - \alpha_n),\\
\mu_n &= \left(1 + \int_0^1 v_0(\beta_n + \theta(\gamma_n - \beta_n))d\theta\right)(\gamma_n - \beta_n)\\
&\quad + \int_0^1 v((1-\theta)(\beta_n - \gamma_n))d\theta(\beta_n - \alpha_n)\\
\alpha_{n+1} &= \beta_n + \left|\frac{a-1}{a^2}\right|(\beta_n - \alpha_n) + \frac{\mu_n}{|a|^2(1 - v_0(\alpha_n))},\\
\delta_{n+1} &= \int_0^1 v((1-\theta)(\alpha_{n+1} - \alpha_n)d\theta)(\alpha_{n+1} - \alpha_n) + (1 + v_0(\alpha_n))(\beta_n - \alpha_n)
\end{aligned}$$

and

$$\beta_{n+1} = \alpha_{n+1} + \frac{\delta_{n+1}}{1 - v_0(\alpha_{n+1})}.$$

The motivational calculations are in turn:

$$\begin{aligned}
z_n - y_n &= F'(x_n)^{-1}F(x_n) + a(y_n - x_n) = (a-1)(y_n - x_n),\\
\|z_n - y_n\| &\leq |a-1|\|y_n - x_n\| \leq |a-1|(\beta_n - \alpha_n) = \gamma_n - \beta_n,\\
F(y_n) &= F(y_n) - F(x_n) - F'(x_n)(y_n - x_n),\\
\|L^{-1}F(y_n)\| &\leq \int_0^1 v((1-\theta)(\beta_n - \alpha_n))d\theta(\beta_n - \alpha_n),\\
F(z_n) &= F(z_n) - F(y_n) + F(y_n),\\
\|L^{-1}F(z_n)^{-1}\| &\leq \left(1 + \int_0^1 v(\beta_n + \theta(\gamma_n - \beta_n))d\theta\right)(\gamma_n - \beta_n)\\
&\quad + \int_0^1 v((1-\theta)(\beta_n - \alpha_n))d\theta(\beta_n - \alpha_n) = \mu_n,\\
x_{n+1} - y_n &= -\frac{a-1}{a^2}F'(x_n)^{-1}F(z_n) - \frac{1}{a^2}F'(x_n)^{-1}F(z_n)\\
&= \frac{a-1}{a^2}(y_n - x_n) - \frac{1}{a^2}F'(x_n)^{-1}F(z_n),\\
\|x_{n+1} - y_n\| &\leq \left|\frac{a-1}{a^2}\right|(\beta_n - \alpha_n) + \frac{1}{|a|^2}\frac{\mu_n}{1 - v_0(\alpha_n)} = \alpha_{n+1} - \gamma_n,\\
F(x_{n+1}) &= F(x_{n+1}) - F(x_n) - F'(x_n)(x_{n+1} - x_n) + F'(x_n)(y_n - x_n),\\
\|L^{-1}F(x_{n+1})\| &\leq \delta_{n+1}
\end{aligned}$$

and

$$\begin{aligned}
\|y_{n+1} - x_{n+1}\| &\leq \|F'(x_{n+1})^{-1}L\|\|L^{-1}F(x_{n+1})\|\\
&\leq \frac{\delta_{n+1}}{1 - v_0(\alpha_{n+1})} = \beta_{n+1} - \alpha_{n+1}.
\end{aligned}$$

The rest of the convergence analysis is as given in previous chapters.

4. References

[1] Campos, B., Cordero, A., Torregrosa, J. R., Vindel, P., *A multidimensional dynamical approach to iterative methods with memory*, Appl. Math. Comput., 271, (2015), 701–715.

[2] Chicharro, F. I., Cordero, A., Garrido, N., *Generalizing Traub's method to a parametric iterative class for solving multidimensional nonlinear problems*, Math. Meth. Appl. Sci., (2023), 1–14.

[3] Chun, C., Neta, B., *Developing high order methods for the solution of systems of nonlinear equations*, Appl. Math. Comput., 342, (2019), 178–190.

[4] Cordero, A., Soleymani, F., Torregrosa, J. R., *Dynamical analysis of iterative methods for nonlinear systems or how to deal with the dimension?*, Appl. Math. Comput., 244, (2014), 398–412.

[5] Traub, J. F., *Iterative methods for the solution of equations*, Prentice-Hall, New York, 1964.

[6] Xiao, X., Yin, H., *Increasing the order of convergence for iterative methods to solve nonlinear systems*, Calcolo, 53, (2016), 285–300.

Chapter 55

Extended Halley-Chebyshev-Type Method for Equations

1. Introduction

The method is defined for $x_0 \in \Omega, a \in \mathbb{R}$ and each $n = 0, 1, 2, \ldots$ by

$$\begin{aligned} u_n &= x_n - M(x_n), v_n = x_n + M(x_n), \\ D_n &= 2[u_n, x_n, v_n; F], \\ B_n &= [u_n, v_n; F], \\ A_n &= B_n^{-1} D_n B_n^{-1} F(x_n) \end{aligned} \tag{55.1}$$

and

$$x_{n+1} = x_n - [I + \frac{1}{2} A_n (I - a A_n)^{-1}] B_n^{-1} F(x_n),$$

where $[\cdot, \cdot; F]$ and $[\cdot, \cdot, \cdot; F]$ are divided differencces of order one and two, respectively and $M : \Omega \to \Omega$ is a continuous operator.

Notice that if $A_n = F'(x_n)^{-1} F''(x_n) F'(x_n)^{-1} F(x_n)$ and $B_n = F'(x_n)$, the method (55.1) reduces to the Halley method [2]-[9]. The semi-local convergence of the special case of method (55.1) for $M(x_n) = \beta_n x_n$ and $\{\beta_n\}$ a controlled scalar sequence is given in [1]. In this chapter, we extend the applicability of the methods in [2]-[9] by using majorizing sequences and weaker convergence criteria.

2. Convergence Analysis

The convergence conditions are:

(C1) There exist an invertible operator L and continuous and nondecreasing functions f, g, ψ_0, ψ and ψ_1 such that for each $x \in \Omega$

$$\begin{aligned}\|u-x_0\| &\leq f(\|x-x_0\|) \leq \|x-x_0\|,\\ \|v-x_0\| &\leq g(\|x-x_0\|) \leq \|x-x_0\|,\\ \|L^{-1}(B-L)\| &\leq \psi_0(\|u-x_0\|, \|v-x_0\|)\\ \|L^{-1}F(x)\| &\leq \psi_1(\|x-x_0\|)\end{aligned}$$

and

$$\|L^{-1}D\| \leq 2\psi(\|u-x_0\|, \|x-x_0\|, \|v-x_0\|)$$

(C2)

$$\begin{aligned}q &= q(x) = \psi_0(\|u-x_0\|, \|v-x_0\|) < 1\\ \lambda &= \lambda(x) = \frac{|a|p(x)\psi_1(\|x-x_0\|)}{1-q(x)} < 1,\end{aligned}$$

where $p = p(x) = 2\psi(\|u-x_0\|, \|x-x_0\|, \|v-x_0\|)$.
Define the scalar sequence $\{\alpha_n\}$ for $x_0 = 0, \alpha_1 \geq \|x_1 - x_0\|$ and each $n = 0, 1, 2, \ldots$ by

$$\begin{aligned}q_n &= \psi_0(f(\alpha_n), g(\alpha_n)),\\ p_n &= 2\psi(f(\alpha_n), \alpha_n, g(\alpha_n)),\\ \lambda_n &= \frac{|a|p_n\psi_1(\alpha_n)}{(1-q_n)^2}\end{aligned} \tag{55.2}$$

and

$$\alpha_{n+1} = \alpha_n + \left(1 + \frac{\lambda_n}{2|a|(1-\lambda_n)}\right)\frac{\psi_1(\alpha_n)}{1-q_n}.$$

(C3) The sequence $\{\alpha_n\}$ is convergent to some $\alpha^* \geq 0$.
Notice that this can happen if $\lambda_n < 1, q_n < 1$ and $\alpha_n \leq \bar{\alpha}$ for each $n = 0, 1, 2, \ldots$ and some $\bar{\alpha} \geq 0$. In this case, we also have $0 \leq \alpha_n \leq \alpha_{n+1}\bar{\alpha} \leq \alpha^*$.

(C4) $U[x_0, \alpha^*] \subset \Omega$.

The motivational calculations are in turn:

$$\begin{aligned}\|L^{-1}(B_n-L)\| &\leq \psi(f(\|x_n-x_0\|), g(\|x_n-x_0\|)) = \bar{q}_n \leq q_n < 1,\\ \|B_n^{-1}L\| &\leq \frac{1}{1-q_n},\\ \|L^{-1}F(x_n)\| &\leq \psi_1(\|x_n-x_0\|) \leq \psi_1(\alpha_n),\\ \|L^{-1}D_n\| &\leq 2\psi(f(\|x_n-x_0\|), \|x_n-x_0\|, g(\|x_n-x_0\|)) = p_n < 1,\\ \|aA_n\| &\leq \frac{|a|p_n\psi_1(\alpha_n)}{(1-q_n)^2} = \lambda_n < 1\end{aligned}$$

and

$$\begin{aligned}\|x_{n+1}-x_0\| &\leq \|B_n^{-1}L\|\|L^{-1}F(x_n)\| \\ &\quad +\frac{1}{2}\|A_n\|\|(I-aA_n)^{-1}\|\|B_n^{-1}L\|\|L^{-1}F(x_n)\| \\ &\leq \left[1+\frac{1}{2|a|}\frac{\lambda_n}{1-\lambda_n}\right]\frac{\psi_1(\alpha_n)}{1-q_n}=\alpha_{n+1}-\alpha_n.\end{aligned}$$

The rest of the analysis appeared in earlier chapters.

3. References

[1] Amat, S., Bermúdez, C., Busquier, S., Mestiri, D., *A family of Halley-Chebyshev iterative schemes for non-Fréchet differentiable operators*, J. Comput. Appl. Math., 228, (2009), 486–493.

[2] Argyros, I. K., *Advances in the efficiency of computational methods and applications*, World Science, Singapore, 2000.

[3] Candela, V. F., Marquina, A., *Recurrence relations for rational cubic methods II: The Chebyshev method*, Computing, 45, (1990), 169–184.

[4] Ezquerro, J. A., Gutiérrez, J. M., Hernández, M. A., Salanova, M. A., *Chebyshev-like methods and quadratic equations*, Revue d'Analyse Numerique et theorie de l'Approximation, 28(1), (1999), 23–35.

[5] Gutiérrez, J. M., Hernández, M. A., *Third-order methods for operators with bounded second derivative*, J. Comput. Appl. Math., 82, (1997), 171–183.

[6] Hernndez, M. A., Romero, N., *On a characterization of some Newton-like methods of R-order at least three*, J. Comput. Appl. Math., 183(1), (2005), 53–66.

[7] Ortega, J. M., Rhienbolt, W. C., *Iterative solution of nonlinear equations in several variables*, Academic Press, New York, 1970.

[8] Potra, F. A., Pták, V., *Nondiscrete induction and iterative processes, in: Research Notes in Mathematics*, vol. 103, Pitman, Boston, 1984.

[9] Werner, W., *Newton-like methods for the computation of fixed points*, Comp. Math. Appl., 10(1), (1984), 77–86.

Chapter 56

Efficient Jarratt-Like Sixth-Order Methods

1. Introduction

The studied methods are defined for $x_0 \in \Omega$ and each $n = 0, 1, 2, \ldots$ by

$$\begin{aligned} y_n &= x_n - \frac{2}{3}F'(x_n)^{-1}F(x_n), A_n = 3F'(y_n) - F'(x_n), \\ z_n &= x_n - \frac{1}{2}A_n^{-1}(3F'(y_n) + F'(x_n))F'(x_n)^{-1}F(x_n), \\ x_{n+1} &= z_n - 2A_n^{-1}F(z_n) \end{aligned} \tag{56.1}$$

and

$$\begin{aligned} y_n &= x_n - \frac{2}{3}F'(x_n), \\ z_n &= x_n - \frac{1}{8}[5I + 3F'(y_n)^{-1}F'(x_n)]F'(x_n)^{-1}F(x_n), \\ x_{n+1} &= z_n + 2[(1+3b)F'(x_n) - (3+5b)F'(y_n)] \\ &\quad \times (F'(x_n) + bF'(y_n))F'(x_n)^{-1}F(z_n), \end{aligned} \tag{56.2}$$

where $b \in \mathbb{R} - \{-1, 0\}$.

The local sixth convergence is determined in [3,8], respectively using the Taylor series expansion approach for $B_1 = B_2 = \mathbb{R}^k$ and provided that $F^{(7)}$ exists and is bounded. Other limitations of this approach are reported in previous chapters. That is why the new local convergence is based only on operators on the methods. Moreover, the more interesting semi-local convergence (not given in [3,8]) is presented based on majorizing sequences and in the setting of a Banach space. The same technique can be used to extend the applicability of other methods [1,2], [4]-[7].

2. Local Convergence

Method (56.1)

The majorant functions $g_i, i = 1,2,3$ are:

$$
\begin{aligned}
g_1(t) &= \frac{\int_0^1 w((1-\theta)t)d\theta + \frac{1}{3}\left(1+\int_0^1 w_0(\theta t)d\theta\right)}{1-w_0(t)}, \\
q(t) &= \frac{1}{2}(3w_0(g_1(t)t)+w_0(t)), \\
\bar{w}(t) &= \begin{cases} w((1+g_1(t))t) \\ w_0(t)+w_0(g_1(t)t) \end{cases}, \\
g_2(t) &= \frac{\int_0^1 w((1-\theta)t)d\theta}{1-w_0(t)} \\
&+ \frac{3\bar{w}(t)\left(1+\int_0^1 w_0(\theta t)d\theta\right)}{4(1-q(t))(1-w_0(t))}, \\
\bar{\bar{w}}(t) &= \begin{cases} w((g_1(t)+g_2(t))t) \\ w_0(g_1(t)t)+w_0(g_2(t)t) \end{cases}
\end{aligned}
$$

and

$$
\begin{aligned}
g_3(t) &= \left[\frac{\int_0^1 w((1-\theta)g_2(t)t)d\theta}{(1-w_0(g_2(t)t))}\right. \\
&+ \left.\frac{(2\bar{\bar{w}}(t)+\bar{w}(t))\left(1+\int_0^1 w_0(\theta g_2(t)t)d\theta\right)}{2(1-w_0(g_2(t)t))(1-q(t))}\right] g_2(t).
\end{aligned}
$$

The motivation is in turn:

$$
\begin{aligned}
y_n - x^* &= x_n - x^* - F'(x_n)^{-1}F(x_n) + \frac{1}{3}F'(x_n)F(x_n), \\
\|y_n - x^*\| &\leq \frac{1}{1-w_0(\|x_n-x^*\|)}\left[\int_0^1 w((1-\theta)\|x_n-x^*\|)d\theta\right. \\
&\left. + \frac{1}{3}\left(1+\int_0^1 w_0(\theta\|x_n-x^*\|)d\theta\right)\right]\|x_n-x^*\| \\
&\leq g_1(\|x_n-x^*\|)\|x_n-x^*\| \leq \|x_n-x^*\| < r, \\
z_n - x^* &= x_n - x^* - F'(x_n)^{-1}F(x_n) \\
&+ \left[I - \frac{1}{2}A_n^{-1}(3F'(y_n)+F'(x_n))\right]F'(x_n)^{-1}F(x_n) \\
&= x_n - x^* - F'(x_n)^{-1}F(x_n) \\
&+ \frac{3}{2}A_n^{-1}(F'(y_n)-F'(x_n))F'(x_n)^{-1}F(x_n),
\end{aligned}
$$

$$\begin{aligned}
\|z_n - x^*\| &\leq \left[\frac{\int_0^1 w((1-\theta)\|x_n - x^*\|)d\theta}{1 - w_0(\|x_n - x^*\|)} \right. \\
&\quad \left. + \frac{3\bar{w}_n \left(1 + \int_0^1 w_0(\theta\|x_n - x^*\|)d\theta\right)}{4(1-q_n)(1 - w_0(\|x_n - x^*\|))} \right] \|x_n - x^*\| \\
&\leq g_2(\|x_n - x^*\|)\|x_n - x^*\| \leq \|x_n - x^*\|, \\
x_{n+1} - x^* &= z_n - x^* - F'(z_n)^{-1}F(z_n) + (F'(z_n)^{-1} - 2A_n^{-1})F(z_n) \\
&= z_n - x^* - F'(z_n)^{-1} + F'(z_n)^{-1}(A_n - 2F'(z_n))A_n^{-1}F(z_n) \\
&\quad + F'(z_n)^{-1}[2(F'(y_n) - F'(z_n)) + (F'(y_n) - F'(x_n))]A_n^{-1}F(z_n)
\end{aligned}$$

and

$$\begin{aligned}
\|x_{n+1} - x^*\| &\leq \left[\frac{\int_0^1 w((1-\theta)\|z_n - x^*\|)d\theta}{1 - w_0(\|z_n - x^*\|)} \right. \\
&\quad \left. + \frac{(2\bar{\bar{w}}_n + \bar{w}_n)\left(1 + \int_0^1 w_0(\theta\|z_n - x^*\|)d\theta\right)}{2(1 - w_0(\|z_n - x^*\|))(1-q_n)} \right] \|z_n - x^*\| \\
&\leq g_3(\|x_n - x^*\|)\|x_n - x^*\| \leq \|x_n - x^*\|.
\end{aligned}$$

Method (56.2)

The function g_1 is as defined in method (56.1). Moreover, the other two functions are:

$$\begin{aligned}
g_2(t) &= \frac{\int_0^1 w((1-\theta)t)d\theta}{1 - w_0(t)} \\
&\quad + \frac{3}{8}\frac{\bar{w}(t)}{1 - w_0(t)}\left(2 + \frac{\bar{w}(t)}{1 - w_0(t)}\right)\frac{\left(1 + \int_0^1 w_0(\theta t)d\theta\right)}{1 - w_0(t)}
\end{aligned}$$

and

$$\begin{aligned}
g_3(t) &= \left[\frac{\int_0^1 w((1-\theta)g_2(t)t)d\theta}{1 - w_0(g_2(t)t)} \right. \\
&\quad + \frac{\bar{\bar{w}}(t)\left(1 + \int_0^1 w_0(\theta g_2(t)t)d\theta\right)}{(1 - w_0(t))(1 - w_0(g_2(t)t))} \\
&\quad \left. + \frac{3\bar{w}(t)\left(1 + \int_0^1 w_0(\theta g_2(t)t)d\theta\right)}{2(1 - q(t))(1 - w_0(t))} \right] g_2(t).
\end{aligned}$$

where

$$q(t) = \frac{1}{2|1+b|}[|1 + 3b|w_0(t) + |3 + 5b|w_0(g_1(t)t)].$$

The calculations are in turn:

$$\begin{aligned}
z_n - x^* &= x_n - x^* - F'(x_n)^{-1}F(x_n) \\
&+\frac{3}{8}(I-(F'(y_n)^{-1}F'(x_n))^2)F'(x_n)^{-1}F(x_n), \\
\|z_n - x^*\| &\leq \left[\frac{\int_0^1 w((1-\theta)\|x_n-x^*\|)d\theta}{1-w_0(\|x_n-x^*\|)}\right. \\
&+\frac{3}{8}\frac{\bar{w}_n}{1-w_0(\|x_n-x^*\|)}\left(2+\frac{\bar{w}_n}{1-w_0(\|x_n-x^*\|)}\right) \\
&\left.\frac{\left(1+\int_0^1 w_0(\theta\|x_n-x^*\|)\right)d\theta}{1-w_0(\|x_n-x^*\|)}\right]\|x_n-x^*\| \\
&\leq g_2(\|x_n-x^*\|)\|x_n-x^*\| \leq \|x_n-x^*\|,
\end{aligned}$$

$$\begin{aligned}
x_{n+1} - x^* &= z_n - x^* - F'(z_n)^{-1}F(z_n) + (F'(z_n)^{-1} - F'(x_n)^{-1})F(z_n) \\
&+ \left(I + 2A_n^{-1}(F'(x_n) + bF'(y_n))\right) F'(x_n)^{-1}F(z_n) \\
&= z_n - x^* - F'(z_n)^{-1}F(z_n) \\
&+F'(z_n)^{-1}(F'(x_n) - F'(z_n))F'(x_n)^{-1}F(z_n) \\
&-3(a+b)A_n^{-1}(F'(y_n) - F'(x_n))F'(x_n)^{-1}F(z_n), \\
\|x_n - x^*\| &\leq \left[\frac{\int_0^1 w((1-\theta)\|z_n-x^*\|)d\theta}{1-w_0(\|z_n-x^*\|)}\right. \\
&+\frac{\bar{\bar{w}}_n\left(1+\int_0^1 w_0(\theta\|z_n-x^*\|)d\theta\right)}{(1-w_0(\|x_n-x^*\|))(1-w_0(\|z_n-x^*\|))} \\
&\left.+\frac{3|1+b|\bar{w}_n\left(1+\int_0^1 w_0(\theta\|z_n-x^*\|)d\theta\right)}{2|1+b|(1-q_n)(1-w_0(\|x_n-x^*\|))}\right]\|z_n-x^*\|. \\
&\leq g_3(\|x_n-x^*\|)\|x_n-x^*\| \leq \|x_n-x^*\|.
\end{aligned}$$

3. Semi-local Convergence

Method (56.1)

The majorant sequence $\{\alpha_n\}$ is defined for $\alpha_0 = 0, \beta_0 \geq \frac{2}{3}\|F'(x_0)^{-1}F(x_0)\|$ and each $n = 0, 1, 2, \ldots$ by

$$\bar{q}_n = \frac{1}{2}(3v_0(\beta_n) + v_0(\alpha_n)),$$

$$\bar{v}_n = \left\{\begin{array}{l} v(\beta_n - \alpha_n) \\ v_0(\alpha_n) + v_0(\beta_n) \end{array}\right.,$$

$$\begin{aligned}
\gamma_n &= \beta_n + \frac{1}{8(1-\bar{q}_n)}(3\bar{v}_n + 4(1+v_0(\alpha_n)))(\beta_n - \alpha_n),\\
\lambda_n &= \left(1 + \int_0^1 v_0(\alpha_n + \theta(\gamma_n - \alpha_n))d\theta\right)(\gamma_n - \alpha_n)\\
&\quad + \frac{3}{2}(1+v_0(\alpha_n))(\beta_n - \alpha_n),\\
\alpha_{n+1} &= \gamma_n + \frac{\lambda_n}{1-\bar{q}_n},\\
\delta_{n+1} &= \left(1 + \int_0^1 v_0(\alpha_n + \theta(\alpha_{n+1} - \alpha_n))d\theta\right)(\alpha_{n+1} - \alpha_n)\\
&\quad + \frac{3}{2}(1+v_0(\alpha_n))(\beta_n - \alpha_n)
\end{aligned}$$

and

$$\beta_{n+1} = \alpha_{n+1} + \frac{2\delta_{n+1}}{3(1-v_0(\alpha_{n+1}))}.$$

The calculations are:

$$\begin{aligned}
z_n - y_n &= \left[\frac{2}{3}I - \frac{1}{2}A_n^{-1}(3F'(y_n) + F'(x_n))\right]\left(-\frac{3}{2}(y_n - x_n)\right)\\
&= -\frac{1}{4}A_n^{-1}[3(F'(y_n)^{-1} - F'(x_n)) - 4F'(x_n)](y_n - x_n),\\
\|z_n - y_n\| &\leq \frac{1}{8(1-\bar{q}_n)}[3\bar{v}_n + 4(1+v_0(\alpha_n))](\beta_n - \alpha_n) = \gamma_n - \beta_n,\\
F(z_n) &= F(z_n) - F(x_n) - \frac{3}{2}F'(x_n)(y_n - x_n),\\
\|L^{-1}F(z_n)\| &\leq \left(1 + \int_0^1 v_0(\alpha_n + \theta(\gamma_n - \alpha_n)d\theta)\right)(\gamma_n - \alpha_n)\\
&\quad + \frac{3}{2}(1+v_0(\alpha_n))(\beta_n - \alpha_n) = \lambda_n,\\
\|x_{n+1} - z_n\| &\leq 2\|A_n^{-1}L\|\|L^{-1}F(z_n)\|\\
&\leq \frac{2\lambda_n}{2(1-\bar{q}_n)} = \alpha_{n+1} - \gamma_n,\\
F(x_{n+1}) &= F(x_{n+1}) - F(x_n) - \frac{3}{2}F'(x_n)(y_n - x_n),\\
\|L^{-1}F(x_{n+1})\| &\leq \left(1 + \int_0^1 v_0(\alpha_n + \theta(\alpha_{n+1} - \alpha_n))d\theta\right)(\alpha_{n+1} - \alpha_n)\\
&\quad + \frac{3}{2}(1+v_0(\alpha_n))(\beta_n - \alpha_n) = \delta_{n+1}
\end{aligned}$$

and

$$\begin{aligned}
\|y_{n+1} - x_{n+1}\| &\leq \frac{2}{3}\|F'(x_{n+1})^{-1}L\|\|L^{-1}F(x_{n+1})\|\\
&\leq \frac{2}{3}\frac{\bar{\delta}_{n+1}}{1-v_0(\|x_{n+1} - x_n\|)} \leq \frac{2}{3}\frac{\delta_{n+1}}{1-v_0(\alpha_{n+1})} = \beta_{n+1} - \alpha_{n+1}.
\end{aligned}$$

Method (56.2)

The majorant sequence $\{\alpha_n\}$ starts as in method (56.2) and for each $n = 0, 1, 2, \ldots$

$$\begin{aligned}
\gamma_n &= \beta_n + \frac{1}{16}\left[8 + 18\left(\frac{\bar{v}_n}{1 - v_0(\beta_n)}\right) + 9\left(\frac{\bar{v}_n}{1 - v_0(\beta_n)}\right)^2\right](\beta_n - \alpha_n),\\
\mu_n &= \left(1 + \int_0^1 v_0(\alpha_n + \theta(\gamma_n - \alpha_n))d\theta\right)(\gamma_n - \alpha_n)\\
&\quad + \frac{3}{2}(1 + v_0(\alpha_n))(\beta_n - \alpha_n),\\
\alpha_{n+1} &= \gamma_n + \frac{[(1 + v_0(\alpha_n)) + |b|(1 + v_0(\beta_n)) + 1 + |b|]\mu_n}{2|1 + b|(1 - \bar{q}_n)(1 - v_0(\alpha_n))}
\end{aligned}$$

whereas the iterate β_{n+1} is as defined in method (56.1).

The calculations are in turn:

$$\begin{aligned}
z_n - y_n &= \left[\frac{2}{3}I - \frac{1}{8}(5I + 3(F'(y_n)^{-1}F'(x_n))^2)\right]\left(-\frac{3}{2}(y_n - x_n)\right)\\
&= \frac{1}{16}[8I + 9(F'(y_n)^{-1}F'(x_n) - I)^2 + 18(F'(y_n)^{-1}F'(x_n) - I)](y_n - x_n),\\
\|z_n - y_n\| &\leq \frac{1}{16}\left(8 + 9\left(\frac{\bar{v}_n}{1 - v_0(\beta_n)}\right)^2 + 18\left(\frac{\bar{v}_n}{1 - v_0(\beta_n)}\right)\right)(\beta_n - \alpha_n)\\
&= \gamma_n - \beta_n,\\
x_{n+1} - z_n &= 2A_n^{-1}(F'(x_n) + bF'(y_n))F'(x_n)^{-1}F(z_n)
\end{aligned}$$

and

$$\|x_{n+1} - z_n\| \leq \frac{[(1 + v_0(\alpha_n)) + |b|(1 + v_0(\beta_n)) + 1 + |b|]\mu_n}{2|1 + b|(1 - \bar{q}_n)(1 - v_0(\alpha_n))},$$

where we also used

$$\begin{aligned}
F(z_n) &= F(z_n) - F(x_n) + F(x_n),\\
\|L^{-1}F(z_n)\| &\leq \left(1 + \int_0^1 v_0(\alpha_n + \theta(\gamma_n - \alpha_n))d\theta\right)(\gamma_n - \alpha_n)\\
&\quad + \frac{3}{2}(1 + v_0(\alpha_n))(\beta_n - \alpha_n) = \mu_n
\end{aligned}$$

and

$$\bar{q}_n = \frac{1}{2|1 + b|}(|1 + 3b|v_0(\alpha_n) + |3 + 5b|v_0(\beta_n)).$$

The rest of the analysis appeared in earlier chapters.

4. References

[1] Argyros, I. K., *Convergence and Application of Newton-Type Iterations*, Springer, 2008.

[2] Argyros, I. K., George, S., *High convergence order Q-step methods for solving equations and systems of equations*, Contemp. Math., 1(3), (2020), 119–126, http://dx.doi.org/10.37256/cm.132020403.

[3] Argyros, I. K., Hilout, S., *Numerical Methods in Nonlinear Analysis*, World Scientific Publ. Comp., New Jersey, 2013.

[4] Behl, R., Srria, I., González, R., Magrenán, A. A., *Highly efficient family of iterative methods for solving nonlinear models*, J. Comput. Appl. Math., 346, (2019), 110–132.

[5] Chicharro, F. I., Cordero, A., Garrido, N., Torregrosa, J. R., *On the improvement of the order of convergence of iterative methods for solving nonlinear systems by means of memory*, Appl. Math. Lett., 104, (2020), 106277, http://dx.doi.org/10.1016/j.aml.2020.106277.

[6] Cordero, A., Hueso, J. L., Martnez, E., Torregrosa, J. R., *A modified Newton-Jarratts composition*, Numer. Algorithms, 55, (2010), 87–99.

[7] Grau-Sánchez, M., Noguera, Á., Amat, S., *On the approximation of derivatives using divided difference operators preserving the local convergence order of iterative methods*, J. Comput. Appl. Math., 237, (2013), 363–372.

[8] Sharma, J. R., Kumar, D., Jántschi, L., *On a reduced cost higher order Traub-Steffensen-Like method for nonlinear systems*, Symmetry, 11(7), (2019), 891, http://dx.doi.org/10.3390/sym11070891.

Chapter 57

Fifth-Order Solvers

1. Introduction

The four three step methods are defined for $x_0 \in \Omega$ and for each $n = 0, 1, 2, \ldots$ by

$$\begin{aligned}
y_n &= x_n - F'(x_n)^{-1}F(x_n), & (57.1)\\
A_n &= 5F'(y_n) - F'(x_n), B_n = F'(y_n) + 3F'(x_n), \\
x_{n+1} &= y_n - A_n^{-1}B_nF'(x_n)^{-1}F(y_n), \\
y_n &= x_n - F'(x_n)^{-1}F(x_n), & (57.2)\\
z_n &= x_n - 2(F'(x_n) + F'(y_n))^{-1}F(x_n), \\
x_{n+1} &= z_n - F'(y_n)^{-1}F(z_n), \\
y_n &= x_n - \frac{1}{2}F'(x_n)^{-1}F(x_n), & (57.3)\\
z_n &= x_n - F'(y_n)^{-1}F(x_n), \\
x_{n+1} &= z_n - [2F'(y_n)^{-1} - F'(x_n)^{-1}]F(z_n), \\
y_n &= x_n - \frac{2}{3}F'(x_n)^{-1}F(x_n), & (57.4)\\
z_n &= x_n - 4(F'(x_n) + 3F'(y_n))^{-1}F(x_n), \\
x_{n+1} &= z_n - (8(F'(x_n) + 3F'(y_n))^{-1} - F'(x_n)^{-1})F(z_n).
\end{aligned}$$

The local fifth convergence order is established in [1,2,3,5], respectively under the Taylor series expansion technique whose limitations are reported in earlier chapters. The new local convergence uses conditions only on the methods and avoids the assumption on the sixth derivative not on these methods. Moreover, the semi-local convergence not studied in [1,2,3,5] is also given in this chapter based on majorizing sequences. The technique can be used to extend the applicability of other methods in a similar way [4].

2. Local Convergence

Method (57.1)

The majorant functions $g_i, i = 1,2$ are:

$$\begin{aligned} g_1(t) &= \frac{\int_0^1 w((1-\theta)t)d\theta}{1-w_0(t)}, \\ q(t) &= \frac{1}{4}(5w_0(g_1(t)t) + w_0(t)), \end{aligned}$$

$$\bar{w}(t) = \begin{cases} w((1+g_1(t))t) \\ w_0(t) + w_0(g_1(t)t) \end{cases},$$

and

$$\begin{aligned} g_2(t) &= \Bigg[\frac{\int_0^1 w((1-\theta)g_1(t)t)d\theta}{1-w_0(g_1(t)t)} + \frac{\bar{w}(t)\left(1+\int_0^1 w_0(\theta g_1(t)t)d\theta\right)}{(1-w_0(t))(1-w_0(g_1(t)t))} \\ &\quad + \frac{\bar{w}(t)\left(1+\int_0^1 w_0(\theta g_1(t)t)d\theta\right)}{(1-w_0(t))(1-q(t))}\Bigg] g_1(t). \end{aligned} \tag{57.5}$$

The motivational calculations are:

$$\begin{aligned} y_n - x^* &= x_n - x^* - F'(x_n)^{-1}F(x_n), \\ \|y_n - x^*\| &\leq \frac{\int_0^1 w((1-\theta)\|x_n - x^*\|)d\theta \|x_n - x^*\|}{1-w_0(\|x_n - x^*\|)} \\ &\leq g_1(\|x_n - x^*\|)\|x_n - x^*\| \leq \|x_n - x^*\| < r, \\ x_{n+1} - x^* &= y_n - x^* - F'(y_n)^{-1}F(y_n) + (F'(y_n)^{-1} - F'(x_n)^{-1})F(y_n) \\ &\quad + A_n^{-1}(A_n - B_n)F'(x_n)^{-1}F(y_n) \end{aligned}$$

and

$$\begin{aligned} \|x_{n+1} - x^*\| &\leq \Bigg[\frac{\int_0^1 w((1-\theta)\|y_n - x^*\|)d\theta}{1-w_0(\|y_n - x^*\|)} \\ &\quad + \frac{\bar{w}_n\left(1+\int_0^1 w_0(\theta\|y_n - x^*\|)d\theta\right)}{(1-w_0(\|x_n - x^*\|))(1-w_0(\|y_n - x^*\|))} \\ &\quad + \frac{\bar{w}_n\left(1+\int_0^1 w_0(\theta\|y_n - x^*\|)d\theta\right)}{(1-q_n)(1-w_0(\|x_n - x^*\|))}\Bigg] \|y_n - x^*\| \\ &\leq g_2(\|x_n - x^*\|)\|x_n - x^*\| \leq \|x_n - x^*\|. \end{aligned}$$

Method (57.2)

The majorant functions $g_i, i = 1,2,3$ are:

$$\begin{aligned} g_1(t) &= \frac{\int_0^1 w((1-\theta)t)d\theta}{1-w_0(t)}, \\ \bar{q}(t) &= \frac{1}{2}(w_0(g_1(t)t)+w_0(t)), \\ g_2(t) &= \frac{\int_0^1 w((1-\theta)t)d\theta}{1-w_0(t)} + \frac{\bar{w}(t)\left(1+\int_0^1 w_0(\theta t)d\theta\right)}{2(1-q(t))(1-w_0(t))} \end{aligned}$$

$$\bar{\bar{w}}(t) = \begin{cases} w((g_1(t)+g_2(t))t) \\ w_0(g_1(t)t)+w_0(g_2(t)t) \end{cases},$$

and

$$g_3(t) = \left[\frac{\int_0^1 w((1-\theta)g_2(t)t)d\theta}{1-w_0(g_2(t)t)} + \frac{\bar{\bar{w}}(t)\left(1+\int_0^1 w_0(\theta g_2(t)t)d\theta\right)}{(1-w_0(g_1(t)t))(1-w_0(g_2(t)t))}\right] g_2(t).$$

The motivational calculation for the function g_1 is given in method (57.1). Moreover, we obtain

$$\begin{aligned} z_n - x^* &= x_n - x^* - F'(x_n)^{-1}F(x_n) \\ &\quad +(F'(x_n)^{-1} - 2(F'(x_n)+F'(y_n))^{-1})F(x_n) \\ &= x_n - x^* - F'(x_n)^{-1}F(x_n) \\ &\quad +(F'(x_n)+F'(y_n))^{-1}(F'(y_n)-F'(x_n))F'(x_n)^{-1}F(x_n), \\ \|z_n - x^*\| &\le \left[\frac{\int_0^1 w((1-\theta)\|x_n-x^*\|)d\theta}{1-w_0(\|x_n-x^*\|)}\right. \\ &\quad \left. + \frac{\bar{w}_n\left(1+\int_0^1 w_0(\theta\|y_n-x^*\|)d\theta\right)}{2(1-q_n)(1-w_0(\|x_n-x^*\|))}\right]\|x_n-x^*\| \\ &\le g_2(\|x_n-x^*\|)\|x_n-x^*\| \le \|x_n-x^*\|, \\ x_{n+1}-x^* &= z_n - x^* - F'(z_n)^{-1}F(z_n) + (F'(z_n)^{-1} - F'(y_n)^{-1})F(z_n) \end{aligned}$$

and

$$\begin{aligned} \|x_{n+1}-x^*\| &\le \left[\frac{\int_0^1 w((1-\theta)\|z_n-x^*\|)d\theta}{1-w_0(\|z_n-x^*\|)}\right. \\ &\quad \left. + \frac{\bar{\bar{w}}_n\left(1+\int_0^1 w_0(\theta\|z_n-x^*\|)d\theta\right)}{(1-w_0(\|y_n-x^*\|))(1-w_0(\|z_n-x^*\|))}\right]\|z_n-x^*\| \\ &\le g_3(\|x_n-x^*\|)\|x_n-x^*\| \le \|x_n-x^*\|. \end{aligned}$$

Method (57.3)

The majorant functions $g_i, i = 1,2,3$ are:

$$g_1(t) = \frac{\int_0^1 w((1-\theta)t)d\theta + \frac{1}{2}\left(1+\int_0^1 w_0(\theta t)d\theta\right)}{1-w_0(t)},$$

$$g_2(t) = \frac{\int_0^1 w((1-\theta)t)d\theta}{1-w_0(t)} + \frac{\bar{w}(t)\left(1+\int_0^1 w_0(\theta t)d\theta\right)}{(1-w_0(t))(1-w_0(g_1(t)t))},$$

$$\bar{\bar{w}}(t) = \begin{cases} w((g_1(t)+g_2(t))t) \\ w_0(g_1(t)t)+w_0(g_2(t)t) \end{cases}$$

and

$$g_3(t) = \left[\frac{\int_0^1 w((1-\theta)g_2(t)t)d\theta}{1-w_0(g_2(t)t)} + \frac{\bar{\bar{w}}(t)\left(1+\int_0^1 w_0(\theta g_2(t)t)d\theta\right)}{(1-w_0(g_1(t)t))(1-w_0(g_2(t)t))}\right] g_2(t).$$

The motivational calculations are in turn:

$$\begin{aligned}
y_n - x^* &= x_n - x^* - F'(x_n)^{-1}F(x_n) + \frac{1}{2}F'(x_n)^{-1}F(x_n),\\
\|y_n - x^*\| &\leq \frac{1}{1-w_0(\|x_n-x^*\|)}\left[\int_0^1 w((1-\theta)\|x_n-x^*\|)d\theta\right.\\
&\quad \left.+\frac{1}{2}\left(1+\int_0^1 w_0(\theta\|x_n-x^*\|)d\theta\right)\right]\|x_n-x^*\|\\
&\leq g_1(\|x_n-x^*\|)\|x_n-x^*\| \leq \|x_n-x^*\| < r,\\
z_n - x^* &= x_n - x^* - F'(x_n)^{-1}F(x_n) + (F'(x_n)^{-1} - F'(y_n)^{-1})F(x_n),\\
\|z_n - x^*\| &\leq \left[\frac{\int_0^1 w((1-\theta)\|x_n-x^*\|)d\theta}{1-w_0(\|x_n-x^*\|)}\right.\\
&\quad \left.+\frac{\bar{w}_n\left(1+\int_0^1 w_0(\theta\|x_n-x^*\|)d\theta\right)}{(1-w_0(\|x_n-x^*\|))(1-w_0(\|y_n-x^*\|))}\right]\|x_n-x^*\|\\
&\leq g_2(\|x_n-x^*\|)\|x_n-x^*\| \leq \|x_n-x^*\|,\\
x_{n+1} - x^* &= z_n - x^* - F'(z_n)^{-1}F(z_n) + (F'(z_n)^{-1} - F'(y_n)^{-1})F(z_n)\\
&\quad -(F'(y_n)^{-1} - F'(x_n)^{-1})F(z_n)
\end{aligned}$$

and

$$\begin{aligned}
\|x_{n+1} - x^*\| &\leq \left[\frac{\int_0^1 w((1-\theta)\|z_n-x^*\|)d\theta}{1-w_0(\|z_n-x^*\|)}\right.\\
&\quad +\frac{\bar{\bar{w}}_n\left(1+\int_0^1 w_0(\theta\|z_n-x^*\|)d\theta\right)}{(1-w_0(\|y_n-x^*\|))(1-w_0(\|z_n-x^*\|))}\\
&\quad \left.+\frac{\bar{w}_n\left(1+\int_0^1 w_0(\theta\|z_n-x^*\|)d\theta\right)}{(1-w_0(\|x_n-x^*\|))(1-w_0(\|y_n-x^*\|))}\right]\|z_n-x^*\|\\
&\leq g_3(\|x_n-x^*\|)\|x_n-x^*\| \leq \|x_n-x^*\|.
\end{aligned}$$

Method (57.4):

The majorant functions $g_i, i = 1,2,3$ are:

$$
\begin{aligned}
g_1(t) &= \frac{\int_0^1 w((1-\theta)t)d\theta + \frac{1}{3}\left(1+\int_0^1 w_0(\theta t)d\theta\right)}{1-w_0(t)},\\
q(t) &= \frac{1}{4}(3w_0(g_1(t)t)+w_0(t)),\\
g_2(t) &= \frac{\int_0^1 w((1-\theta)t)d\theta}{1-w_0(t)} + \frac{3\bar{w}(t)\left(1+\int_0^1 w_0(\theta t)d\theta\right)}{4(1-q(t))(1-w_0(t))},
\end{aligned}
$$

and

$$
\begin{aligned}
g_3(t) &= \left[\frac{\int_0^1 w((1-\theta)g_2(t)t)d\theta}{1-w_0(g_2(t)t)}\right.\\
&\quad + \frac{(\bar{\bar{w}}(t)+3(1+w_0(g_1(t)t)))\left(1+\int_0^1 w_0(\theta g_2(t)t)d\theta\right)}{4(1-q(t))(1-w_0(g_2(t)t))}\\
&\quad + \frac{3(1+w_0(g_1(t)t))\left(1+\int_0^1 w_0(\theta g_2(t)t)d\theta\right)}{4(1-q(t))(1-w_0(t))}\\
&\quad \left. + \frac{3\left(1+\int_0^1 w_0(\theta g_2(t)t)d\theta\right)}{2(1-q(t))}\right] g_2(t).
\end{aligned}
$$

The motivational calculations are in turn:

$$
\begin{aligned}
y_n - x^* &= x_n - x^* - F'(x_n)^{-1}F(x_n) + \frac{1}{3}F'(x_n)F(x_n)\\
\|y_n - x^*\| &\leq \frac{1}{1-w_0(\|x_n-x^*\|)}[\int_0^1 w((1-\theta)\|x_n-x^*\|)d\theta\\
&\quad + \frac{1}{3}\left(1+\int_0^1 w_0(\theta\|x_n-x^*\|)\right)d\theta]\|x_n-x^*\|\\
&\leq g_1(\|x_n-x^*\|)\|x_n-x^*\| \leq \|x_n-x^*\| < r,\\
z_n - x^* &= x_n - x^* - F'(x_n)^{-1}F(x_n) - (F'(x_n)+3F'(y_n))^{-1}\\
&\quad \times(4F'(x_n)-F'(x_n)-3F'(y_n))F'(x_n)^{-1}F(x_n)\\
&= x_n - x^* - F'(x_n)^{-1}F(x_n)\\
&\quad +3F'(y_n)^{-1}(F'(y_n)-F'(x_n))F'(x_n)^{-1}F(x_n),
\end{aligned}
$$

$$\begin{aligned}
\|z_n - x^*\| &\leq \left[\frac{\int_0^1 w((1-\theta)\|x_n - x^*\|)d\theta}{1 - w_0(\|x_n - x^*\|)}\right. \\
&\left. + \frac{3\bar{w}\left(1 + \int_0^1 w_0(\theta\|y_n - x^*\|)d\theta\right)}{4(1-q_n)(1 - w_0(\|x_n - x^*\|))}\right]\|x_n - x^*\| \\
&\leq g_2(\|x_n - x^*\|)\|x_n - x^*\| \leq \|x_n - x^*\|, \\
x_{n+1} - x^* &= z_n - x^* - F'(z_n)^{-1}F(z_n) \\
&+ (F'(z_n)^{-1} - (F'(x_n) + 3F'(y_n))^{-1})F(z_n) \\
&- ((F'(x_n) + 3F'(y_n))^{-1} - F'(x_n)^{-1})F(z_n) \\
&- 6(F'(x_n) + 3F'(y_n))^{-1}F(z_n)
\end{aligned}$$

and

$$\begin{aligned}
\|x_{n+1} - x^*\| &\leq \left[\frac{\int_0^1 w((1-\theta)\|z_n - x^*\|)d\theta}{1 - w_0(\|z_n - x^*\|)}\right. \\
&+ \frac{(\bar{\bar{w}}_n + 3(1 + w_0(\|y_n - x^*\|)))\left(1 + \int_0^1 w_0(\theta\|z_n - x^*\|)d\theta\right)}{4(1-q_n)(1 - w_0(\|z_n - x^*\|))} \\
&+ \frac{3(1 + w_0(\|y_n - x^*\|))\left(1 + \int_0^1 w_0(\theta\|z_n - x^*\|)d\theta\right)}{4(1-q_n)(1 - w_0(\|x_n - x^*\|))} \\
&\left. + \frac{3\left(1 + \int_0^1 w_0(\theta\|z_n - x^*\|)d\theta\right)}{2(1-q_n)}\right]\|z_n - x^*\| \\
&\leq g_3(\|x_n - x^*\|)\|x_n - x^*\| \leq \|x_n - x^*\|.
\end{aligned}$$

3. Semi-local Convergence

The role of x^*,"w" is exchanged by x_0,"v" in the calculations.

Method (57.1):

The majorant sequence $\{\alpha_n\}$ is defined for $\alpha_0 = 0, \beta_0 \geq \|F'(x_0)^{-1}F(x_0)\|$ and each $n = 0, 1, 2, \ldots$ by

$$\begin{aligned}
\lambda_n &= \int_0^1 v((1-\theta))(\beta_n - \alpha_n)d\theta(\beta_n - \alpha_n), \\
\bar{q}_n &= \frac{1}{4}(5w_0(\beta_n) + w_0(\alpha_n)), \\
\alpha_{n+1} &= \beta_n + \frac{\left(3 + \frac{1+v_0(\beta_n)}{1-v_0(\alpha_n)}\right)\lambda_n}{4(1-\bar{q}_n)} \\
\delta_{n+1} &= \int_0^1 v((1-\theta)(\alpha_{n+1} - \alpha_n))d\theta(\alpha_{n+1} - \alpha_n) + (1 + v_0(\alpha_n))(\beta_n - \alpha_n)
\end{aligned}$$

and

$$\beta_{n+1} = \alpha_{n+1} + \frac{\delta_{n+1}}{1 - v_0(\alpha_{n+1})}.$$

The motivational calculations are in turn:

$$\begin{aligned}
F(y_n) &= F(y_n)-F(x_n)-F'(x_n)(y_n-x_n),\\
\|L^{-1}F(y_n)\| &\leq \lambda_n,\\
x_{n+1}-y_n &= -A_n^{-1}B_nF'(x_n)^{-1}F(y_n),\\
\|x_{n+1}-y_n\| &\leq \frac{\left(3+\frac{1+v_0(\beta_n)}{1-v_0(\alpha_n)}\right)\lambda_n}{4(1-\bar{q}_n)}=\alpha_{n+1}-\beta_n,\\
\|y_{n+1}-x_{n+1}\| &\leq \|F'(x_{n+1})^{-1}L\|\|L^{-1}F(x_{n+1})\|\\
&\leq \frac{\delta_{n+1}}{1-v_0(\alpha_{n+1})}=\beta_{n+1}-\alpha_{n+1}.
\end{aligned}$$

Method (57.2)

The majorant sequence $\{\alpha_n\}$ is defined for $\alpha_0=0,\beta_0\geq\|F'(x_0)^{-1}F(x_0)\|$ and each $n=0,1,2,\ldots$ by

$$\bar{v}_n=\begin{cases} v(\beta_n-\alpha_n)\\ v_0(\alpha_n)+v_0(\beta_n),\end{cases}$$

$$\begin{aligned}
\bar{q}_n &= \frac{1}{2}(v_0(\beta_n)+v_0(\alpha_n)),\\
\gamma_n &= \beta_n+\frac{\bar{v}_n(\beta_n-\alpha_n)}{2(1-\bar{q}_n)},\\
\mu_n &= \left(1+\int_0^1 v_0(\beta_n+\theta(\gamma_n-\beta_n)d\theta)\right)(\gamma_n-\beta_n)\\
&\quad+\int_0^1 v((1-\theta)(\beta_n-\alpha_n))d\theta(\beta_n-\alpha_n),\\
\alpha_{n+1} &= \gamma_n+\frac{\mu_n}{1-v_0(\beta_n)},\\
\delta_{n+1} &= \int_0^1 v((1-\theta)(\alpha_{n+1}-\alpha_n))d\theta(\alpha_{n+1}-\alpha_n)+(1+v_0(\alpha_n))(\beta_n-\alpha_n)
\end{aligned}$$

and

$$\beta_{n+1} = \alpha_{n+1}+\frac{\delta_{n+1}}{1-v_0(\alpha_{n+1})}.$$

The motivational calculations are in turn:

$$\begin{aligned}
\|z_n - y_n\| &\leq \frac{\bar{v}_n(\beta_n - \alpha_n)}{2(1-\bar{q}_n)} = \gamma_n - \beta_n, \\
F(z_n) &= F(z_n) - F(y_n) + F(y_n), \\
\|L^{-1}F(z_n)\| &\leq \left(1 + \int_0^1 v_0(\beta_n + \theta(\gamma_n - \beta_n))d\theta\right)(\gamma_n - \beta_n) \\
&\quad + \int_0^1 v((1-\theta)(\beta_n - \alpha_n))d\theta(\beta_n - \alpha_n) = \mu_n, \\
\|x_{n+1} - z_n\| &\leq \|F'(y_n)^{-1}L\|\|L^{-1}F(z_n)\| \\
&\leq \frac{\mu_n}{1 - v_0(\beta_n)} = \alpha_{n+1} - \gamma_n, \\
\|y_{n+1} - x_{n+1}\| &\leq \|F'(x_{n+1})^{-1}L\|\|L^{-1}F(x_{n+1})\| \\
&\leq \frac{\delta_{n+1}}{1 - v_0(\alpha_{n+1})} = \beta_{n+1} - \alpha_{n+1}.
\end{aligned}$$

Method (57.3)

The majorant sequence $\{\alpha_n\}$ is defined for $\alpha_0 = 0, \beta_0 \geq \frac{1}{2}\|F'(x_0)^{-1}F(x_0)\|$ and each $n = 0, 1, 2, \ldots$ by

$$\begin{aligned}
\gamma_n &= \beta_n + \frac{1}{1 - v_0(\beta_n)}(1 + v_0(\alpha_n) + \bar{v}_n)(\beta_n - \alpha_n), \\
\xi_n &= \left(1 + \int_0^1 v_0(\alpha_n + \theta(\gamma_n - \alpha_n))d\theta\right)(\gamma_n - \alpha_n) \\
&\quad + 2(1 + v_0(\alpha_n))(\beta_n - \alpha_n), \\
\alpha_{n+1} &= \gamma_n + \frac{1}{1 - v_0(\beta_n)}\left(1 + \frac{\bar{v}_n}{1 - v_0(\alpha_n)}\right)\xi_n, \\
p_{n+1} &= \left(1 + \int_0^1 v_0(\alpha_n + \theta(\alpha_{n+1} - \alpha_n))d\theta\right)(\alpha_{n+1} - \alpha_n) \\
&\quad + 2(1 + v_0(\alpha_n))(\beta_n - \alpha_n)
\end{aligned}$$

and

$$\beta_{n+1} = \alpha_{n+1} + \frac{1}{2}\frac{p_{n+1}}{1 - v_0(\alpha_{n+1})}.$$

The motivational calculations are in turn:

$$\begin{aligned}
z_n - y_n &= \left[\frac{1}{2}F'(x_n)^{-1} - F'(y_n)^{-1}\right]F(x_n)\\
&= \frac{1}{2}F'(y_n)^{-1}[F'(x_n) + (F'(x_n) - F'(y_n))](y_n - x_n),\\
\|z_n - y_n\| &\leq \frac{1}{1-v_0(\beta_n)}(1 + v_0(\alpha_n) + \bar{v}_n)(\beta_n - \alpha_n) = \gamma_n - \beta_n,\\
x_{n+1} - z_n &= -F'(y_n)^{-1}F(z_n) - F'(y_n)^{-1}(F'(x_n) - F'(y_n))F'(x_n)^{-1}F(z_n),\\
F(z_n) &= F(z_n) - F(x_n) + F(x_n)\\
&= F(z_n) - F(x_n) - 2F'(x_n)(y_n - x_n),\\
\|L^{-1}F(z_n)\| &\leq \xi_n,\\
\|x_{n+1} - z_n\| &\leq \left[\frac{1}{1-v_0(\beta_n)} + \frac{\bar{v}_n}{(1-v_0(\beta_n))(1-v_0(\alpha_n))}\right]\xi_n\\
&= \alpha_{n+1} - \gamma_n,\\
F(x_{n+1}) &= F(x_{n+1}) - F(x_n) - 2F'(x_n)(y_n - x_n),\\
\|L^{-1}F(x_{n+1})\| &\leq p_{n+1}
\end{aligned}$$

and

$$\begin{aligned}
\|y_{n+1} - x_{n+1}\| &\leq \frac{1}{2}\|F'(x_{n+1})^{-1}L\|\|L^{-1}F(x_{n+1})\|\\
&\leq \frac{1}{2}\frac{p_{n+1}}{1-v_0(\alpha_{n+1})} = \beta_{n+1} - \alpha_{n+1}.
\end{aligned}$$

Method (57.4)

The majorant sequence $\{\alpha_n\}$ is defined for $\alpha_0 = 0, \beta_0 \geq \frac{2}{3}\|F'(x_0)^{-1}F(x_0)\|$ and each $n = 0, 1, 2, \ldots$ by

$$\begin{aligned}
\bar{q}_n &= \frac{1}{4}(3v_0(\beta_n) + v_0(\alpha_n)),\\
\gamma_n &= \beta_n + \frac{(3\bar{v}_n + 2(1 + v_0(\alpha_n)))(\beta_n - \alpha_n)}{4(1-\bar{q}_n)},\\
a_n &= \left(1 + \int_0^1 v_0(\alpha_n + \theta(\gamma_n - \alpha_n))d\theta\right)(\gamma_n - \alpha_n)\\
&\quad + \frac{3}{2}(1 + v_0(\alpha_n))(\beta_n - \alpha_n),\\
\alpha_{n+1} &= \gamma_n + \left[\frac{2}{1-\bar{q}_n} + \frac{1}{1-v_0(\alpha_n)}\right]a_n,\\
b_{n+1} &= \left(1 + \int_0^1 v_0(\alpha_n + \theta(\alpha_{n+1} - \alpha_n))d\theta\right)(\alpha_{n+1} - \alpha_n)\\
&\quad + \frac{3}{2}(1 + v_0(\alpha_n))(\beta_n - \alpha_n)
\end{aligned}$$

and

$$\beta_{n+1} = \alpha_{n+1} + \frac{2}{3}\frac{b_{n+1}}{1-v_0(\alpha_{n+1})}.$$

The motivational calculations are in turn:

$$
\begin{aligned}
z_n - y_n &= \left[\frac{2}{3}F'(x_n)^{-1} - 4(F'(x_n) - 3F'(y_n))^{-1}\right] F(x_n) \\
&= (F'(x_n) + 3F'(y_n))^{-1}[2F'(x_n) + 3(F'(x_n) - F'(y_n))](y_n - x_n),
\end{aligned}
$$

$$
\begin{aligned}
\|z_n - y_n\| &\leq \frac{(3\bar{v}_n + 2(1 + v_0(\alpha_n)))(\beta_n - \alpha_n)}{4(1 - \bar{q}_n)} = \gamma_n - \beta_n, \\
x_{n+1} - z_n &= [8(F'(x_n) + 3F'(y_n))^{-1} - F'(x_n)^{-1}]F(z_n), \\
\|x_{n+1} - z_n\| &\leq \left[\frac{8}{4(1 - \bar{q}_n)} + \frac{1}{1 - v_0(\alpha_n)}\right] a_n \\
&= \alpha_{n+1} - \gamma_n, \\
F(x_{n+1}) &= F(x_{n+1}) - F(x_n) - \frac{3}{2}F'(x_n)(y_n - x_n), \\
\|L^{-1}F(x_{n+1})\| &\leq b_{n+1}
\end{aligned}
$$

and

$$
\begin{aligned}
\|y_{n+1} - x_{n+1}\| &\leq \frac{2}{3}\|F'(x_{n+1})L\|\|L^{-1}F(x_{n+1})\| \\
&\leq \frac{2}{3}\frac{b_{n+1}}{1 - v_0(\alpha_{n+1})} = \beta_{n+1} - \alpha_{n+1}.
\end{aligned}
$$

The rest of the convergence analysis is given in earlier chapters.

4. References

[1] Cordero, A., Heuso, J. L., Martnez, E., Torregrosa, J. R., *Increasing the convergence order of an iterative method for nonlinear systems*, Applied Mathematics Letters, 25(12), (2012), 2369–2374. https://doi.org/10.1016/j.aml.2012.07.005.

[2] Ham, Y., Chun, C., *A fifth-order iterative method for solving nonlinear equations*, Applied Mathematics and Computation, 194, (2007), 287–290.

[3] Sharma, J. R., Gupta, P., *An efficient fifth order method for solving systems of nonlinear equations*, Computers and Mathematics with Applications, 67(3), (2014), 591–601. https://doi.org/10.1016/j.camwa.2013.12.004.

[4] Xiao, X. Y., Yin, H. W., *A new class of methods with higher order of convergence for solving systems of nonlinear equations*, Applied Mathematics and Computation, 264, (2015), 300–309. https://doi.org/10.1016/j.amc.2015.04.094.

[5] Xiao, X. Y., Yin, H. W., *Increasing the order of convergence for iterative methods to solve nonlinear systems*, Calcolo, 53(3), (2016), 285–300. https://doi.org/10.1007/s10092-0150149-9.

Chapter 58

The Parametric Family of Fourth-Order Methods

1. Introduction

The two step method is defined for $x_0 \in \Omega$ and each $n = 0, 1, 2, \ldots$ by

$$\begin{aligned} y_n &= x_n - \frac{2}{3}F'(x_n)^{-1}F(x_n), \\ A_n &= F'(x_n)^{-1}F'(y_n), a \neq 0 \end{aligned}$$

and

$$x_{n+1} = y_n - \left[I - \frac{3}{4}(A_n - I) + \frac{9}{8}(A_n - I)^2 + a(A_n - I)^3\right] F'(x_n)^{-1}F(x_n).$$

The local fourth convergence order four is established in [7] when $B_1 = B_2 = \mathbb{R}^k$ under the Taylor expansion series approach and provided that $F^{(5)}$ is bounded and continuous about the solution x^*. Other limitations with this approach have been reported in previous chapters. The new local convergence analysis uses only conditions on the operators of the method. Moreover, the more challenging semi-local convergence analysis (not given in [7]) is presented in this chapter. This is how we extend the applicability of the method. Furthermore, the same approach can be used to extend the applicability of other methods [1]-[6], [8, 9].

2. Local Convergence

The majorizing sequences $g_i, i = 1, 2$ are:

$$g_1(t) = \frac{\int_0^1 w((1-\theta)t)d\theta + \frac{1}{3}\left(1 + \int_0^1 w_0(\theta t)d\theta\right)}{1 - w_0(t)}$$

and

$$\bar{w}(t) = \begin{cases} w((1+g_1(t))t) \\ \text{or} \\ w_0(t) + w_0(g_1(t)t) \end{cases},$$

$$\begin{aligned} g_2(t) &= \frac{\int_0^1 w((1-\theta)t)d\theta}{1-w_0(t)} + \frac{2}{3}\frac{1+\int_0^1 w_0(\theta t)d\theta}{1-w_0(t)} \\ &+ \left(\frac{3}{4}\frac{\bar{w}(t)}{1-w_0(t)} + \frac{9}{8}\left(\frac{\bar{w}(t)}{1-w_0(t)}\right)^2 + |a|\left(\frac{\bar{w}(t)}{1-w_0(t)}\right)^2\right) \\ &\times \frac{\left(1+\int_0^1 w_0(\theta t)d\theta\right)}{1-w_0(t)}. \end{aligned}$$

The motivational calculations are in turn:

$$\begin{aligned} y_n - x^* &= x_n - x^* - F'(x_n)^{-1} + \frac{1}{3}F'(x_n)^{-1}F(x_n), \\ \|y_n - x^*\| &\leq \left[\frac{\int_0^1 w((1-\theta)\|x_n-x^*\|)d\theta}{1-w_0(\|x_n-x^*\|)} + \frac{1+\int_0^1 w_0(\theta\|x_n-x^*\|)d\theta}{3(1-w_0(\|x_n-x^*\|))}\right]\|x_n-x^*\| \\ &\leq g_1(\|x_n-x^*\|)\|x_n-x^*\| \leq \|x_n-x^*\| < r, \\ x_{n+1}-x^* &= x_n - x^* - F'(x_n)^{-1}F(x_n) - \frac{2}{3}F'(x_n)^{-1}F(x_n) \\ &- \left[-\frac{3}{4}(A_n-I) + \frac{9}{8}(A_n-I)^2 + a(A_n-I)^3\right]F'(x_n)^{-1}F(x_n) \end{aligned}$$

and

$$\begin{aligned} \|x_{n+1}-x^*\| &\leq \left[\frac{\int_0^1 w((1-\theta)\|x_n-x^*\|)d\theta}{1-w_0(\|x_n-x^*\|)} + \frac{2}{3}\frac{\left(1+\int_0^1 w_0(\theta\|x_n-x^*\|)d\theta\right)}{1-w_0(\|x_n-x^*\|)}\right. \\ &+ \left(\frac{3}{4}\frac{\bar{w}_n}{1-w_0(\|x_n-x^*\|)} + \frac{9}{8}\left(\frac{\bar{w}_n}{1-w_0(\|x_n-x^*\|)}\right)\right. \\ &\left. + |a|\left(\frac{\bar{w}_n}{1-w_0(\|x_n-x^*\|)}\right)^2\right) \\ &\left.\times \frac{\left(1+\int_0^1 w_0(\theta\|x_n-x^*\|)d\theta\right)}{1-w_0(\|x_n-x^*\|)}\right]\|x_n-x^*\| \\ &\leq g_2(\|x_n-x^*\|)\|x_n-x^*\| \leq \|x_n-x^*\|. \end{aligned}$$

3. Semi-local Convergence

The majorant sequence $\{\alpha_n\}$ is defined for $\alpha_0 = 0, \beta_0 \geq \frac{2}{3}\|F'(x_0)^{-1}F(x_0)\|$ and each $n = 0, 1, 2, \ldots$ by

$$\bar{v}_n = \begin{cases} v(\beta_n - \alpha_n) \\ v_0(\alpha_n) + v_0(\beta_n) \end{cases},$$

$$
\begin{aligned}
\lambda_n &= \frac{3}{2}\left(1+\frac{3}{4}\frac{\bar{v}_n}{1-v_0(\alpha_n)}+\frac{9}{8}\left(\frac{\bar{v}_n}{1-v_0(\alpha_n)}\right)^2+|a|\left(\frac{\bar{v}_n}{1-v_0(\alpha_n)}\right)^3\right),\\
\alpha_{n+1} &= \beta_n+\lambda_n(\beta_n-\alpha_n),\\
\delta_{n+1} &= \left(1+\int_0^1 v_0(\alpha_n+\theta(\alpha_{n+1}-\alpha_n))d\theta\right)(\alpha_{n+1}-\alpha_n)\\
&\quad +\frac{3}{2}(1+v_0(\alpha_n))(\beta_n-\alpha_n)
\end{aligned}
$$

and

$$
\beta_{n+1} = \alpha_{n+1}+\frac{2}{3}\frac{\delta_{n+1}}{1-v_0(\alpha_{n+1})}.
$$

$$
\begin{aligned}
x_{n+1}-y_n &= -\left[I-\frac{3}{4}(A_n-I)+\frac{9}{8}(A_n-I)^2+|a|(A_n-I)^2\right]\\
&\quad \left(-\frac{3}{2}(y_n-x_n)\right)\\
\|x_{n+1}-y_n\| &\leq \frac{3}{2}\left[1+\frac{3}{4}\frac{\bar{v}_n}{1-v_0(\alpha_n)}+\frac{9}{8}\left(\frac{\bar{v}_n}{1-v_0(\alpha_n)}\right)^2\right.\\
&\quad \left.+|a|\left(\frac{\bar{v}_n}{1-v_0(\alpha_n)}\right)^3\right](\beta_n-\alpha_n)\\
&= \alpha_{n+1}-\beta_n,\\
F(x_{n+1}) &= F(x_{n+1})-F(x_n)-\frac{3}{2}F'(x_n)(y_n-x_n)\\
\|L^{-1}F(x_{n+1})\| &\leq \left(1+\int_0^1 v_0(\alpha_n+\theta(\alpha_{n+1}-\alpha_n))d\theta\right)(\alpha_{n+1}-\alpha_n)\\
&\quad +\frac{3}{2}(1+v_0(\alpha_n))(\beta_n-\alpha_n)=\delta_{n+1}
\end{aligned}
$$

and

$$
\begin{aligned}
\|y_{n+1}-x_{n+1}\| &\leq \frac{2}{3}\|F'(x_{n+1})^{-1}L\|\|L^{-1}F(x_{n+1})\|\\
&\leq \frac{2}{3}\frac{\delta_{n+1}}{1-v_0(\alpha_{n+1})}=\beta_{n+1}-\alpha_{n+1}.
\end{aligned}
$$

The rest of the analysis is given in previous chapters.

4. References

[1] Artidiello, S., Chicharro, F., Cordero, A., Torregrosa, J.R., *Local convergence and dynamical analysis of a new family of optimal fourth-order iterative methods*, Int. J. Comput. Math., 90, (2013), 2049–2060.

[2] Cordero, A., Torregrosa, J.R., Vindel, P., *Dynamics of a family of Chebyshev-Halley type methods*, Appl. Math. Comput., 219, (2013), 8568–8583.

[3] Hueso, J.L., Martnez, E., Teruel, C., *Convergence, efficiency and dynamics of new fourth and sixth order families of iterative methods for nonlinear systems*, Comput. Appl. Math., 275, (2015), 412–420.

[4] Jarratt, P., *Some fourth order multipoint iterative methods for solving equations*, Math. Comput., 20, (1966), 434–437.

[5] Kung, H.T., Traub, J.F., *Optimal order of one-point and multi-point iteration*, J. Assoc. Comput. Math., 21, (1974), 643–651.

[6] Kou, J., Li, Y., *A family of new Newton-like methods*, Appl. Math. Comput., 192, (2007), 162–167.

[7] Li, W., Pang, Y., *Application of Adomian decomposition method to nonlinear systems*, Adv. Differ. Equ., 2020, 67.

[8] Padilla, J. J., Chicharro, F. I., Cordero, A., Torregrosa, J.R., *Parametric family of root-finding iterative methods: Fractals of the basins of attraction*, Fractal Fract., 6, (2022), 572, https://doi.org/10.3390/fractalfract6100572.

[9] Sharma, D., Argyros, I.K., Parhi, S.K., Sunanda, S.K., *Local Convergence and Dynamical Analysis of a Third and Fourth Order Class of Equation Solvers*, Fractal Fract., 5, 27, (2021).

Chapter 59

Efficient Chebyshev-Secant and Chebyshev-Kurchatov Derivative-Free Methods

1. Introduction

The methods are defined for $x_{-1}, x_0 \in \Omega$, $a, b, c \in \mathbb{R}$ and each $n = 0, 1, 2, \ldots$ by

$$\begin{aligned} A_n &= [x_{n-1}, x_n; F], \\ y_n &= x_n - A_n^{-1} F(x_n), \\ z_n &= x_n + a(y_n - x_n), \\ x_{n+1} &= x_n - A_n^{-1}(bF(x_n) + cF(z_n)), \end{aligned} \tag{59.1}$$

and

$$\begin{aligned} B_n &= [2x_n - x_{n-1}; F], \\ y_n &= x_n - B_n^{-1} F(x_n), \\ z_n &= x_n + a(y_n - x_n), \\ x_{n+1} &= x_n - B_n^{-1}(bF(x_n) + cF(z_n)). \end{aligned} \tag{59.2}$$

These methods are called Chebyshev-Secant and Chebyshev-Kurchatov, respectively and are found to be very efficient [4]. In this chapter we revisit the local convergence under weaker conditions and also develop the semi-local convergence not given in [1]-[3], [5]-[20].

2. Local Convergence

Method (59.1):

The majorant functions $g_i, i = 1,2,3$ are

$$\begin{aligned} g_1(t) &= \frac{\psi(t,t)}{1-\psi_0(t,t)}, \\ g_2(t) &= |1-a|+|a|g_1(t) \end{aligned}$$

and

$$g_3(t) = g_1(t) + \frac{|1-b|(1+\psi_1(t))}{1-\psi_0(t,t)} + \frac{|c|(1+\psi_1(g_2(t)))g_2(t)}{1-\psi_0(t,t)}.$$

The motivational estimate are int turn:

$$\begin{aligned} y_n - x^* &= x_n - x^* - A_n^{-1}F(x_n) = A_n^{-1}(A_n - [x_n,x^*;F])(x_n - x^*), \\ \|y_n - x^*\| &\leq \frac{\psi(\|x_{n-1}-x^*\|, \|x_n-x^*\|)\|x_n-x^*\|}{1-\psi_0(\|x_{n-1}-x^*\|, \|x_n-x^*\|)} \leq g_1(\|x_n-x^*\|)\|x_n-x^*\| \\ &\leq \|x_n-x^*\| < r, \\ z_n - x^* &= x_n - x^* + a(y_n-x^*) + a(x^*-x_n) = (1-a)(x_n-x^*) + a(y_n-x^*), \\ \|z_n-x^*\| &\leq (|1-a|+|a|g_1(\|x_n-x^*\|))\|x_n-x^*\| \\ &= g_2(\|x_n-x^*\|)\|x_n-x^*\| \leq \|x_n-x^*\|, \\ F(x_n) &= [x_n,x^*;F](x_n-x^*), \\ \|L^{-1}F(x_n)\| &\leq \|L^{-1}([x_n,x^*;F]-L+L)(x_n-x^*)\| \\ &\leq (1+\psi_1(\|x_n-x^*\|))\|x_n-x^*\|, \\ x_{n+1}-x^* &= x_n - x^* - bA_n^{-1}F(x_n) - cA_n^{-1}F(z_n) \\ &= x_n - x^* - A_n^{-1}F(x_n) - (1-b)A_n^{-1}F(x_n) - cA_n^{-1}F(z_n) \end{aligned}$$

and

$$\begin{aligned} \|x_{n+1}-x^*\| &\leq \left[g_1(\|x_n-x^*\|) + \frac{|1-b|(1+\psi_1(\|x_n-x^*\|))}{1-\psi_0(\|x_{n-1}-x^*\|, \|x_n-x^*\|)} \right. \\ &\quad \left. + \frac{|c|(1+\psi_1(g_2(\|x_n-x^*\|)))g_2(\|x_n-x^*\|)}{1-\psi_0(\|x_{n-1}-x^*\|, \|x_n-x^*\|)} \right] \|x_n-x^*\| \\ &\leq g_3(\|x_n-x^*\|)\|x_n-x^*\| \leq \|x_n-x^*\|. \end{aligned}$$

Method (59.2):

The majorant functions $g_i, i = 1,2,3$ are:

$$\begin{aligned} g_1(t) &= \frac{\psi(3t,t)}{1-\psi_0(3t,t)}, \\ g_2(t) &= |1-a|+|a|g_1(t) \end{aligned}$$

and

$$g_3(t) = g_1(t) + \frac{|1-b|(1-\psi_1(t))}{1-\psi_0(3t,t)} + \frac{|c|(1+\psi_1(g_2(t)))g_2(t)}{1-\psi_0(3t,t)}.$$

The motivational estimates are in turn:

$$\begin{aligned} y_n - x^* &= B_n^{-1}(B_n - [x_n,x^*;F])(x_n-x^*), \\ \|y_n-x^*\| &\leq \frac{\psi(2\|x_n-x^*\| + \|x_{n-1}-x^*\|, \|x_n-x^*\|)\|x_n-x^*\|}{1-\psi_0(2\|x_n-x^*\| + \|x_{n-1}-x^*\|, \|x_n-x^*\|)} \\ &\leq g_1(\|x_n-x^*\|)\|x_n-x^*\| \leq \|x_n-x^*\| < r. \end{aligned}$$

The calculations for the function g_2 are given in method (59.1). Moreover, we have

$$x_{n+1}-x^* = x_n-x^*-B_n^{-1}F(x_n)+(1-b)B_n^{-1}F(x_n)-cB_n^{-1}F(z_n)$$

and

$$\begin{aligned}\|x_{n+1}-x^*\| &\leq \Big[g_1(\|x_n-x^*\|)+\frac{|1-b|(1+\psi_1(\|x_n-x^*\|))}{1-\psi_0(2\|x_n-x^*\|+\|x_{n-1}-x^*\|,\|x_n-x^*\|)}\\ &\quad+\frac{|c|(1+\psi_1(g_2(\|x_n-x^*\|)))g_2(\|x_n-x^*\|)}{1-\psi_0(2\|x_n-x^*\|+\|x_{n-1}-x^*\|,\|x_n-x^*\|)}\Big]\|x_n-x^*\|\\ &\leq g_3(\|x_n-x^*\|)\|x_n-x^*\|\leq\|x_n-x^*\|.\end{aligned}$$

3. Semi-local Convergence

Method (59.1):

The majorant sequence $\{\alpha_n\}$ is defined for $\alpha_{-1}=0, \alpha_0=\alpha$, $\beta_0\geq\|A_0^{-1}F(x_0)\|$ and each $n=0,1,2,\ldots$ by

$$\begin{aligned}\gamma_n &= \beta_n+|1-a|(\beta_n-\alpha_n),\\ \lambda_n &= (1+\Psi_0(\alpha_n,\gamma_n))(\gamma_n-\alpha_n)+(1+\Psi_0(\alpha_{n-1}-\alpha_n))(\beta_n-\alpha_n),\\ \alpha_{n+1} &= \gamma_n+|b-a|(\beta_n-\alpha_n)+\frac{|c|\lambda_n}{1-\Psi_0(\alpha_{n-1},\alpha_n)},\\ \delta_{n+1} &= (1+\Psi_0(\alpha_n,\alpha_{n+1}))(\alpha_{n+1}-\alpha_n)+(1+\Psi_0(\alpha_{n-1},\alpha_n))(\beta_n-\alpha_n)\end{aligned}$$

and

$$\beta_{n+1} = \alpha_{n+1}+\frac{\delta_{n+1}}{1-\Psi_0(\alpha_n,\alpha_{n+1})}.$$

The motivational calculations are in turn:

$$\begin{aligned}z_n-y_n &= A_n^{-1}F(z_n)+a(y_n-x_n)=-(y_n-x_n)+a(y_n-x_n)=(a-1)(y_n-x_n),\\ \|z_n-y_n\| &= |1-a|\|y_n-x_n\|\leq|1-a|(\beta_n-\alpha_n)=\gamma_n-\beta_n,\\ x_{n+1}-z_n &= -a(y_n-x_n)-bA_n^{-1}F(z_n)\\ &= -a(y_n-x_n)+b(y_n-x_n)-cA_n^{-1}F(z_n)\\ &= (b-a)(y_n-x_n)-cA_n^{-1}F(z_n),\\ F(z_n) &= F(z_n)-F(x_n)+F(x_n)=F(z_n)-F(x_n)-A_n(y_n-x_n),\\ \|L^{-1}F(z_n)\| &\leq (1+\Psi_0(\|x_n-x_0\|,\|z_n-x_0\|))\|z_n-x_n\|\\ &\quad+(1+\Psi_0(\|x_{n-1}-x_0\|,\|x_n-x_0\|))\|y_n-x_n\|=\bar{\lambda}_n\\ &\leq (1+\Psi_0(\alpha_n,\gamma_n))(\gamma_n-\alpha_n)+(1+\Psi_0(\alpha_{n-1},\alpha_n))(\beta_n-\alpha_n)=\lambda_n,\\ \|x_{n+1}-z_n\| &\leq |b-a|(\beta_n-\alpha_n)+\frac{|c|\lambda_n}{1-\Psi_0(\alpha_{n-1},\alpha_n)}=\alpha_{n+1}-\gamma_n,\\ F(x_{n+1}) &= F(x_{n+1})-F(x_n)-A_n(y_n-x_n),\\ \|L^{-1}F(x_{n+1})\| &\leq (1+\Psi_0(\alpha_n,\alpha_{n-1}))(\alpha_{n+1}-\alpha_n)\\ &\quad+(1+\Psi_0(\alpha_{n-1},\alpha_n))(\beta_n-\alpha_n)=\delta_{n+1}\end{aligned}$$

and

$$\|y_{n+1}-x_{n+1}\| \leq \|A_{n+1}^{-1}L\|\|L^{-1}F(x_{n+1})\|\leq\frac{\delta_{n+1}}{1-\Psi_0(\alpha_n,\alpha_{n+1})}=\beta_{n+1}-\alpha_{n+1}.$$

Method (59.2):

This time the iterate γ_n is the same as in method (59.1),

$$\begin{aligned}
\mu_n &= (1+\Psi_0(\alpha_n,\gamma_n))(\gamma_n-\alpha_n) \\
&\quad +(1+\Psi_0(2\alpha_n+\alpha_{n-1},\alpha_n))(\beta_n-\alpha_n), \\
\alpha_{n+1} &= \gamma_n+|b-a|(\beta_n-\alpha_n)+\frac{|c|\mu_n}{1-\Psi_0(2\alpha_n+\alpha_{n-1},\alpha_n)}, \\
\zeta_{n+1} &= (1+\Psi_0(\alpha_n,\alpha_{n+1}))(\alpha_{n+1}-\alpha_n)+(1+\Psi_0(2\alpha_n+\alpha_{n-1},\alpha_n))(\beta_n-\alpha_n)
\end{aligned}$$

and

$$\beta_{n+1} = \alpha_{n+1}+\frac{\zeta_{n+1}}{1-\Psi_0(2\alpha_{n+1}+\alpha_n,\alpha_{n+1})}.$$

The motivational calculations are in turn:

$$\begin{aligned}
x_{n+1}-z_n &= (b-a)(y_n-x_n)-cB_n^{-1}F(z_n), \\
F(z_n) &= F(z_n)-F(x_n)-B_n(y_n-x_n), \\
\|L^{-1}F(z_n)\| &\leq (1+\Psi_0(\alpha_n,\gamma_n))(\gamma_n-\alpha_n) \\
&\quad +(1+\Psi_0(2\alpha_n+\alpha_{n+1},\alpha_n))(\beta_n-\alpha_n)=\mu_n, \\
\|x_{n+1}-z_n\| &\leq |b-a|(\beta_n-\alpha_n)+\frac{|c|\mu_n}{1-\Psi_0(2\alpha_n+\alpha_{n-1},\alpha_n)}, \\
F(x_{n+1}) &= F(x_{n+1})-F(x_n)-B_n(y_n-x_n), \\
\|L^{-1}F(x_{n+1})\| &\leq (1+\Psi_0(\alpha_n,\alpha_{n+1}))(\alpha_{n+1}-\alpha_n) \\
&\quad +(1+\Psi_0(2\alpha_n+\alpha_{n-1},\alpha_n))(\beta_n-\alpha_n)=\zeta_{n+1}
\end{aligned}$$

and

$$\begin{aligned}
\|y_{n+1}-x_{n+1}\| &\leq \|B_{n+1}^{-1}L\|\|L^{-1}F(x_{n+1})\|\leq\frac{\zeta_{n+1}}{1-\Psi_0(2\alpha_{n+1}+\alpha_n,\alpha_{n+1})} \\
&= \beta_{n+1}-\alpha_{n+1}.
\end{aligned}$$

The rest of the convergence analysis has appeared in earlier chapters.

4. References

[1] Ahmad, F., Soleymani, F., Haghani, F., Argyros, I. K., *The theory and applications of iteration methods*, 2nd Edition, CRC Press, Engineering Series, Taylor and Francis Group, 2022.

[2] Argyros, C. I., Regmi, S., Argyros, I. K., George, S., *Contemporary Algorithms: Theory and Applications*, Volume III, NOVA Publishers, (2023).

[3] Argyros, I. K., George, S., *Local convergence of two competing third order methods in Banach space*, Applicationes Mathematicae 4 (41), (2014), 341-350.

[4] Chun, C., Neta, B., *An Efficient Derivative-Free Method for the Solution of Systems of Equations*, Numerical Functional Analysis and Optimization, 42,7, (2021), 834-848, DOI: 10.1080/01630563.2021.1931313.

[5] Chun, C., Neta, B., *Developing high order methods for the solution of systems of nonlinear equations*, Appl. Math. Comput., 342, (2019), 178-190, DOI: 10.1016/j.amc.2018.09.032.

[6] Cordero, A., Maim, J. G., Torregrosa, J. R., Vassileva, M. P., *Iterative methods with memory for solving systems of nonlinear equations using a second order approximation*, Mathematics, 7(11), (2019), 1069, DOI: 10.3390/math7111069.

[7] Ezquerro, J. A., Hernndez, M. A., Romero, N., Velasco, A. I., *On Steffensens method on Banach spaces*, J. Comput. Appl. Math., 249, (2013), 9-23, DOI: 10.1016/j.cam.2013.02.004.

[8] Hueso, J. L., Martinez, E., Teruel, C., *Convergence, efficiency and dynamics of new fourth and sixth order families of iterative methods for nonlinear systems*, J. Comput. Appl. Math., 275, (2015), 412-420, DOI: 10.1016/j.cam.2014.06.010.

[9] Jangveladze, T., Kiguradze, Z., Neta, B., *Large time behavior of solutions and finite difference scheme to a nonlinear integro-differential equation*, Comput. Math. Appl., 57(5), (2009), 7991-811, DOI: 10.1016/j.camwa.2008.09.055.

[10] Kumar, A., Gupta, D. K., Martinez, E., Hueso, J. L., *Convergence and dynamics of improved Chebyshev-Secant-type methods for non differentiable operators*, Numer. Algor., 86(3), (2021), 1051-1070, DOI: 10.1007/s11075-020-00922-9.

[11] Liu, Z., Zheng, Q., Zhao, P., *A variant of Steffensens method of fourth-order convergence and its applications*, Appl. Math. Comput., 216, (2010), 1978-1983.

[12] Montazeri, H., Soleymani, F., Shateyi, S., Motsa, S. S., *On a new method for computing the numerical solution of systems of nonlinear equations*, J. Appl. Math., 2012, (2012), 751975, DOI: 10.1155/2012/751975.

[13] Narang, M., Bhatia, S., Kanwar, V., *New two parameter Chebyshev-Halley like family of fourth and sixth-order methods for systems of nonlinear equations*, Appl. Math. Comput., 275, (2016), 394-403, DOI: 10.1016/j.amc.2015.11.063.

[14] Narang, M., Bhatia, S., Alshomrani, A., Kanwar, V., *General efficient class of Steffensen type methods with memory for solving systems of nonlinear equations*, J. Comput. Appl. Math., 352, (2019), 23-39, DOI: 10.1016/j.cam.2018.10.048.

[15] Petkovi, M. S., Neta, B., Petkovi, L. D., Dzuni, J., *Multipoint Methods for Solving Nonlinear Equations*, Waltham: Elsevier, (2013).

[16] Ramandeep, B., Cordero, A., Motsa, S. S., Torregrosa, J. R., *Stable high-order iterative methods for solving nonlinear models*, Appl. Math. Comput., 303, (2017), 70-88, DOI: 10.1016/j.amc.2017.01.029.

[17] Sharma, J. R., Arora, H., *An efficient derivative-free iterative method for solving systems of nonlinear equations*, Appl. Anal. Discrete Math., 7(2), (2013), 390-403, DOI: 10.2298/AADM130725016S.

[18] Sharma, J. R., Arora, H., *Efficient Jarratt-like methods for solving systems of nonlinear equations*, Calcolo, 51(1), (2014), 193-210, DOI: 10.1007/s10092-013-0097-1.

[19] Sharma, J. R., Sharma, R., Kalra, N., *A novel family of composite Newton-Traub methods for solving systems of nonlinear equations*, Appl. Math. Comput., 269, (2015), 520-535, DOI: 10.1016/j.amc.2015.07.092.

[20] Traub, J. F., *Iterative Methods for the Solution of Equations*, Englewood Cliffs: Prentice-Hall, (1964).

Chapter 60

Stable Sixth-Order Methods for Equations

1. Introduction

The three step method is defined for $x_0 \in \Omega$ and each $n = 0, 1, 2, ...$ by

$$\begin{aligned} y_n &= x_n - c_1 F'(x_n)^{-1} F(x_n), \\ A_n &= c_3 F'(x_n) + c_4 F'(y_n), \\ z_n &= y_n - (c_2 F'(x_n)^{-1} + A_n^{-1}) F(z_n), \\ B_n &= c_6 F'(x_n) + c_7 F'(y_n) \\ x_{n+1} &= z_n - (c_5 F'(x_n)^{-1} + B_n^{-1}) F(z_n) \end{aligned} \tag{60.1}$$

where $c_i,\ i = 1, 2, ..., 7$ are real parameters. In particular, if $c_1 = \frac{2}{3}, c_2 = -\frac{1}{6}, c_3 = -1, c_4 = 3, c_6 = -\frac{2c_5 + 1}{2(c_5 - 1)^2}$ and $c_7 = \frac{3}{2(c_5 - 1)^2}$ for $c_5 \neq 1$ the local sixth order is established in [6] if $B_1 = B_2 = \mathbb{R}$ provided that $F^{(7)}$ is bounded. Other limitations of the Taylor series expansion approach have been reported in earlier chapters. In this chapter conditions only on F' which appears on the method are used in the local convergence. Moreover, the semi-local convergence (not given in [6]) is presented using majorizing sequences. In both cases, the convergence is established in the more general setting of a Banach space. The new approach is also applicable to other methods [1]-[5], [7]-[13].

Method(60.1):

The majorant functions $g_i,\ i=1,2,3$ are in turn:

$$\begin{aligned}
g_1(t) &= \frac{\int_0^1 w((1-\theta)t)d\theta+|1-c_1|(1+\int_0^1 w_0(\theta t)d\theta)}{1-w_0(t)},\\
p(t) &= \frac{1}{|c_3+c_4|}(|c_3|w_0(t)+|c_4|w_0(g_1(t)t)),\\
g_2(t) &= \frac{\int_0^1 w((1-\theta)g_1(t)t)d\theta g_1(t)}{1-w_0(g_1(t)t)}\\
&+\frac{[|1-c_3|(1+w_0(t))+|c_4|(1+w_0(g_1(t)t))](1+\int_0^1 w_0(\theta t)d\theta)}{|c_3+c_4|(1-p(t))(1-w_0(t))}\\
&+\frac{(1+\int_0^1 w_0(\theta g_1(t)t)d\theta)g_1(t)}{1-w_0(g_1(t)t)}+|1+c_2|(1+\int_0^1 w_0(\theta t)d\theta)\\
q(t) &= \frac{1}{|c_6+c_7|}(|c_6|w_0(t)+|c_7|w_0(g_1(t)t))
\end{aligned}$$

and

$$\begin{aligned}
g_3(t) &= \left[\frac{\int_0^1 w((1-\theta)g_2(t)t)d\theta}{1-w_0(g_2(t)t)}+\frac{\bar{w}(t)(1+\int_0^1 w_0(\theta g_2(t)t)d\theta)}{(1-w_0(t))(1-w_0(g_2(t)t))}\right.\\
&\left.+\frac{|1-c_5|(1+\int_0^1 w_0(\theta g_2(t)t)d\theta}{1-w_0(t)}+\frac{1+\int_0^1 w_0(\theta g_2(t)t)d\theta}{|c_6+c_7|(1-q(t))}\right]g_2(t).
\end{aligned}$$

The motivational calculations are in turn:

$$\begin{aligned}
y_n-x^* &= x_n-x^*-F'(x_n)^{-1}F(x_n)+(1-c_1)F'(x_n)^{-1}F(x_n),\\
\|y_n-x^*\| &\leq \frac{1}{1-w_0(\|x_n-x^*\|)}\left[\int_0^1 w((1-\theta)\|x_n-x^*\|)d\theta\right.\\
&\left.+|1-c_1|(1+\int_0^1 w_0(\theta\|x_n-x^*\|)d\theta\right]\|x_n-x^*\|\\
&\leq g_1(\|x_n-x^*\|)\|x_n-x^*\| \leq \|x_n-x^*\| < 1,\\
z_n-x^* &= y_n-x^*-F'(y_n)^{-1}F(y_n)-A_n^{-1}(F'(x_n)-A_n)F'(x_n)^{-1}F(x_n)\\
&+F'(y_n)F(y_n)-(1+c_2)F'(x_n)^{-1}F(x_n),
\end{aligned}$$

$$\begin{aligned}
&\|((c_3+c_4)L)^{-1}[c_3(F'(x_n)-L)+c_4(F'(y_n)-L)]\|\\
\leq\ & \frac{1}{|c_3+c_4|}(|c_3|w_0(\|x_n-x^*\|)+|c_4|w_0(\|y_n-x^*\|))=p_n<1,\ (c_3+c_4\neq 0)
\end{aligned}$$

$$
\begin{aligned}
\|A_n^{-1}L\| &\leq \frac{1}{|c_3+c_4|(1-p_n)},\\
\|z_n-x^*\| &\leq \frac{\int_0^1 w((1-\theta)\|y_n-x^*\|)d\theta\|y_n-x^*\|}{1-w_0(\|y_n-x^*\|)}\\
&\leq \frac{1}{|c_3+c_4|(1-p_n)(1-w_0(\|x_n-x^*\|))}\Big[|1-c_3|(1+w_0(\|x_n-x^*\|))\\
&\quad +|c_4|(1+w_0(\|y_n-x^*\|))\Big](1+\int_0^1 w_0(\theta\|x_n-x^*\|)d\theta)\|x_n-x^*\|\\
&\quad +\frac{(1+\int_0^1 w_0(\theta\|y_n-x^*\|)d\theta)\|y_n-x^*\|}{1-w_0(\|y_n-x^*\|)}\\
&\quad +\frac{|1+c_2|(1+\int_0^1 w_0(\theta\|x_n-x^*\|)d\theta)\|x_n-x^*\|}{1-w_0(\|x_n-x^*\|)}\\
&\leq g_2(\|x_n-x^*\|)\|x_n-x^*\| \leq \|x_n-x^*\|,\\
x_{n+1}-x^* &= z_n-x^*-F'(z_n)^{-1}F(z_n)+(F'(z_n)^{-1}-F'(x_n)^{-1})F(z_n)\\
&\quad +(1-c_5)F'(x_n)^{-1}F(z_n)-B_n^{-1}F(z_n)
\end{aligned}
$$

and

$$
\begin{aligned}
\|x_{n+1}-x^*\| &\leq \Bigg[\frac{\int_0^1 w((1-\theta)\|z_n-x^*\|)d\theta}{1-w_0(\|z_n-x^*\|)}\\
&\quad +\frac{\bar{w}_n(1+\int_0^1 w_0(\theta\|z_n-x^*\|)d\theta)}{(1-w_0(\|x_n-x^*\|))(1-w_0(\|z_n-x^*\|))}\\
&\quad +\frac{|1-c_5|(1+\int_0^1 w_0(\theta\|z_n-x^*\|)d\theta)}{1-w_0(\|x_n-x^*\|)}\\
&\quad +\frac{1+\int_0^1 w_0(\theta\|z_n-x^*\|)d\theta}{|c_6+c_7|(1-q_n)}\Bigg]\|z_n-x^*\|\\
&\leq g_3(\|x_n-x^*\|)\|x_n-x^*\| \leq \|x_n-x^*\|.
\end{aligned}
$$

2. Semi-local Convergence

The majorant sequence $\{\alpha_n\}$ is defined for $\alpha_0=0, \beta_0 \geq |c_1|\|F'(x_0)^{-1}F(x_0)\|$ and each $n = 0,1,2,...$ by

$$
\begin{aligned}
\bar{p}_n &= \frac{1}{|c_3+c_4|}(v_0(\alpha_n)+v_0(\beta_n)),\\
\gamma_n &= \beta_n+\left[\left|\frac{c_2}{c_1}\right|\frac{1+v_0(\alpha_n)}{|c_1||c_3+c_4|(1-\bar{p}_n)}\right](\beta_n-\alpha_n)\\
&\quad (c_1\neq 0, c_3+c_4\neq 0),\\
\lambda_n &= (1+\int_0^1 v_0(\alpha_n+\theta(\gamma_n-\alpha_n))d\theta)(\gamma_n-\alpha_n)+(1+v_0(\alpha_n))(\beta_n-\alpha_n)
\end{aligned}
$$

$$\begin{aligned}
\bar{q}_n &= \frac{1}{|c_6+c_7|}(v_0(\alpha_n)+v_0(\beta_n)),\ c_6+c_7\neq 0,\\
\alpha_{n+1} &= \left[\frac{|c_5|}{1-v_0(\alpha_n)}+\frac{1}{|c_6+c_7|(1-\bar{q}_n)}\right]\lambda_n\\
\delta_{n+1} &= (1+\int_0^1 v_0(\alpha_n+\theta(\alpha_{n+1}-\alpha_n))d\theta)(\alpha_{n+1}-\alpha_n)+\frac{1}{|c_1|}(1+v_0(\alpha_n))(\beta_n-\alpha_n)
\end{aligned}$$

and

$$\lambda_{n+1} = \alpha_{n+1}+\frac{|c_1|\delta_{n+1}}{1-v_0(\alpha_{n+1})}.$$

The calculations are in turn

$$\begin{aligned}
z_n-y_n &= -\frac{c_2}{c_1}(y_n-x_n)+\frac{1}{c_1}A_n^{-1}F'(x_n)(y_n-x_n),\\
\|z_n-y_n\| &\leq \left[\left|\frac{c_2}{c_1}\right|+\frac{1+v_0(\alpha_n)}{|c_1||c_3+c_4|(1-\bar{p}_n)}\right](\beta_n-\alpha_n)=\gamma_n-\beta_n,\\
F(z_n) &= F(z_n)-F(x_n)-F'(x_n)(y_n-x_n),\\
\|L^{-1}F(z_n)\| &\leq (1+\int_0^1 v_0(\alpha_n+\theta(\gamma_n-\alpha_n))d\theta)(\gamma_n-\alpha_n)+(1+v_0(\alpha_n))(\beta_n-\alpha_n)\\
&= \lambda_n,
\end{aligned}$$

$$\begin{aligned}
\|x_{n+1}-z_n\| &\leq \left[\frac{|c_5|}{1-v_0(\alpha_n)}+\frac{1}{|c_6+c_7|(1-\bar{q}_n)}\right]\lambda_n=\alpha_{n+1}-\gamma_n,\\
F(x_{n+1}) &= F(x_{n+1})-F(x_n)-\frac{1}{c_1}F'(x_n)(y_n-x_n),\\
\|L^{-1}F(x_{n+1})\| &\leq (1+\int_0^1 v_0(\alpha_n+\theta(\alpha_{n+1}-\alpha_n))d\theta)(\alpha_{n+1}-\alpha_n)\\
&\quad +\frac{1}{|c_1|}(1+v_0(\alpha_n))(\beta_n-\alpha_n)=\delta_{n+1}
\end{aligned}$$

and $\|y_{n+1}-x_{n+1}\|\leq |c_1|\|F'(x_{n+1})^{-1}L\|\|L^{-1}F(x_{n+1})\|\leq \beta_{n+1}-\alpha_{n+1}$.
The rest of the analysis can be found in earlier chapters.

3. References

[1] Abad, M. F., Cordero, A., Torregrosa, J. R., *A family of seventh-order schemes for solving nonlinear systems*, Bull. Math. Soc. Sci. Math. Roum, 57, (105)(2), (2014), 133-145.

[2] Argyros, I. K., Magrenn, A. A., *A study on the local convergence and the dynamics of Chebyshev-Halley-type methods free from second derivative*, Numer. Algor., 71(1), (2016), 1-23.

[3] Argyros, I. K., Magrenn, A. A., *On the convergence of an optimal fourth-order family of methods and its dynamics*, Appl. Math. Comput., 252, (2015), 336-346.

[4] Artidiello, S., Cordero, A., Torregrosa, J. R., Vassileva, M. P., *Multidimensional generalization of iterative methods for solving nonlinear problems by means of weight-function procedure*, Appl. Math. Comput., 268, (2015), 1064-1071.

[5] Behl, R., Cordero, A., Motsa, S. S., Torregrosa, J. R., *Stable high-order iterative methods for solving nonlinear models*, Appl. Math. Comput., 303, (2017), 70-88.

[6] Behl, R., Kanwar, V., Sharma, K. K., *Optimal equi-scaled families of Jarratts method*, Int. J. Comput. Math., 90, (2013), 408-422.

[7] Chicharro, F., Cordero, A., Gutirrez, J. M., Torregrosa, J. R., *Complex dynamics of derivative-free methods for nonlinear equations*, Appl. Math. Comput., 219, (2013), 7023-7035.

[8] Hueso, J. L., Martnez, E., Teruel, C., *Convergence, efficiency and dynamics of new fourth and sixth order families of iterative methods for nonlinear systems*, Comput. Appl. Math., 275, (2015), 412-420.

[9] Jarratt, P., *Some fourth order multipoint iterative methods for solving equations*, Math. Comput., 20, (1966), 434-437.

[10] Lin, Y., Bao, L., Jia, X., *Convergence analysis of a variant of the Newton method for solving nonlinear equations*, Comput. Math. Appl., 59, (2010), 2121-2127.

[11] Sharma, J. R., Arora, H., *A novel derivative free algorithm with seventh order convergence for solving systems of nonlinear equations*, Numer. Algorithm, 67, (2014), 917-933.

[12] Sharma, J. R., Guna, R. K., Sharma, R., *An efficient fourth order weighted-newton method for systems of nonlinear equations*, Numer. Algorithm., 2, (2013), 307-323.

[13] Wang, X., Zhang, T., *A family of Steffensen type methods with seventh-order convergence*, Numer. Algor., 62, (2013), 429-444.

Chapter 61

Efficient and Comparable Fourth-Order Methods for Equations

1. Introduction

The methods are defined for $x_0 \in \Omega$ and each $n = 0, 1, 2, ...$ by

$$\begin{aligned} y_n &= x_n - \frac{2}{3}F'(x_n)^{-1}F(x_n) \\ x_{n+1} &= x_n - \frac{1}{2}[-I + \frac{9}{4}F'(y_n)^{-1}F'(x_n) + \frac{3}{4}F'(x_n)^{-1}F'(y_n)]F'(x_n)^{-1}F(x_n) \end{aligned} \tag{61.1}$$

and

$$\begin{aligned} y_n &= x_n - \frac{2}{3}F'(x_n)^{-1}F(x_n) \\ x_{n+1} &= x_n - [I - \frac{3}{8}(I - F'(y_n)^{-1}F'(x_n))^2]F'(x_n)^{-1}F(x_n). \end{aligned} \tag{61.2}$$

The local convergence order four is determined in [5, 6], respectively using the Taylor expansion series approach for $B_1 = B_2 = \mathbb{R}^k$ and by assuming the boundedness and existence of at least $F^{(5)}$. Other limitations with this approach have been reported in earlier chapters. This is the reason why the new local convergence uses only F' that actually appears on these methods. Moreover, the more interesting semi-local convergence not studied is also presented in this chapter. Both analyses are developed in a Banach space setting. The new approach can be used to extend other methods [1]-[4], [7] analogously.

2. Local Convergence

Method(61.1)

The majorant sequences $g_i,\ i = 1, 2$ are

$$g_1(t) = \frac{\int_0^1 w((1-\theta)t)d\theta + \frac{1}{3}(1 + \int_0^1 w_0(\theta t)d\theta)}{1 - w_0(t)}$$

$$\bar{w}_n = \begin{cases} w((1+g_1(t))t) \\ or \\ w_0(t) + w_0(g_1(t)t), \end{cases}$$

and

$$\begin{aligned} g_2(t) &= \frac{\int_0^1 w((1-\theta)t)d\theta}{1-w_0(t)} \\ &+\frac{3}{8}\frac{\bar{w}(t)}{1-w_0(g_1(t)t)}\left(2+\frac{\bar{w}(t)}{1-w_0(t)}\right)\frac{(1+\int_0^1 w_0(\theta t)d\theta)}{1-w_0(t)}. \end{aligned}$$

The motivational calculations are in turn:

$$\begin{aligned} y_n - x^* &= x_n - x^* - F'(x_n)^{-1}F(x_n) + \frac{1}{3}F'(x_n)^{-1}F(x_n), \\ \|y_n - x^*\| &\leq \frac{[\int_0^1 w((1-\theta)\|x_n-x^*\|)d\theta + \frac{1}{3}(1+\int_0^1 w_0(\theta\|x_n-x^*\|)d\theta)]\|x_n-x^*\|}{1-w_0(\|x_n-x^*\|)} \\ &\leq g_1(\|x_n-x^*\|)\|x_n-x^*\| \leq \|x_n-x^*\| < r, \\ x_{n+1}-x^* &= x_n - x^* - F'(x_n)^{-1}F(x_n) \\ &+\left[\frac{3}{2}I - \frac{9}{8}I - \frac{3}{8} - \frac{9}{8}F'(y_n)^{-1}(F'(x_n)-F'(y_n))\right. \\ &-\frac{3}{8}F'(x_n)^{-1}(F'(y_n)-F'(x_n))]F'(x_n)^{-1}F(x_n) \\ &= x_n - x^* - F'(x_n)^{-1}F(x_n) \\ &-\frac{3}{8}[2F'(y_n)^{-1}(F'(x_n)-F'(y_n)) \\ &+(F'(y_n)^{-1}-F'(x_n)^{-1})(F'(x_n)-F'(y_n))]F'(x_n)^{-1}F(x_n) \end{aligned}$$

$$\begin{aligned} \|x_{n+1}-x^*\| &\leq \left[\frac{\int_0^1 w((1-\theta)\|x_n-x^*\|)d\theta}{1-w_0(\|x_n-x^*\|)}\right. \\ &+\frac{3}{8}\left(\frac{2\bar{w}_n}{1-w_0(\|y_n-x^*\|)} + \frac{\bar{w}_n^2}{(1-w_0(\|x_n-x^*\|))(1-w_0(\|y_n-x^*\|))}\right) \\ &\left.\times\frac{(1+\int_0^1 w_0(\theta\|x_n-x^*\|)d\theta)}{1-w_0(\|x_n-x^*\|)}\right]\|x_n-x^*\| \\ &\leq g_2(\|x_n-x^*\|)\|x_n-x^*\| \leq \|x_n-x^*\|. \end{aligned}$$

Method(61.2)

The function g_1 is the same as in method (61.1). Then, we also have

$$x_{n+1}-x^* = x_n - x^* - F'(x_n)^{-1}F(x_n) + \frac{3}{8}(I - F'(y_n)^{-1}F'(x_n))^2F'(x_n)^{-1}F(x_n),$$

and

$$\begin{aligned} \|x_{n+1}-x^*\| &\leq \left[\frac{\int_0^1 w((1-\theta)\|y_n-x^*\|)d\theta}{1-w_0(\|x_n-x^*\|)}\right. \\ &\left.+\frac{3}{8}\frac{\bar{w}_n^2(1+\int_0^1 w_0(\theta\|x_n-x^*\|)d\theta)}{(1-w_0(\|y_n-x^*\|))^2(1-w_0(\|x_n-x^*\|))}\right]\|x_n-x^*\| \\ &\leq g_2(\|x_n-x^*\|)\|x_n-x^*\| \leq \|x_n-x^*\|. \end{aligned}$$

Thus, we can choose

$$\begin{aligned} g_2(t) &= \frac{\int_0^1 w((1-\theta)t)d\theta}{1-w_0(t)} \\ &+\frac{3}{8}\frac{\bar{w}(t)^2(1+\int_0^1 w_0(\theta t)d\theta)}{(1-w_0(g_1(t)t))^2(1-w_0(t))}. \end{aligned}$$

3. Semi-local Convergence

Method(61.1)

The majorant sequence $\{\alpha_n\}$ is defined for $\alpha_0 = 0, \beta_0 \geq \frac{2}{3}\|F'(x_0)^{-1}F(x_0)\|$ and each $n = 0,1,2,...$ by

$$\begin{aligned}
\bar{v}_n &= \begin{cases} v(\beta_n - \alpha_n) \\ or \\ v_0(\alpha_n) + v_0(\beta_n), \end{cases} \\
\alpha_{n+1} &= \beta_n + \frac{3}{2}\left[\frac{1}{3} + \frac{3\bar{v}_n}{4(1-v_0(\beta_n))} + \frac{3}{8}\frac{\bar{v}_n^2}{(1-v_0(\alpha_n))(1-v_0(\beta_n))}\right](\beta_n - \alpha_n), \\
\delta_{n+1} &= (1 + \int_0^1 v_0(\alpha_n + \theta(\alpha_{n+1} - \alpha_n))d\theta)(\alpha_{n+1} - \alpha_n) \\
&\quad + \frac{3}{2}(1 + v_0(\alpha_n))(\beta_n - \alpha_n)
\end{aligned}$$

and

$$\beta_{n+1} = \alpha_{n+1} + \frac{2}{3}\frac{\delta_{n+1}}{1 - v_0(\alpha_{n+1})}.$$

The calculation are in turn as in the local case:

$$\begin{aligned}
x_{n+1} - y_n &= -\frac{3}{2}\left[-\frac{1}{3} + \frac{3}{8}(F'(y_n)^{-1} - F'(x_n)^{-1})(F'(y_n) - F'(x_n))\right. \\
&\quad \left. -\frac{3}{4}F'(y_n)^{-1}(F'(x_n) - F'(y_n))\right](y_n - x_n), \\
\|x_{n+1} - y_n\| &\leq \frac{3}{2}\left[\frac{1}{3} + \frac{3\bar{v}_n}{4(1-v_0(\beta_n))} + \frac{3}{8}\frac{\bar{v}_n^2}{(1-v_0(\alpha_n))(1-v_0(\beta_n))}\right](\beta_n - \alpha_n), \\
&= \alpha_{n+1} - \beta_n, \\
F(x_{n+1}) &= F(x_{n+1}) - F(x_n) - \frac{3}{2}F'(x_n)(y_n - x_n)
\end{aligned}$$

4. References

[1] Darvishi, M. T., Shin, B. C., *High-order Newton-Krylov methods to solve systems of nonlinear equation*, *Journal of the Korean Society for Industrial and Applied Mathematics*, 15, (2011), 19-30.

[2] Ezquerro, J. A., Grau-Snchez, M., Grau, A., Hernndez, M. A., Noguera, M., Romero, N., *On iterative methods with accelerated convergence for solving systems of nonlinear equations*, *Journal of Optimization Theory and Applications*, 151, (2011), 163-174.

[3] Grau-Snchez, M., Grau, A., Noguera, M., *On the computational efficiency index and some iterative methods for solving systems of nonlinear equations*, *Journal of Computational and Applied Mathematics*, 236, 6, (2011), 1259-1266.

[4] Sharma, J. R., Guha, R. K., Sharma, R., *An efficient fourth order weighted-Newton method for systems of nonlinear equations*, *Numerical Algorithms*. In press.

[5] Shin, B. C., Darvishi, M. T., Kim, C. H., *A comparison of the Newton-Krylov method with high order Newton-like methods to solve nonlinear systems*, Applied Mathematics and Computation, 217, 7, (2010), 3190-3198.

[6] Soleymani, F., *Regarding the accuracy of optimal eighth-order methods*, Mathematical and Computer Modelling, 53, 5-6, (2011), 1351-1357.

[7] Turan, E., Ecder, A., *Set reduction in nonlinear equations*, In press, http://arxiv.org/abs/1203.3059v1.

Index